Lecture Notes in Physics

Volume 878

For further volumes:
www.springer.com/series/5304

The Lecture Notes in Physics

The series Lecture Notes in Physics (LNP), founded in 1969, reports new developments in physics research and teaching—quickly and informally, but with a high quality and the explicit aim to summarize and communicate current knowledge in an accessible way. Books published in this series are conceived as bridging material between advanced graduate textbooks and the forefront of research and to serve three purposes:

- to be a compact and modern up-to-date source of reference on a well-defined topic
- to serve as an accessible introduction to the field to postgraduate students and nonspecialist researchers from related areas
- to be a source of advanced teaching material for specialized seminars, courses and schools

Both monographs and multi-author volumes will be considered for publication. Edited volumes should, however, consist of a very limited number of contributions only. Proceedings will not be considered for LNP.

Volumes published in LNP are disseminated both in print and in electronic formats, the electronic archive being available at springerlink.com. The series content is indexed, abstracted and referenced by many abstracting and information services, bibliographic networks, subscription agencies, library networks, and consortia.

Proposals should be sent to a member of the Editorial Board, or directly to the managing editor at Springer:

Christian Caron
Springer Heidelberg
Physics Editorial Department I
Tiergartenstrasse 17
69121 Heidelberg/Germany
christian.caron@springer.com

Wilhelm von Waldenfels

A Measure Theoretical Approach to Quantum Stochastic Processes

Springer

Wilhelm von Waldenfels
Himmelpfort, Germany

ISSN 0075-8450 ISSN 1616-6361 (electronic)
Lecture Notes in Physics
ISBN 978-3-642-45081-5 ISBN 978-3-642-45082-2 (eBook)
DOI 10.1007/978-3-642-45082-2
Springer Heidelberg New York Dordrecht London

Library of Congress Control Number: 2013956252

Printed on acid-free paper

Springer is part of Springer Science+Business Media (www.springer.com)

Preface

Let us start by considering a finite set of operators a_x, called annihilation operators, and a_x^+, called creation operators, indexed by x in the finite set X. They have the commutation relations, for $x, y \in X$,

$$\left[a_x, a_y^+\right] = \delta_{x,y}$$
$$[a_x, a_y] = \left[a_x^+, a_y^+\right] = 0.$$

First we realize these operators in a purely algebraic way. We define them as generators of a complex associative algebra with the above commutation relations as defining relations. We denote this algebra by $\mathfrak{W}(X)$. It is a special form of a *Weyl algebra*. A normal ordered monomial of the $a_x, a_x^+, x \in X$ is what we call a monomial of the form

$$a_{x_1}^+ \cdots a_{x_m}^+ a_{y_1} \cdots a_{y_n}.$$

The normal ordered monomials form a basis of $\mathfrak{W}(X)$. This means any element of $\mathfrak{W}(X)$ can be represented in a unique way according to the formula

$$\sum K(x_1, \ldots, x_m; y_1, \ldots, y_n) a_{x_1}^+ \cdots a_{x_m}^+ a_{y_1} \cdots a_{y_n},$$

where K is a function symmetric both in the x_i and in the y_i.

We can then move on to consider a continuous set of annihilation and creation operators, e.g., $a_x, a_x^+, x \in \mathbb{R}$, with the commutation relations

$$\left[a_x, a_y^+\right] = \delta(x - y)$$
$$[a_x, a_y] = \left[a_x^+, a_y^+\right] = 0$$

where $\delta(x - y)$ is Dirac's δ-function. These operators are harder to define rigorously. One possibility is to use the integrals

$$a(\varphi) = \int \mathrm{d}x\, \overline{\varphi}(x) a_x$$

$$a^+(\psi) = \int \mathrm{d}x\ \psi(x) a_x^+,$$

where the arguments φ and ψ are square-integrable functions. Then the non-vanishing commutation relations read

$$\left[a(\varphi), a^+(\psi)\right] = \int \mathrm{d}x\ \overline{\varphi}(x)\psi(x).$$

Everything in this context can be well defined using what is called Fock space.

Another way to approach the problem was chosen by Obata [35]. He uses an infinite system of nested Hilbert spaces, first defines a_x, and then the adjoint a_x^+ in the dual system.

In quantum field theory, one uses for operators the representation developed by Berezin [8]

$$\sum_{m,n} \int \cdots \int \mathrm{d}x_1 \cdots \mathrm{d}x_m \mathrm{d}y_1 \cdots \mathrm{d}y_n K_{m,n}(x_1, \ldots, x_m; y_1, \ldots, y_n) \\ \times a_{x_1}^+ \cdots a_{x_m}^+ a_{y_1} \cdots a_{y_n}, \qquad (*)$$

where $K_{m,n}$ might be quite irregular generalized functions. The multiplication of these operators can be performed by using the commutation relations. Berezin provides for that purpose an attractive functional integral.

Another way to perform the multiplication of these operators is to define a convolution for the coefficients K, using the commutation relations formally, and then to forget about the a_x and a_x^+ and work only with the convolution. This can be done in a rigorous way. This is the theory of *kernels* introduced by Hans Maassen [31] and continued by Paul-André Meyer [34]. These kernels are therefore called *Maassen-Meyer-kernels*. The theory works for Lebesgue measurable kernels [41].

We now mention the usual way of defining $a(\varphi)$ and $a^+(\varphi)$. Denote by

$$\mathfrak{R} = \{\emptyset\} + \mathbb{R} + \mathbb{R}^2 + \cdots$$

the space of all finite sequences of real numbers, where we use the $+$ sign for union of disjoint sets. Equip it with the measure

$$\hat{\mathrm{e}}(\lambda)(f) = f(\emptyset) + \sum_{n=1}^{\infty} \frac{1}{n!} \int \cdots \int \mathrm{d}x_1 \cdots \mathrm{d}x_n f(x_1, \ldots, x_n),$$

where the function $f(x_1, \ldots, x_n)$ is supposed to be symmetric in the x_i. The notation $\hat{\mathrm{e}}(\lambda)$ is used because this is essentially the exponential of the Lebesgue measure λ. Then Fock space is defined to be

$$L_{\mathrm{s}}^2(\mathfrak{R}, \hat{\mathrm{e}}(\lambda)),$$

where the letter s stands for symmetric. If $L^2_s(\mathbb{R}^n) = L(n)$ is the space of symmetric Lebesgue square-integrable functions on $\mathbb{R}^n$, then

$$a(\varphi) : L(n+1) \to L(n),$$

$$\big(a(\varphi)f\big)(x_1, \ldots, x_n) = \int \mathrm{d}x_0 \overline{\varphi}(x_0) f(x_0, x_1, \ldots, x_n)$$

and

$$a^+(\varphi) : L(n) \to L(n+1),$$

$$\big(a^+(\varphi)f\big)(x_0, x_1, \ldots, x_n)$$
$$= \varphi(x_0) f(x_1, \ldots, x_n) + \varphi(x_1) f(x_0, x_2, \ldots, x_n) + \cdots$$
$$+ \varphi(x_n) f(x_0, x_1, \ldots, x_{n-1}).$$

Thus $a(\varphi)$ and $a^+(\varphi)$ can be defined on the pre-Hilbert space

$$\bigoplus_{n=0,\mathrm{f}}^{\infty} L(n) \subset L^2_s\big(\mathfrak{R}, \hat{\mathrm{e}}(\lambda)\big),$$

where the suffix f means, that any element $f = (f_0, f_1, \ldots, f_n, \ldots)$ has components $f_n = 0$ for sufficiently large n.

This approach is based on the duality of the Hilbert space $L^2_s(\mathfrak{R}, \mathrm{e}(\lambda))$ with itself. We use Bourbaki's measure theory [10] and employ the duality between measures and functions. The space $\mathfrak{R}$ is locally compact when provided with the obvious topology. Use the notation $\mathscr{M}_s(\mathfrak{R})$ for the space of symmetric measures and $\mathscr{K}_s(\mathfrak{R})$ for the space of symmetric continuous functions of compact support. We can now define, for a measure ν on $\mathbb{R}$ and a symmetric function $f \in \mathscr{K}_s(\mathfrak{R})$,

$$a(\nu) : \mathscr{K}_s(\mathfrak{R}) \to \mathscr{K}_s(\mathfrak{R}),$$

$$\big(a(\nu)f\big)(x_1, \ldots, x_n) = \int \overline{\nu}(\mathrm{d}x_0) f(x_0, x_1, \ldots, x_n)$$

and for a continuous function φ with compact support in $\mathbb{R}$

$$a^+(\varphi) : \mathscr{K}_s(\mathfrak{R}) \to \mathscr{K}_s(\mathfrak{R}),$$

$$\big(a^+(\varphi)f\big)(x_0, x_1, \ldots, x_n)$$
$$= \varphi(x_0) f(x_1, \ldots, x_n) + \varphi(x_1) f(x_0, x_2, \ldots, x_n) + \cdots$$
$$+ \varphi(x_n) f(x_0, x_1, \ldots, x_{n-1})$$

which is essentially the same formula as above.

By making use of the δ-function we have raised both a conceptual and a semantic problem. Denote the point measure at the point x by ε_x, with

$$\int \varepsilon_x(\mathrm{d}y)\varphi(y) = \varphi(x).$$

In the physical literature, the δ-function can have three different meanings corresponding to the different differentials with which it is combined:

$$\begin{aligned}
\delta(x-y)\mathrm{d}y &= \varepsilon_x(\mathrm{d}y)\\
\delta(x-y)\mathrm{d}x &= \varepsilon_y(\mathrm{d}x)\\
\delta(x-y)\mathrm{d}x\mathrm{d}y &= \Lambda(\mathrm{d}x,\mathrm{d}y),
\end{aligned}$$

where Λ is the measure on $\mathbb{R}^2$ concentrated on the diagonal and given by

$$\int \Lambda(\mathrm{d}x,\mathrm{d}y)\varphi(x,y) = \int \mathrm{d}x\varphi(x,x).$$

We will use both types of notation: one is mathematically clearer, the other one is often more convenient for calculations. In mathematics one very often uses δ_x for the point measure ε_x. We tend to avoid this notation.

Now we can define easily

$$\begin{aligned}
&a(x) = a(\varepsilon_x) : \mathcal{K}_s(\mathfrak{R}) \to \mathcal{K}_s(\mathfrak{R}),\\
&\big(a(x)f\big)(x_1,\dots,x_n) = f(x,x_1,\dots,x_n).
\end{aligned}$$

The definition of the creation operator is more difficult. Consider the measure-valued function

$$x \to \varepsilon_x$$

and define

$$\begin{aligned}
&a^+(\mathrm{d}x) = a^+\big(\varepsilon(\mathrm{d}x)\big) : \mathcal{K}_s(\mathfrak{R}) \to \mathcal{M}(\mathbb{R}),\\
&\big(a^+(\,dx)f\big)(x_0,x_1,\dots,x_n)\\
&\quad = \varepsilon_{x_0}(\mathrm{d}x)f(x_1,\dots,x_n) + \varepsilon_{x_1}(\mathrm{d}x)f(x_0,x_2,\dots,x_n) + \cdots\\
&\qquad + \varepsilon_{x_n}(\mathrm{d}x)f(x_0,x_1,\dots,x_{n-1}),
\end{aligned}$$

where the result is a sum of point measures on $\mathbb{R}$. With the help of these operators it is possible to establish a *quantum white noise* calculus.

We have the commutation relation

$$\big[a(x),a^+(\mathrm{d}y)\big] = \varepsilon_x(\mathrm{d}y).$$

There is an important operator called the number operator informally given as

$$N = \int_{\mathbb{R}} \mathrm{d}x \; a^+(x)a(x).$$

The differential of the number operator can be defined rigorously by

$$\mathfrak{n}(\mathrm{d}x) = a^+(\mathrm{d}x)a(x),$$

$$\big(\mathfrak{n}(\mathrm{d}x)f\big)(x_1, \ldots, x_n) = \sum_{i=1}^{n} \varepsilon_{x_i}(\mathrm{d}x) f(x_1, \ldots, x_n).$$

The normal ordered monomials have the form

$$\begin{aligned} M_{lmn} &= M(s_1, \ldots, s_l; t_1, \ldots, t_m; u_1, \ldots, u_n) \\ &= a^+(\mathrm{d}s_1) \cdots a^+(\mathrm{d}s_l) a^+(\mathrm{d}t_1) \cdots a^+(\mathrm{d}t_m) a(t_1) \cdots a(t_m) a(u_1) \cdots \\ &\quad \times a(u_m) \mathrm{d}u_1 \cdots \mathrm{d}u_n. \end{aligned}$$

We define a measure on $\mathfrak{R}^5$ by

$$\begin{aligned} \mathfrak{m}_{plmnq} &= \mathfrak{m}(x_1, \ldots, x_p; s_1, \ldots, s_l; t_1, \ldots, t_m; u_1, \ldots, u_n; y_1, \ldots, y_q) \\ &= \langle \emptyset | a(x_1) \cdots a(x_p) \mathrm{d}x_1 \cdots \mathrm{d}x_p M_{lmn}(s_1, \ldots, s_l; t_1, \ldots, t_m; u_1, \ldots, u_n) \\ &\quad a^+(\mathrm{d}y_1) \cdots a^+(\mathrm{d}y_q) | \emptyset \rangle. \end{aligned}$$

Fix a Hilbert space $\mathfrak{k}$, and denote by $B(\mathfrak{k})$ the space of bounded operators on it. Consider a Lebesgue locally integrable function

$$F = (F_{lmn})_{lmn \in \mathbb{N}^3} : \mathfrak{R}^3 \to B(\mathfrak{k})$$

$$F_{lmn} = F_{lmn}(s_1, \ldots, s_l; t_1, \ldots, t_m; u_1, \ldots, u_n)$$

which is symmetric in the variables s_i, t_i and u_i, and two functions $f, g \in \mathscr{K}_s(\mathfrak{R}, \mathfrak{k})$,

$$f = f_p(x_1, \ldots, x_p)$$

$$g = g_q(y_1, \ldots, y_q).$$

We associate with F the sesquilinear form $\mathscr{B}(F)$ given by

$$\langle f | \mathscr{B}(F) | g \rangle = \sum \frac{1}{p!l!m!n!q!} \int \mathfrak{m}_{plmnq} f_p^+ F_{lmn} g_q$$

where f^+ denotes the adjoint vector to f. This formula may look terrifying, but it becomes more manageable by using multi-indices. It gives to Berezin's formula $(*)$ above a rigorous mathematical meaning, and it has the big advantage that it is a classical integral, so that we have all the tools of classical measure theory available.

These considerations can easily be generalized from $\mathbb{R}$ to any locally compact space X, and to an arbitrary measure λ on X instead of the Lebesgue measure. We will need that in Example 2 below.

The δ-function, or equivalently the point measure ε_0, can be approximated by measures continuous with respect to the Lebesgue measure. If $\varphi \geq 0$ is a continuous function of compact support on $\mathbb{R}$, with $\int \mathrm{d}x\, \varphi(x) = 1$, put

$$\varphi_\zeta(x) = \frac{1}{\zeta}\varphi\left(\frac{x}{\zeta}\right)$$

and

$$\varphi_\zeta^x(y) = \varphi_\zeta(x-y).$$

Then for $\zeta \downarrow 0$

$$\varphi_\zeta^x(y)\mathrm{d}x = \varphi_\zeta(x-y)\mathrm{d}x \to \varepsilon_y(\mathrm{d}x) = \delta(x-y)\mathrm{d}x$$

and

$$\varphi_\zeta^x(y)\mathrm{d}y = \varphi_\zeta(x-y)\mathrm{d}y \to \varepsilon_x(\mathrm{d}y) = \delta(x-y)\mathrm{d}y.$$

Recall

$$a^+(\varphi) = \int \varphi(x)a^+(\mathrm{d}x), \qquad a(\varphi) = \int \mathrm{d}x\, \varphi(x)a_x.$$

These were the operators defined above. We have

$$a^+\left(\varphi_\zeta^x\right)\mathrm{d}x \to a^+(\mathrm{d}x), \qquad a\left(\varphi_\zeta^x\right) \to a_x$$

since

$$\begin{aligned}&\left(a^+\left(\varphi_\zeta^x\right)\mathrm{d}x f\right)(x_0, x_1, \ldots, x_n)\\ &\quad= \left(\varphi_\zeta(x-x_0)f(x_1,\ldots,x_n) + \cdots + \varphi_\zeta(x-x_n)f(x_0,x_1,\ldots,x_{n-1})\right)\mathrm{d}x\\ &\quad\to \varepsilon_{x_0}(\mathrm{d}x)f(x_1,\ldots,x_n) + \cdots + \varepsilon_{x_n}(\mathrm{d}x)f(x_0,x_1,\ldots,x_{n-1}),\end{aligned}$$

and

$$\left(a\left(\varphi_\zeta^x\right)f\right)(x_1,\ldots,x_n) = \int \mathrm{d}x_0 \varphi_\zeta^x(x_0)f(x_0,x_1,\ldots,x_n) \to f(x,x_1,\ldots,x_n).$$

In this context the operators $a^+(\varphi_\zeta^x)$ and $a(\varphi_\zeta^x)$ are called *coloured noise* operators, and the transition $\zeta \downarrow 0$ is called, for historical reasons, the *singular coupling limit*.

Without introducing any heavy apparatus we can treat four examples, where we restrict ourselves to the zero-particle case and to the one-particle case, i.e. just to the vacuum $|\emptyset\rangle$ and $L(1) = L^1(\mathbb{R}, \mathfrak{k})$, and do not need the whole Fock space.

1. *A two-level atom coupled to a heat bath of oscillators, or equivalently the damped oscillator*

   We restrict to the one-excitation case: We have either all oscillators in the ground state and the atom in the upper level, or one oscillator is in the first state and the atom is in the lower state. In the rotating wave approximation the Hamiltonian can be reduced to

$$H = \int \omega a^+(\mathrm{d}\omega)a(\omega) + E_{10}a^+(\varphi) + E_{01}a(\varphi),$$

   where

$$E_{01} = \begin{pmatrix} 0 & 0 \\ 1 & 0 \end{pmatrix}, \qquad E_{10} = \begin{pmatrix} 0 & 1 \\ 0 & 0 \end{pmatrix}, \qquad E_{11} = \begin{pmatrix} 1 & 0 \\ 0 & 0 \end{pmatrix}$$

   and φ is a continuous function ≥ 0, with compact support in $\mathbb{R}$, and $\int \mathrm{d}t\varphi(t) = 1$. We consider $a^+(\varphi)$ and $a^+(\varphi)$ as coloured noise operators, replace φ by φ_ς, calculate the resolvent and perform the singular coupling limit. This means, in frequency space, that φ approaches 1 and not δ. Then the resolvent converges to the resolvent of a one-parameter strongly continuous unitary group on the space

$$\mathfrak{H} = \left(\mathbb{C}\begin{pmatrix} 1 \\ 0 \end{pmatrix} \otimes \mathbb{C}|\emptyset\rangle\right) \oplus \left(\mathbb{C}\begin{pmatrix} 0 \\ 1 \end{pmatrix} \otimes L(1)\right).$$

   The one-parameter group can be calculated explicitly, then we obtain the Hamiltonian as a singular operator, and calculate the spectral decomposition of the Hamiltonian explicitly.

   After establishing a more general theory on the entire Fock space we recognize the interaction representation $V(t)$ of the time-development operator in the formal time representation as the restriction of U_0^t to $\mathfrak{H}$, where U_s^t is the solution of the quantum stochastic differential equation (QSDE)

$$\mathrm{d}_t U_s^t = -\mathrm{i}\sqrt{2\pi}\, E_{01}a^+(\mathrm{d}t)U_s^t - \mathrm{i}\sqrt{2\pi}\, E_{10}U_s^t a(t)\mathrm{d}t - \pi E_{11}\mathrm{d}t$$

   with $U_s^s = 1$; so U_s^t is an operator on

$$L^2(\mathfrak{R}, \mathbb{C}^2) \supset \mathfrak{H}.$$

2. *A two-level atom interacting with polarized radiation*

   This is very similar to the first example, but we have to consider not only the frequency but also the direction and the polarization of the photons. So for the photons we are concerned with the space

$$X = L^2(\mathbb{R} \times \mathbb{S}^2 \times \{1, 2, 3\}),$$

   where the first factor stands for the formal time (replacing the frequency via Fourier transform), the second one for the direction and the third one for the polarization. We have added a fictional longitudinal polarization in order to make the calculations easier. We provide X with the measure

$$\langle \lambda | f \rangle = \iint \mathrm{d}t\omega_0^2 \mathrm{d}\mathbf{n} \sum_{i=1,2,3} f(t, \mathbf{n}, i),$$

where $\mathrm{d}\mathbf{n}$ is the surface element on the unit sphere such that

$$\int_{\mathbb{S}^2} \mathrm{d}\mathbf{n} = 4\pi$$

and ω_0 is the transition frequency. Define

$$\mathfrak{X} = \{\emptyset\} + X + X^2 + \cdots$$

and consider

$$\Gamma = L^2(\mathfrak{X}, \mathbb{C}^2).$$

Denote by $\Pi(\mathbf{n})$ the projector on the plane perpendicular to $\mathbf{n}$,

$$\Pi(\mathbf{n})_{ij} = \delta_{ij} - \mathbf{n}_i \mathbf{n}_j.$$

After some approximations we obtain the Hamiltonian

$$H = \int \mathrm{d}\mathbf{n}\omega_0^2\omega \sum_{i,l} \Pi(\mathbf{n})_{i,l} a^+(\mathrm{d}\omega, \mathbf{n}, i) a(\omega, \mathbf{n}, l)$$
$$+ \int \mathrm{d}\mathbf{n}\omega_0^2\varphi(\omega) \sum_{i,l} \Pi(\mathbf{k})_{i,l} \big(E_{10} q_i a(\omega, \mathbf{n}, l)\mathrm{d}\omega + E_{01}\overline{q}_i a^+(\mathrm{d}\omega, \mathbf{n}, l)\big)$$

where (q_1, q_2, q_3) is a vector proportional to the dipole moment. We perform the singular coupling limit via the resolvent, and arrive at a strongly continuous unitary one-parameter group on

$$\mathfrak{H} = \left(\mathbb{C}\begin{pmatrix}1\\0\end{pmatrix} \otimes \mathbb{C}|\emptyset\rangle\right) \oplus \left(\mathbb{C}\begin{pmatrix}0\\1\end{pmatrix} \otimes L^2(X, \lambda)\right).$$

We calculate the time evolution explicitly, calculate the Hamiltonian as a singular operator and give its spectral decomposition. If $V(t)$ is the interaction representation of the time evolution in a formal time representation, then $V(t)$ turns out to be the restriction of U_s^t to $\mathfrak{H}$. Here U_s^t is the solution of the differential equation

$$\mathrm{d}_t U_s^t = -\mathrm{i}\sqrt{2\pi} \int_{\mathbb{S}^2} \sum_{il} \Pi(\mathbf{n})_{il} \big(E_{01}\overline{q}_i a^+(\mathrm{d}(t, \mathbf{n}), l) U_s^t$$
$$+ E_{10} U_s^t q_i a(t, \mathbf{n}, l)\omega_0^2 \mathrm{d}\mathbf{n}\mathrm{d}t\big) - \pi\gamma E_{11} U_s^t \mathrm{d}t$$

with

$$\gamma = \frac{8\pi}{3}|\mathbf{q}|^2.$$

This is a new type of QSDE and should be investigated further.

3. *The Heisenberg equation of the amplified oscillator*

In the coloured noise approximation the Hamiltonian reads

$$H = \int \omega a^+(\mathrm{d}\omega)a(\omega) + \int b^+ a^+(\varphi) + \int ba(\varphi)$$

where b and b^+ are the usual oscillator operators with the non-vanishing commutator $[b, b^+] = 1$. Whereas the evolution corresponding to H is difficult and will be treated in Chap. 9, the Heisenberg evolution is very easy. Define

$$\mathfrak{H} = \mathbb{C}b^+ \oplus \{a(\psi) : \psi \in L^2(\mathbb{R})\},$$

then $\mathfrak{H}$ stays invariant under the mapping

$$A \mapsto \mathrm{e}^{\mathrm{i}Ht} A \mathrm{e}^{-\mathrm{i}Ht}.$$

Hence we obtain a one-parameter group on the space $\mathfrak{H}$. We perform the weak coupling limit via the resolvent and obtain, similarly to the first example, that evolution forms a strongly continuous one-parameter group on $\mathfrak{H}$. We identify $\mathfrak{H}$ with the $\mathfrak{H}$ of Example 1 and define E_{ij} accordingly. Then the interaction representation $V(t)$ of the evolution is the restriction to $\mathfrak{H}$ of the solution U_s^t to the QSDE

$$\mathrm{d}_t U_s^t = \mathrm{i}\sqrt{2\pi}\, a^+(\mathrm{d}t) E_{01} U_s^t - \mathrm{i}\sqrt{2\pi}\, E_{10} U_s^t a(t)\mathrm{d}t + \pi E_{11} U_s^t \mathrm{d}t.$$

We calculate the evolution on $\mathfrak{H}$ explicitly, determine the Hamiltonian and its spectral decomposition. Whereas this example looks algebraically very similar to the first one, it is analytically very different. The evolution is not unitary, but it does leave invariant the hermitian form

$$(c, f) \mapsto |c|^2 - \|f\|^2.$$

The spectrum of the Hamiltonian consists of the real line and the points $\pm \mathrm{i}\pi$.

4. *The pure number process*

We consider the coloured noise Hamiltonian

$$H = \int \omega a^+(\mathrm{d}\omega)a(\omega) + a^+(\varphi)a(\varphi).$$

The one-particle space $L(1) = L^2(\mathbb{R})$ stays invariant. We calculate on this subspace the resolvent, and determine the weak coupling limit. We again compute the unitary one-parameter group, the Hamiltonian and its spectral decomposition. The interaction representation is the restriction of the solution of the QSDE

$$d_t U_s^t = \frac{-i2\pi}{1+i\pi} a^+(dt) U_s^t a(t).$$

After using coloured noise we establish a *white noise theory*. Then we attack *the general Hudson-Parthasarathy differential equation*, i.e., the QSDE

$$dU_s^t = A_1 a^+(dt) U_s^t + A_0 a^+(dt) U_s^t a(t) + A_{-1} U_s^t a(t) dt + B dt$$

with $U_s^s = 1$. The solution can be given as an infinite power series in normal ordered monomials. The coefficients A_i, B are in $B(\mathfrak{k})$ for some Hilbert space $\mathfrak{k}$. If the coefficients satisfy some well-known conditions, the evolution is unitary. We give an explicit formula for the Hamiltonian. In Chap. 10 we show how this differential equation can be approximated by coloured noise.

In order to treat *the amplified oscillator* we investigate the QSDE

$$d_t U_s^t = -ia^+(dt) b^+ U_s^t - ibU_s^t a(t) dt - \frac{1}{2} bb^+.$$

This is an example of a QSDE with unbounded coefficients. For this we need the white noise theory, and establish an infinite power series in normal ordered polynomials. Using an algebraic theorem due to Wick, we sum the series and obtain an a priori estimate. We prove unitarity, strong continuity and the Heisenberg evolution of Example 3. With the help of the Heisenberg evolution we get estimates which allow the calculation of the Hamiltonian.

I would like to express my sincere thanks to my good friend and colleague, Patrick D.F. Ion. He spent weeks reading and discussing the present work with me, finding a number of mathematical errors and providing good advice. Last but not least, he improved my clumsy English as well as the LaTeX layout of the mathematical formulae. This book could never have been completed without the untiring help of Hartmut Krafft, a fellow citizen of our village, rescuing me on all computer and LaTeX issues. I owe a great deal to the continuous moral support of my dear friend Sigrun Stumpf.

Contents

Chapter 1
Weyl Algebras

Abstract We define creation and annihilation operators as generators of an associative algebra with the commutation relations as defining relations. This is a special case of a Weyl algebra. We discuss Weyl algebras, show that ordered monomials form a basis, introduce multisets and their notation. The vacuum and the scalar product are defined in a natural way. We prove an algebraic theorem due to Wick.

1.1 Definition of a Weyl Algebra

By an algebra we understand, if not stated otherwise, a complex associative algebra with unit element denoted by 1. We will define the quantum mechanical momentum and position operators in an algebraic way following the ideas of Hermann Weyl [45]. They are elements of a special *Weyl algebra*. Weyl algebras are defined as quotients of a free algebra. The complex *free algebra* with indeterminates X_i, $i \in I$, is the associative algebra of all *noncommutative* polynomials in the X_i. So, for instance, $X_1 X_2 \neq X_2 X_1$. The algebra is denoted by $\mathfrak{F} = \mathbb{C}\langle X_i, i \in I\rangle$. A basis for it is the collection of monomials or words W formed out of X_i, $i \in I$

$$W = X_{i_k} \cdots X_{i_2} X_{i_1}.$$

Assume given a skew-symmetric matrix $H = (H_{ij})_{i,j \in I}$, and divide the algebra $\mathbb{C}\langle X_i, i \in I\rangle$ by the ideal generated by the elements

$$X_i X_j - X_j X_i - H_{ij}, \quad i, j \in I.$$

The resulting algebra is generated by the canonical images x_i, $i \in I$, and has the relations

$$x_i x_j - x_j x_i = H_{ij}.$$

It is called the *Weyl algebra* generated by the x_i with the defining relations $x_i x_j - x_j x_i = H_{ij}$.

The canonical commutation relations provide the best known example: the quantities p_i and q_i, with $i = 1, \ldots, n$, generate a Weyl algebra with the defining rela-

W. von Waldenfels, *A Measure Theoretical Approach to Quantum Stochastic Processes*, Lecture Notes in Physics 878, DOI 10.1007/978-3-642-45082-2_1,

tions

$$p_i q_j - q_j p_i = -\mathrm{i}\delta_{ij},$$
$$p_i p_j - p_j p_i = q_i q_j - q_j q_i = 0.$$

1.2 The Algebraic Tensor Product

We introduce the tensor product in a coordinate-free way, following Bourbaki [12]. Assume we have n vector spaces $V_1, \dots, V_n$, and consider the space C of formal linear combinations of the n-tuples

$$(x_1, \dots, x_n) \in V_1 \times \cdots \times V_n.$$

Then define the subspace $D \subset C$ generated by

$$(x_1, \dots, x_{i-1}, x_i + y_i, x_{i+1}, \dots, x_n)$$
$$- (x_1, \dots, x_{i-1}, x_i, x_{i+1}, \dots, x_n) - (x_1, \dots, x_{i-1}, y_i, x_{i+1}, \dots, x_n),$$
$$(x_1, \dots, x_{i-1}, c x_i, x_{i+1}, \dots, x_n) - c(x_1, \dots, x_{i-1}, x_i, x_{i+1}, \dots, x_n)$$

for $i = 1, \dots, n$; $x_i, y_i \in V_i$; $c \in \mathbb{C}$.

The tensor product is the quotient C/D,

$$C/D = V_1 \otimes \cdots \otimes V_n = \bigotimes_{i=1}^{n} V_i.$$

The canonical image of $(x_1, \dots, x_n)$ is written

$$x_1 \otimes \cdots \otimes x_n.$$

Definition 1.2.1 A mapping

$$F : V_1 \times \cdots \times V_n \to U,$$

where U is a vector space, is called multilinear, if

$$F(x_1, \dots, x_{i-1}, x_i + y_i, x_{i+1}, \dots, x_n)$$
$$= F(x_1, \dots, x_{i-1}, x_i, x_{i+1}, \dots, x_n) + F(x_1, \dots, x_{i-1}, y_i, x_{i+1}, \dots, x_n),$$
$$F(x_1, \dots, x_{i-1}, c x_i, x_{i+1}, \dots, x_n) = c F(x_1, \dots, x_{i-1}, x_i, x_{i+1}, \dots, x_n)$$

for $i = 1, \dots, n$; x_i, y_i, in V_i; $c \in \mathbb{C}$.

A direct consequence of the definition of the tensor product is the following proposition.

Proposition 1.2.1 *One has*

- *The mapping*

$$(x_1, \ldots, x_n) \in V_1 \times \cdots \times V_n \mapsto x_1 \otimes \cdots \otimes x_n \in V_1 \otimes \cdots \otimes V_n$$

is multilinear.

- *If*

$$F : V_1 \times \cdots \times V_n \to U$$

is a multilinear mapping into a complex vector space U, then there exists a unique linear mapping

$$\tilde{F} : V_1 \otimes \cdots \otimes V_n \to U$$

such that

$$\tilde{F}(x_1 \otimes \cdots \otimes x_n) = F(x_1, \ldots, x_n).$$

For completeness we prove the following proposition.

Proposition 1.2.2 *Assume that $B_i \subset V_i$ is a basis for each V_i, $i = 1, \ldots, n$. Then the set*

$$\{b_1 \otimes \cdots \otimes b_n : b_i \in B_i\}$$

forms a basis of $V_1 \otimes \cdots \otimes V_n$.

Proof It is clear, that the $b_1 \otimes \cdots \otimes b_n$, $b_i \in B_i$, generate $V_1 \otimes \cdots \otimes V_n$. We have to show that they are independent. Recall the space C of formal linear combinations of the $(x_1, \ldots, x_n)$, and consider the subspace U spanned by the $(b_1, \ldots, b_n)$, $b_i \in B_i$. If $x_i \in V_i$, then

$$x_i = \sum_{b \in B_i} x_i(b) b$$

where $x_i(b)$ is the component of x_i along $b \in B$. Recall that only finitely many $x_i(b)$ are not equal to 0. The mapping

$$F : V_1 \times \cdots \times V_n \to U$$
$$(x_1, \ldots, x_n) \mapsto \sum_{k_1, \ldots, k_n} x_1(b_{1,k_1}) \cdots x_n(b_{n,k_n})(b_{1,k_1}, \ldots, b_{n,k_n})$$

with $b_{i,k_i} \in B_i$, is multilinear. Hence there exists a unique linear mapping

$$\tilde{F} : V_1 \otimes \cdots \otimes V_n \to U$$

with

$$\tilde{F}(x_1 \otimes \cdots \otimes x_n) = F(x_1, \ldots, x_n).$$

In particular,

$$\tilde{F}(b_{1,k_1} \otimes \cdots \otimes b_{n,k_n}) = (b_{1,k_1}, \ldots, b_{n,k_n}).$$

As the elements on the right-hand side are independent, the tensor products

$$(b_{1,k_1} \otimes \cdots \otimes b_{n,k_n})$$

have to be independent too. □

If $\mathfrak{A}$ is an algebra, the multiplication mapping

$$m : f \otimes g \in \mathfrak{A} \otimes \mathfrak{A} \mapsto fg \in \mathfrak{A}$$

is bilinear, and hence well defined.

Assume we have n algebras, and define a product in their tensor product in the following way:

$$\begin{aligned}
(f_1 \otimes \cdots \otimes f_n) \otimes (g_1 \otimes \cdots \otimes g_n) &\in (\mathfrak{A}_1 \otimes \cdots \otimes \mathfrak{A}_n) \otimes (\mathfrak{A}_1 \otimes \cdots \otimes \mathfrak{A}_n) \\
&\mapsto (f_1 \otimes g_1) \otimes \cdots \otimes (f_n \otimes g_n) \in (\mathfrak{A}_1 \otimes \mathfrak{A}_1) \otimes \cdots \otimes (\mathfrak{A}_n \otimes \mathfrak{A}_n) \\
&\mapsto m_1(f_1 \otimes g_1) \otimes \cdots \otimes m_n(f_n \otimes g_n) \in \mathfrak{A}_1 \otimes \cdots \otimes \mathfrak{A}_n.
\end{aligned}$$

So finally

$$(f_1 \otimes \cdots \otimes f_n)(g_1 \otimes \cdots \otimes g_n) = f_1 g_1 \otimes \cdots \otimes f_n g_n.$$

We imbed $\mathfrak{A}_i$ into $\bigotimes_i \mathfrak{A}_i$ by putting

$$u_1 : \mathfrak{A}_1 \ni f_1 \mapsto u_1(f_1) = f_1 \otimes 1 \otimes \cdots \otimes 1 \in \bigotimes_i \mathfrak{A}_i$$

$$\vdots$$

$$u_n : \mathfrak{A}_n \ni f_n \mapsto u_n(f_n) = 1 \otimes \cdots \otimes 1 \otimes f_n \in \bigotimes_i \mathfrak{A}_i.$$

The images $u_i(f_i)$ commute for different i. Conversely we have the following proposition [12].

Proposition 1.2.3 *If $\mathfrak{A}$ is an algebra and $\mathfrak{A}_i$ are subalgebras, commuting for different i, then $\mathfrak{A}$ is isomorphic to $\bigotimes_i \mathfrak{A}_i$. We write*

$$\mathfrak{A} \cong \bigotimes_i \mathfrak{A}_i.$$

Proposition 1.2.4 *If $\mathfrak{W}$ is the Weyl algebra generated by $x_1, \ldots, x_n$, with defining relations $[x_i, x_j] = H_{i,j}$ (where $[x_i, x_j]$ denotes the commutator as usual), and H*

is the direct sum of a $p \times p$ submatrix H_1 and $(n-p) \times (n-p)$ submatrix H_2, so

$$H = \begin{pmatrix} H_1 & 0 \\ 0 & H_2 \end{pmatrix},$$

then

$$\mathfrak{W} \cong \mathfrak{W}_1 \otimes \mathfrak{W}_2,$$

where $\mathfrak{W}_1$ is the Weyl algebra generated by $x_1, \ldots, x_p$ with the defining relations $[x_i, x_j] = (H_1)_{ij}$, and $\mathfrak{W}_2$ is the Weyl algebra generated by $x_{p+1}, \ldots, x_n$ with the defining relations $[x_i, x_j] = (H_2)_{ij}$.

For the proof consider that groups of generators $x_1, \ldots, x_p$ and $x_{p+1}, \ldots, x_n$ commute, hence the algebras generated by them commute, and we apply the Proposition 1.2.3.

1.3 Wick's Theorem

We cite a well-known theorem in quantum field theory from Jauch-Rohrlich's book [27].

Assume given two linearly ordered sets A and B, a ring $\mathfrak{A}$, and a function $f : A \times B \to \mathfrak{A}$. Define

$$C(\alpha, \beta; \alpha', \beta') = [f(\alpha, \beta), f(\alpha', \beta')](\mathbf{1}\{\alpha > \alpha'\} - \mathbf{1}\{\beta > \beta'\}),$$

where [,] denotes the commutator as usual, and $\mathbf{1}\{\alpha > \alpha'\}$ has the value 1 when $\alpha > \alpha'$ and 0 otherwise. Consider a finite family $(\alpha_i, \beta_i)_{i \in I}$, $\alpha_i \in A$, $\beta_i \in B$ and $f_i = f(\alpha_i, \beta_i)$. Assume, e.g., $I = [1, n]$, then the sequence

$$(f_{i_n}, \ldots, f_{i_1})$$

is called A *-ordered* if $\alpha_{i_n} \geq \cdots \geq \alpha_{i_1}$, and the sequence

$$(f_{j_n}, \ldots, f_{j_1})$$

is called *B-ordered* if $\beta_{j_n} \geq \cdots \geq \beta_{j_1}$. Assume

- $[[f_i, f_j], f_k] = 0$
- $[f_i, f_j] = 0$ if $\alpha_i = \alpha_j$ or $\beta_i = \beta_j$.

Then the A-product

$$A(f_1 \ldots f_n) = \mathbb{O}_A f_1 \cdots f_n := f_{i_n} \cdots f_{i_1}$$

is independent of the choice of the order of the sequence $f_1, \ldots, f_n$. So the elements f_i can supposed to commute on the right side of $\mathbb{O}_A$ and the A-product is commutative. A similar assertion holds for the B-product.

Denote by $\mathfrak{P}(n)$ the set of partitions of $[1,n]$ into singletons and pairs. So $\mathfrak{p} \in \mathfrak{P}(n)$ is of the form

$$\mathfrak{p} = \{\{t_1\}, \ldots, \{t_l\}, \{r_1, s_1\}, \ldots, \{r_m, s_m\}\}.$$

Define

$$B(\mathfrak{p}) = B(\mathfrak{p}; f_1, \ldots, f_n) = B(f_{t_1} \cdots f_{t_l}) C_{r_1,s_1} \cdots C_{r_m,s_m}$$

with

$$C_{rs} = C(\alpha_r, \beta_r; \alpha_s, \beta_s) = C_{sr}.$$

Then we have

Theorem 1.3.1

$$\mathbb{O}_A f_1 \cdots f_n = \sum_{\mathfrak{p} \in \mathfrak{P}(n)} B(\mathfrak{p}; f_1, \ldots, f_n).$$

We start with a lemma.

Lemma 1.3.1 *Assume given n elements in $\mathfrak{A}$ indexed by α_i, β_i,*

$$g_i = g(\alpha_i, \beta_i) \in \mathfrak{A}, \quad i \in [1,n]$$

and assume $\beta_n \geq \cdots \geq \beta_1$, and that there is an element $h = h(\alpha, \beta) \in \mathfrak{A}$ such that $\alpha \geq \alpha_i$, $i \in [1,n]$ and furthermore $[[h, g_i], g_j] = 0$, and if $\alpha = \alpha_i$ then $[h, g_i] = 0$. We have

$$hB(g_1 \cdots g_n) = B(hg_1 \cdots g_n) + \sum_{i=1}^{n} C(\alpha, \beta; \alpha_i, \beta_i) B\Bigg(\prod_{j \in [1,n] \setminus \{i\}} g_j\Bigg). \qquad (*)$$

Proof Assume $\beta \leq \beta_i$, $i = n, \ldots, k$ and $\beta > \beta_i$, $i = k-1, \ldots, 1$. Then

$$\begin{aligned} hB(g_1 \cdots g_n) &= hg_n \cdots g_1 = g_n \cdots g_k h g_{k-1} \cdots g_1 + [h, g_n \cdots g_k] g_{k-1} \cdots g_1 \\ &= g_n \cdots g_k h g_{k-1} \cdots g_1 + \sum_{i=k}^{n} [h, g_i] g_n \cdots g_{i+1} g_{i-1} \cdots g_k g_{k-1} \cdots g_1. \end{aligned}$$

As

$$[h, g_i] \mathbf{1}\{\alpha = \alpha_i\} = 0$$

we have for $i \in [k,n]$

$$[h, g_i] = [h, g_i] \mathbf{1}\{\beta \leq \beta_i\} = [h, g_i] \big(\mathbf{1}\{\alpha > \alpha_i\} - \mathbf{1}\{\beta > \beta_i\}\big) = C(\alpha, \beta; \alpha_i, \beta_i).$$

For $i \in [1, k-1]$, one has anyway

$$C(\alpha, \beta; \alpha_i, \beta_i) = 0. \qquad \square$$

With the lemma in hand, we finish the proof of the theorem.

Proof We prove the theorem by induction from $n-1$ to n. For $n=1$ the theorem is clear. Assume it for $n-1$. Define a mapping $\varphi : \mathfrak{P}(n) \to \mathfrak{P}(n-1)$ by erasing the letter n. Assume again

$$\mathfrak{p} = \big\{\{t_1\}, \dots, \{t_l\}, \{r_1, s_1\}, \dots, \{r_m, s_m\}\big\}.$$

Then

$$\varphi\mathfrak{p} = \begin{cases} \{\{t_2\}, \dots, \{t_l\}, \{r_1, s_1\}, \dots, \{r_m, s_m\}\} & \text{for } t_1 = n, \\ \{\{s_1\}, \{t_1\}, \dots, \{t_l\}, \{r_2, s_2\}, \dots, \{r_m, s_m\}\} & \text{for } r_1 = n. \end{cases}$$

Assume now, with different l and m such that $l + 2m = n - 1$,

$$\mathfrak{q} \in \mathfrak{P}(n-1) = \big\{\{t_1\}, \dots, \{t_l\}, \{r_1, s_1\}, \dots, \{r_m, s_m\}\big\}.$$

Then

$$\varphi^{-1}(\mathfrak{q}) = \big\{\mathfrak{p}^0, \mathfrak{p}^1, \dots, \mathfrak{p}^l\big\}$$

is a set of $l+1$ partitions of $[1, n]$ with

$$\mathfrak{p}^i = \begin{cases} \{\{n\}, \{t_1\}, \dots, \{t_l\}, \{r_1, s_1\}, \dots, \{r_m, s_m\}\} & \text{for } i = 0, \\ \{\{t_1\}, \dots, \{t_{i-1}\}, \{t_{i+1}\}, \dots, \{t_l\}, \{n, t_i\}, \{r_1, s_1\}, \dots, \{r_m, s_m\}\} & \text{for } i > 0. \end{cases}$$

Without loss of generality, we may assume that the f_i are A-ordered. We have by our hypothesis of induction

$$A(f_n \cdots f_1) = f_n A(f_{n-1} \cdots f_1) = f_n \sum_{\mathfrak{q} \in \mathfrak{P}(n-1)} B(\mathfrak{q}; f_1, \dots, f_{n-1}).$$

Now

$$\begin{aligned} f_n B(\mathfrak{q}) &= f_n B(f_{t_1} \cdots f_{t_l}) C_{r_1, s_1} \cdots C_{r_m, s_l} \\ &= \left(B(f_n, f_{t_1}, \dots, f_{t_l}) + \sum_{i=1}^{l} B(f_{t_1} \cdots f_{t_{i-1}} f_{t_{i+1}} \cdots f_{t_{i_l}}) C(n, t_i) \right) \\ &\quad \times C_{r_1, s_1} \cdots C_{r_m, s_l} \\ &= \sum_{\mathfrak{p} \in \varphi^{-1}(\mathfrak{q})} B(\mathfrak{p}; f_1 \cdots f_n) \end{aligned}$$

using our lemma (∗). Finally

$$A(f_n \cdots f_1) = \sum_{\mathfrak{q}\in\mathfrak{P}(n-1)} f_n B(\mathfrak{q}; f_1, \dots, f_{n-1})$$

$$= \sum_{\mathfrak{q}\in\mathfrak{P}(n-1)} \sum_{\mathfrak{p}\in\varphi^{-1}(\mathfrak{q})} B(\mathfrak{p}; f_1 \cdots f_n) = \sum_{\mathfrak{p}\in\mathfrak{P}(n)} B(\mathfrak{p}; f_1, \dots, f_n).$$ □

Now consider a Weyl algebra $\mathfrak{W}$ generated by the elements $x_i,\ i \in I$, with the defining relations $[x_i, x_j] = H_{i,j}$. Assume given a linearly ordered set Γ and a mapping $\gamma : I \to \Gamma$ with the property that $H_{i,j} = 0$ for $\gamma(i) = \gamma(j)$. Then a monomial $W = x_{i_n} \cdots x_{i_1}$ can be Γ-ordered. Denote the Γ-ordering by $\mathbb{O}_\Gamma(W)$. We use Wick's theorem in order to calculate $\mathbb{O}_\Gamma(W)$. The A-ordering of our formulation of Wick's theorem is the natural ordering of factors in W, the B-ordering is the Γ-ordering. Then

$$C_{r,s} = [x_{i_r}, x_{i_s}]\big(\mathbf{1}\{r > s\} - \mathbf{1}\{\gamma(i_r) > \gamma(i_s)\}\big).$$

Define, for $\mathfrak{p} \in \mathfrak{P}(n)$ with

$$\mathfrak{p} = \big\{\{t_1\}, \dots, \{t_l\}, \{r_1, s_1\}, \dots, \{r_m, s_m\}\big\},$$

the expression

$$\lfloor W \rfloor_\mathfrak{p} = \mathbb{O}_\Gamma(f_{t_1} \cdots f_{t_l}) C_{r_1,s_1} \cdots C_{r_m,s_m}.$$

Theorem 1.3.2

$$\mathbb{O}_\Gamma(W) = \sum_{\mathfrak{p}\in\mathfrak{P}(n)} \lfloor W \rfloor_\mathfrak{p}.$$

Proof This is a corollary of the last theorem in the notation just discussed. □

1.4 Basis of a Weyl Algebra

Assume the index set I to be totally ordered. We want to show, that the ordered monomials make up a basis for the Weyl algebra $\mathfrak{W}$ generated by $x_i,\ i \in I$, with the defining relations $[x_i, x_j] = H_{i,j}$. By the last theorem, it is clear that they generate the Weyl algebra. We have to prove their independence. This problem is related to the Poincaré-Birkhoff-Witt Theorem and we shall borrow some ideas from Bourbaki's proof of that [13].

We begin with the special case of $H = 0$. The Weyl algebra is then the algebra $\mathfrak{K} = \mathbb{C}[x_i,\ i \in I]$ of commutative polynomials, with complex coefficients, in the indeterminates $x_i,\ i \in I$.

We shall use the following notation: if $A : [1, k] \to I$ is a mapping, then

$$x_A = x_{A(k)} \cdots x_{A(1)},$$

so A may be called the ordering map for the monomial X_A.

Proposition 1.4.1 *The ordered monomials form a basis of* $\mathfrak{K} = \mathbb{C}[x_i,\ i \in I]$.

Proof If W is a monomial in the free algebra $\mathfrak{F} = \mathbb{C}\langle X_i, i \in I\rangle$ with $W = X_A = X_{A(k)} \cdots X_{A(1)}$ and $\sigma \in \mathfrak{S}_k$, the symmetric group on k elements, then define

$$\sigma W = X_{A(\sigma^{-1}(k))} \cdots X_{A(\sigma^{-1}(1))} = X_{A\circ\sigma^{-1}}$$

and

$$\mathbf{s}W = \frac{1}{k!} \sum_{\sigma\in\mathfrak{S}_k} \sigma W;$$

thus a mapping $\mathbf{s} : \mathfrak{F} \to \mathfrak{F}$ is defined.

The algebra $\mathfrak{K}$ is defined as the quotient $\mathfrak{F}/\mathfrak{J}$, where $\mathfrak{J}$ is the ideal generated by the $X_j X_i - X_i X_j$. An element of $\mathfrak{J}$ is a linear combination of elements of the form

$$\begin{aligned} W(X_j X_i - X_i X_j)W' &= X_{A(k)} \cdots X_{A(l+1)}(X_j X_i - X_i X_j)X_{A(l-2)} \cdots X_{A(1)} \\ &= (1-\tau)X_{A(k)} \cdots X_{A(l+1)} X_j X_i X_{A(l-2)} \cdots X_{A(1)}, \end{aligned}$$

where $\tau = (l-1, l)$ denotes the operator interchanging the indeterminates in places $l-1$ and l. As $\mathbf{s}\tau = \mathbf{s}$ we have

$$\mathbf{s}\big(W(X_j X_i - X_i X_j)W'\big) = 0$$

and $\mathbf{s}$ vanishes on $\mathfrak{J}$.

We want to prove, that $\sum c_i x_{A_i} = 0$ implies $c_i = 0$ for finite sums, if the A_i are different ordering maps $A_i : [1, k] \to I$. This means

$$\sum c_i X_{A_i} \in \mathfrak{J}$$

and

$$\sum c_i \mathbf{s}(X_{A_i}) = 0.$$

As the words for different A_i on the left-hand side are different, the c_i must vanish. □

Theorem 1.4.1 *If* $\mathfrak{W}$ *is a Weyl algebra generated by* $x_1, \dots, x_n$, *with defining relations* $x_i x_j - x_j x_i = H_{i,j}$, *then the ordered monomials form a basis of* $\mathfrak{W}$.

Proof We define the commutative polynomial algebra $\mathfrak{K} = \mathbb{C}[z_1, \dots, z_n]$ with indeterminates $z_1, \dots, z_n$ and denote by $L(\mathfrak{K})$ the algebra of linear maps $\mathfrak{K} \to \mathfrak{K}$. Set

$$m_i \in L(\mathfrak{K}) : m_i(f) = z_i f; \qquad d_i(f) = \sum_{i<l} H_{i,l} \frac{\partial f}{\partial z_l}.$$

Then

$$[m_i + d_i, m_j + d_j] = [d_i, m_j] - [d_j, m_i] = H_{i,j} \mathbf{1}_{i<j} - H_{j,i} \mathbf{1}_{j<i} = H_{i,j}$$

since $H_{i,j} = -H_{j,i}$ and $H_{i,i} = 0$. Here

$$\mathbf{1}_{i<j} = \begin{cases} 1 & \text{for } i < j, \\ 0 & \text{for } i \not< j. \end{cases}$$

We use this kind of notation often. Define a homomorphism $\eta : \mathfrak{F} \to L(\mathfrak{K})$ by $\eta(X_i) = m_i + d_i$. This means that in any polynomial we have to replace X_i by $m_i + d_i$. If $X_A = X_{A(k)} \cdots X_{A(1)}$, with $A(k) \geq \cdots \geq A(1)$, is an ordered monomial, then

$$\begin{aligned} \eta(X_A)(1) &= (m_{A(k)} + d_{A(k)}) \cdots (m_{A(1)} + d_{A(1)})(1) \\ &= (m_{A(k)} \cdots m_{A(1)})(1) = z_{A(k)} \cdots z_{A(1)} = z_A. \end{aligned}$$

The algebra $\mathfrak{W} = \mathfrak{F}/\mathfrak{J}$, where $\mathfrak{J}$ is the ideal generated by $[X_i, X_j] - H_{i,j}$. It is clear, that η vanishes on $\mathfrak{J}$. Assume X_{A_i} to be ordered monomials, with ordering maps A_i as above, and $\sum c_i x_{A_i} = 0$ in $\mathfrak{W}$, so $\sum c_i X_{A_i} \in \mathfrak{J}$ in $\mathfrak{F}$. Then

$$0 = \eta\Big(\sum c_i X_{A_i}\Big)(1) = \sum c_i z_{A_i},$$

hence $c_i = 0$, as the ordered monomials form a basis in $\mathfrak{K}$. □

1.5 Gaussian Functionals

If Q is a complex $n \times n$-matrix, we define the linear functional $\gamma_Q : \mathfrak{F} = \mathbb{C}\langle X_1, \dots, X_n \rangle \to \mathbb{C}$ in the following way. If $k = 2m$ is even, we define the set $\mathfrak{P}$ of partitions of $[1, k]$ into pairs; we will always write the pairs with the first component greater than the second:

$$\mathfrak{P} \ni \mathfrak{p} = \{\mathfrak{p}_1, \dots, \mathfrak{p}_m\}, \quad \mathfrak{p}_i = (r_i, s_i), \; r_i > s_i.$$

Put $\gamma_Q(1) = 1$, and $A : [1, k] \to [1, n]$ with $k = 2m$, and define

$$\lfloor X_A \rfloor_{\mathfrak{p}} = \prod_{i=1}^{m} Q\big(A(\mathfrak{p}_i)\big), \quad \text{with}$$

$$Q\bigl(A(\mathfrak{p}_i)\bigr) = Q\bigl(A(r_i), A(s_i)\bigr) \text{ for } \mathfrak{p}_i = (r_i, s_i), r_i > s_i.$$

Then, for $A : [1, k] \to [1, n]$, we define the Gaussian functional

$$\gamma_Q(X_A) = \begin{cases} 0 & \text{for } k = 2m + 1, \\ \sum_{\mathfrak{p} \in \mathfrak{P}} \lfloor X_A \rfloor_{\mathfrak{p}} & \text{for } k = 2m. \end{cases}$$

Proposition 1.5.1 *The functional γ_Q vanishes on the ideal generated by the polynomials*

$$X_i X_j - X_j X_i - (Q_{i,j} - Q_{j,i}).$$

Proof Consider a monomial, and a specific $l \in [1, k]$,

$$W = X_A = X_{A(k)} \cdots X_{A(l+1)} X_{A(l)} X_{A(l-1)} X_{A(l-2)} \cdots X_{A(1)}$$

and divide the set $\mathfrak{P}$, for the given l, into the subsets

$$M_0 = \bigl\{\mathfrak{p} \in \mathfrak{P} : (l, l-1) \in \mathfrak{p}\bigr\}, \qquad M_{rs} = \bigl\{\mathfrak{p} \in \mathfrak{P} : \mathfrak{p}'_r, \mathfrak{p}''_s \in \mathfrak{p}\bigr\},$$

where $\mathfrak{p}'_r = (r, l)$ resp. $\mathfrak{p}'_r = (l, r)$, if $r > l$ or $r < l$, and $\mathfrak{p}''_s = (s, l-1)$ resp. $\mathfrak{p}''_s = (l-1, s)$. Then when we define

$$W_0 = X_{A(k)} \cdots X_{A(l+1)} X_{A(l-2)} \cdots X_{A(1)}$$

we have

$$\gamma_Q(W) = Q\bigl(A(l), A(l-1)\bigr)\gamma_Q(W_0) + \sum_{r,s \notin \{l,l-1\}, r \neq s} Q(\mathfrak{p}'_r) Q(\mathfrak{p}''_s) \sum_{\mathfrak{p} \in M_{rs}} \prod_{\mathfrak{q} \in \mathfrak{p} \setminus \{\mathfrak{p}'_r, \mathfrak{p}''_s\}} Q(\mathfrak{q}).$$

Now consider

$$W' = X_{A'} = X_{A(k)} \cdots X_{A(l+1)} X_{A(l-1)} X_{A(l)} X_{A(l-2)} \cdots X_{A(1)}.$$

Then $\mathfrak{p}'_r$ and $\mathfrak{p}''_s$ exchange roles, and we obtain

$$\gamma_Q(W) - \gamma_Q(W') = \bigl(Q\bigl(A(l), A(l-1)\bigr) - Q\bigl(A(l-1), A(l)\bigr)\gamma_Q(W_0).$$

From there one obtains the result immediately. □

Corollary 1.5.1 *Consider the Weyl algebra $\mathfrak{W}$ with defining relations*

$$[x_i, x_j] = Q_{i,j} - Q_{j,i}$$

and let $\kappa : \mathfrak{F} \to \mathfrak{W}$ be the canonical homomorphism; then there exists a well defined mapping $\tilde{\gamma}_Q : \mathfrak{W} \to \mathbb{C}$ with $\gamma_Q = \tilde{\gamma}_Q \circ \kappa$.

Example Consider a Weyl algebra $\mathfrak{W}$ with defining relations $[x_i, x_j] = H_{i,j}$, and define a linear mapping by taking the coefficient of the unit in the basis of ordered monomials, cf. Theorem 1.4.1, then by Theorem 1.3.2, under this mapping

$$x_A \mapsto \sum_{\mathfrak{p}\in\mathfrak{P}} Q(\mathfrak{p})$$

with

$$Q_{i,j} = H_{i,j}\mathbf{1}_{i<j}.$$

1.6 Multisets

Let us recall some basic notions. If X is a set, a *list* of n elements of X is typically written, with $x_i \in X, i = 1, 2, \dots, n$, as an n-tuple

$$(x_1, \dots, x_n).$$

It can be defined as a mapping from the interval $[1, n] = (1, 2, \dots, n)$ of the natural numbers into X. We may write

$$(x_1, \dots, x_n) = x_{[1,n]}.$$

More generally, if $A = (a_1, \dots, a_n)$ is an ordered set, and x is seen as a map from A to some target space,

$$x_A = (x_{a_1}, \dots, x_{a_n}).$$

The ordinary *set* defined by x_A is the set

$$\{x_a : a \in A\}.$$

We shall use the notion of *multisets*. A multiset based on a set X is a mapping

$$\mathfrak{m} : X \to \mathbb{N} = \{0, 1, 2, 3, \dots\}.$$

The cardinality of $\mathfrak{m}$ is $\sharp\mathfrak{m} = |\mathfrak{m}| = \sum_{x\in X} \mathfrak{m}(x)$, showing different notations for the same cardinality. The set of multisets is $\mathbb{N}^X$, the set of all mappings $X \to \mathbb{N}$. It forms an additive monoid. A multiset is finite if its cardinality is finite. The commutative, ordered monoid of all finite multisets is denoted $\mathfrak{M}(X)$, and its ordered monoid structure comes from defining

$$(\mathfrak{m}_1 + \mathfrak{m}_2)(x) = \mathfrak{m}_1(x) + \mathfrak{m}_2(x)$$

and

$$\mathfrak{m}_1 \le \mathfrak{m}_2 \Longleftrightarrow \mathfrak{m}_1(x) \le \mathfrak{m}_2(x) \quad \text{for all } x \in X.$$

We denote by $\mathbf{1}_x$ the multiset $\mathfrak{m}(y) = \delta_{x,y}$, and obtain

$$\mathfrak{m} = \sum_{x \in X} \mathfrak{m}(x) \mathbf{1}_x.$$

We associate to a sequence $x = (x_1, \dots, x_n)$ the multiset

$$x^\bullet = (x_1, \dots, x_n)^\bullet = \kappa x = \sum_{i=1}^{n} \mathbf{1}_{x_i}.$$

So κ is the map that associates to a sequence its multiset. If $x = x_{[1,n]} = (x_1, \dots, x_n)$ is a sequence and σ is a permutation, then

$$\sigma x = (x_{\sigma^{-1}(1)}, \dots, x_{\sigma^{-1}(n)}).$$

If x and x' are two sequences, then there exists a permutation σ with $x' = \sigma x$ if and only if $\kappa x = \kappa x'$.

If $\mathfrak{m} = (x_1, \dots, x_n)^\bullet$, then $\mathfrak{m}(y)$ is the number of times that y occurs in the sequence $(x_1, \dots, x_n)$, so $\mathfrak{m}(y)$ is also known as the multiplicity of y in $x^\bullet$. Hence the number of sequences defining the same multiset is

$$\#\big(\kappa^{-1}(\mathfrak{m})\big) = \frac{|\mathfrak{m}|!}{\mathfrak{m}!}$$

with

$$\mathfrak{m}! = \prod_{x \in X} \mathfrak{m}(x)!.$$

We denote by $\mathfrak{X}$ the set of all finite sequences of elements of X

$$\mathfrak{X} = \{\emptyset\} + X + X^2 + \cdots.$$

We use the plus sign to denote the union of *disjoint* sets. A function $f : \mathfrak{X} \to \mathbb{C}$ is called symmetric if for $x \in X^n$ we have $f(\sigma x) = f(x)$ for all permutations σ. If f is a symmetric function and $\mathfrak{M}_n(X)$ is the set of multisets of cardinality n, then there exists a unique function $\tilde{f} : \mathfrak{M}_n(X) \to \mathbb{C}$ such that $f = \tilde{f} \circ \kappa$.

Assume that X is finite and that f vanishes on X^n for sufficiently big n; then we have the formula

$$\sum_{x \in \mathfrak{X}} \frac{1}{(\#x)!} f(x) = \sum_{\mathfrak{m} \in \mathfrak{M}(X)} \frac{1}{\mathfrak{m}!} \tilde{f}(\mathfrak{m}).$$

If

$$\alpha = \{a_1, \dots, a_n\},$$

is a set without a prescribed ordering, we define X^α as the set of all mappings $x_\alpha : \alpha \to X$. Supplement α with an ordering ω so that then the pair (α, ω) is given, e.g., by the sequence

$$(\alpha, \omega) = (a_1, \ldots, a_n).$$

If ω' is another ordering, then

$$(\alpha, \omega') = (a'_1, \ldots, a'_n),$$

where $(a'_1, \ldots, a'_n)$ is a permutation of $(a_1, \ldots, a_n)$. The mapping x_α is represented in the order ω by the sequence

$$(x_{a_1}, \ldots, x_{a_n}).$$

The multiset

$$x^\bullet_\alpha = (x_{a_1}, \ldots, x_{a_n})^\bullet = \sum_i \mathbf{1}_{x_{a_i}}$$

is independent of the ordering of α and hence is well defined. If $f : X^\alpha \to \mathbb{C}$ is a symmetric function, then $f(x_\alpha) = f((x_{a_1}, \ldots, x_{a_n}))$ is well defined, regardless of the ordering of α. If $\beta \subset \alpha$, and x_α is given, then we use the notation for restriction $x_\beta = x_\alpha \restriction \beta$ and $x_{\alpha \setminus \beta} = x_\alpha \restriction (\alpha \setminus \beta)$. If $x_\alpha \in X^\alpha$ and $x_\beta \in X^\beta$ are given, and α and β are disjoint, then there exists a unique $x_{\alpha+\beta} \in X^{\alpha+\beta}$, such that x_α and x_β are the restrictions of $x_{\alpha+\beta}$, and we have

$$x^\bullet_{\alpha+\beta} = x^\bullet_\alpha + x^\bullet_\beta.$$

If $x^\bullet_\alpha = (x_{a_1}, \ldots, x_{a_n})^\bullet$ and $x_\beta = (x_{b_1}, \ldots, x_{b_m})^\bullet$, then

$$x^\bullet_{\alpha+\beta} = (x_{a_1}, \ldots, x_{a_n}, x_{b_1}, \ldots, x_{b_m})^\bullet,$$

regardless of the orderings chosen in α, β, and $\alpha + \beta$.

If X is finite and $f : X^n \to \mathbb{C}$ is symmetric, and also α has n elements then

$$\sum_{x \in X^n} f(x) = \sum_{x_\alpha \in X^\alpha} f(x_\alpha) = \sum_{(x_{a_1}, \ldots, x_{a_n})} f(x_{a_1}, \ldots, x_{a_n}) = \sum_{\mathfrak{m} \in \mathfrak{M}_n} \frac{n!}{\mathfrak{m}!} \tilde{f}(\mathfrak{m}).$$

Assume $f : \mathfrak{X} \to \mathbb{C}$ is a symmetric function, such that f vanishes on X^n for n sufficiently large, and there is a sequence $\alpha = (\alpha_0, \alpha_1, \alpha_2, \ldots)$ of finite sets with $\#\alpha_n = n$. Then

$$\sum_{n=0}^{\infty} \frac{1}{n!} \sum_{x \in X^n} f(x) = \sum_{n=0}^{\infty} \frac{1}{n!} \sum_{x_{\alpha_n} \in X^{\alpha_n}} f(x_{\alpha_n}) = \sum_{\mathfrak{m} \in \mathfrak{M}(X)} \frac{1}{\mathfrak{m}!} \tilde{f}(\mathfrak{m}).$$

We write for short

$$\sum_{n=0}^{\infty} \frac{1}{n!} \sum_{x \in X^n} f(x) = \sum_\alpha f(x_\alpha) \Delta\alpha$$

with

$$\Delta\alpha = \frac{1}{\#\alpha!}.$$

Assume, for example,

$$\begin{aligned}
\alpha_0 &= \emptyset, \\
\alpha_1 &= \{1\}, \\
\alpha_2 &= \{1, 2\}, \\
\alpha_3 &= \{1, 2, 3\}, \\
&\dots
\end{aligned}$$

and

$$\begin{aligned}
x_{\alpha_0} &= \emptyset, \\
x_{\alpha_1} &= x_1, \\
x_{\alpha_2} &= (x_1, x_2), \\
x_{\alpha_3} &= (x_1, x_2, x_3), \\
&\dots.
\end{aligned}$$

Then

$$\begin{aligned}
\sum_{\alpha} f(x_\alpha)\Delta\alpha = f(\emptyset) + \sum_{x_1} f(x_1) + \frac{1}{2!}\sum_{x_1,x_2} f(x_1, x_2) \\
+ \frac{1}{3!}\sum_{x_1,x_2,x_3} f(x_1, x_2, x_3) + \cdots.
\end{aligned}$$

If $\mathbb{C}[X]$ is a free commutative polynomial algebra generated by the elements $x \in X$, we set

$$x_\alpha = x_{a_1} \cdots x_{a_n},$$

so that

$$x_\alpha x_\beta = x_{\alpha+\beta}.$$

If $\partial_{x_0} = \mathrm{d}/(\mathrm{d}x_0)$, then

$$\partial_{x_0} x_\alpha = \sum_{c\in\alpha} \delta(x_0, x_c) x_{\alpha\setminus c},$$

where δ is Kronecker's symbol and $\alpha \setminus c$ stands for $\alpha \setminus \{c\}$.

1.7 Finite Sets of Creation and Annihilation Operators

Assume X to be a finite set and consider the Weyl algebra $\mathfrak{W}(X)$ generated by a_x, a_x^+ for all $x \in X$, with the defining relations

$$\left[a_x, a_y^+\right] = \delta_{x,y}, \qquad [a_x, a_y] = \left[a_x^+, a_y^+\right] = 0$$

for $x, y \in X$. This implies, that the a_x commute with each other and so do the a_x^+. Using Proposition 1.2.4 we obtain

$$\mathfrak{W}(X) = \bigotimes_{x \in X} \mathfrak{W}\left(a_x, a_x^+\right),$$

where $\mathfrak{W}(a_x, a_x^+)$ is the subalgebra generated by a_x, a_x^+. The elements a_x are called *annihilation operators*, the elements a_x^+ *creation operators*.

In $\mathfrak{W}(X)$ we define the anti-isomorphism given by

$$a_x \mapsto a_x^\top, \qquad a_x^+ \mapsto \left(a_x^+\right)^\top = a_x.$$

Consider a monomial

$$M = a_{x_n}^{\vartheta_n} \cdots a_{x_1}^{\vartheta_1}$$

with $\vartheta_i = \pm 1$ and

$$a_x^\vartheta = \begin{cases} a_x^+ & \text{for } \vartheta = +1, \\ a_x & \text{for } \vartheta = -1. \end{cases}$$

Then

$$M^\top = a_{x_1}^{-\vartheta_1} \cdots a_{x_n}^{-\vartheta_n}.$$

A monomial is called *normal ordered*, if the creators precede the annihilators, i.e., if the monomial is of the form

$$a_{x_m}^+ \cdots a_{x_1}^+ a_{y_n} \cdots a_{y_1}.$$

Proposition 1.7.1 *The normal ordered monomials form a basis of* $\mathfrak{W}(X)$.

Proof In order to apply Theorem 1.4.1, we order the generators of $\mathfrak{W}(X)$. We assume that X has N elements, order the elements of X and define $\xi_i = a_{x_i}^+$ and $\xi_{N+i} = a_{x_i}$ for $i = 1, \ldots, N$. □

Consider a monomial

$$M = a_{x(n)}^{\vartheta(n)} \cdots a_{x(1)}^{\vartheta(1)}.$$

As the a_x commute and the a_x^+ commute among themselves, normal ordering is defined, see Sect. 1.3, and

$$:M: = \mathbb{O}_a M = \prod_{i:\vartheta(i)=+1} a^+_{x(i)} \prod_{i:\vartheta(i)=-1} a_{x(i)}.$$

Denote by $\mathfrak{P}(n)$ the set of partitions of $[1,n]$ into singletons $\{t_i\}$ and pairs $\{r_j, s_j\}, r_j > s_j$. So we have a typical partition

$$\mathfrak{p} = \big\{\{t_1\},\dots,\{t_l\},\{r_1,s_1\},\dots,\{r_m,s_m\}\big\}.$$

A direct consequence of Wick's theorem, Theorem 1.3.2, is

Proposition 1.7.2 (Wick's theorem) *Define*

$$\lfloor M \rfloor_{\mathfrak{p}} = :a_{x(t_1)}^{\vartheta(t_1)} \cdots a_{x_{t(l)}}^{\vartheta(t_l)}: C(r_1,s_1)\cdots C(r_m,s_m)$$

with

$$C(r,s) = \begin{cases} 1 & \textit{for } x(r)=x(s), \vartheta(r)=-1, \vartheta(s)=+1, \\ 0 & \textit{otherwise.} \end{cases}$$

Then

$$M = \sum_{\mathfrak{p}\in\mathfrak{P}} \lfloor M \rfloor_{\mathfrak{p}}.$$

If $\mathfrak{m} \in \mathfrak{M}(X)$ is a multiset, $\mathfrak{m} = m_1 \mathbf{1}_{x_1} + \cdots + m_k \mathbf{1}_{x_k}$, then

$$\left(a^+\right)^{\mathfrak{m}} = \left(a^+_{x_1}\right)^{m_1} \cdots \left(a^+_{x_k}\right)^{m_k}, \qquad a^{\mathfrak{m}} = (a_{x_1})^{m_1}\cdots(a_{x_k})^{m_k}.$$

The general form of a normally ordered monomial is

$$\left(a^+\right)^{\mathfrak{m}_1} a^{\mathfrak{m}_2},$$

with $\mathfrak{m}_1, \mathfrak{m}_2 \in \mathfrak{M}(X)$.

We want to define the 'right vacuum' Φ. It is characterized by the property that $a_x \Phi = 0$ for all $x \in X$. We define the left ideal $\mathfrak{I}_l \subset \mathfrak{W}(X)$ generated by the elements $a_x, x \in X$. A normal ordered monomial is in $\mathfrak{I}_l$ if it is of the form $(a^+)^{\mathfrak{m}} a^{\mathfrak{m}'}$ with $\mathfrak{m}' \neq 0$. These elements form a basis of $\mathfrak{I}_l$. The quotient space $\mathfrak{W}(X)/\mathfrak{I}_l$ has the basis $(a^+)^{\mathfrak{m}} + \mathfrak{I}_l$, where $\mathfrak{m}$ runs through all multisets in $\mathfrak{M}(X)$. Denote the zero element $0 + \mathfrak{I}_l$ of $\mathfrak{W}(X)/\mathfrak{I}_l$ by 0, and call

$$\Phi = 1 + \mathfrak{I}_l,$$

then

$$a_x \Phi = \mathfrak{I}_l = 0.$$

This is a natural algebraic definition of Φ. We have

$$(a^+)^{\mathfrak{m}} + \mathfrak{J}_l = (a^+)^{\mathfrak{m}} \Phi.$$

The quotient space $\mathfrak{W}(X)/\mathfrak{J}_l$ is a $\mathfrak{W}(X)$ left module. The action of $\mathfrak{W}(X)$ on $\mathfrak{W}(X)/\mathfrak{J}_l$ is denoted by T_l.

$$f \in \mathfrak{W}(X) \mapsto T_l(f) : \mathfrak{W}(X)/\mathfrak{J}_l \to \mathfrak{W}(X)/\mathfrak{J}_l,$$
$$T_l(f)(g + \mathfrak{J}_l) = fg + \mathfrak{J}_l.$$

As $T_l(fg) = T_l(f)T_l(g)$, the mapping T_l is a homomorphism.

Use Dirac's notation $(a^+)^{\mathfrak{m}}\Phi = |\mathfrak{m}\rangle$, then $\Phi = |0\rangle$ and

$$a_x^+|\mathfrak{m}\rangle = |\mathfrak{m} + \mathbf{1}_x\rangle, \qquad a_x|\mathfrak{m}\rangle = \sum_{y\in X} \delta_{x,y}|\mathfrak{m} - \mathbf{1}_x\rangle.$$

If $\mathfrak{l} \in \mathfrak{M}(X)$, then

$$a^{\mathfrak{l}}|\mathfrak{m}\rangle = \frac{\mathfrak{m}!}{(\mathfrak{m} - \mathfrak{l})!}|\mathfrak{m} - \mathfrak{l}\rangle,$$

recalling $\mathfrak{m}! = \prod_{x\in X} \mathfrak{m}(x)!$. So $a^{\mathfrak{l}}|\mathfrak{m}\rangle \neq 0$ iff $\mathfrak{m} \geq \mathfrak{l}$. Especially

$$a^{\mathfrak{m}}|\mathfrak{m}\rangle = \mathfrak{m}!\Phi.$$

In an analogous way we define the left vacuum Ψ. Consider the right ideal $\mathfrak{J}_r$ generated by the a_x^+, $x \in X$. The elements of the form $(a^+)^{\mathfrak{m}}\, a^{\mathfrak{m}'}$ with $\mathfrak{m} \neq 0$ form a basis of $\mathfrak{J}_r$. The quotient space $\mathfrak{W}(X)/\mathfrak{J}_r$ has the basis $a^{\mathfrak{m}} + \mathfrak{J}_r$, where $\mathfrak{m}$ runs through all multisets in $\mathfrak{M}(X)$. The quotient space $\mathfrak{W}(X)/\mathfrak{J}_r$ is a $\mathfrak{W}(X)$ right module under the action T_r

$$f \in \mathfrak{W}(X) \mapsto T_r(f) : \mathfrak{W}(X)/\mathfrak{J}_r \to \mathfrak{W}(X)/\mathfrak{J}_r,$$
$$T_r(f)(g + \mathfrak{J}_r) = gf + \mathfrak{J}_r.$$

As $T_r(fg) = T_r(g)T_r(f)$, the mapping T_l is an anti-homorphism. Use the notation $\Psi = 1 + \mathfrak{J}_r$, then

$$a^{\mathfrak{m}} + \mathfrak{J}_r = \Psi a^{\mathfrak{m}}.$$

Again use Dirac's notation $\Psi a^{\mathfrak{m}} = \langle\mathfrak{m}|$ and $\Psi = \langle 0|$. Then

$$\Psi a_x^+ = \langle 0|a_x^+ = 0,$$
$$\langle\mathfrak{m}|a_x = \langle\mathfrak{m} + \mathbf{1}_x|,$$
$$\langle\mathfrak{m}|a_x^+ = \sum_{y\in X} \delta_{x,y}\langle\mathfrak{m} - \mathbf{1}_x|,$$

$$\langle\mathfrak{m}|(a^+)^{\mathfrak{l}} = \frac{\mathfrak{m}!}{(\mathfrak{m} - \mathfrak{l})!}\langle\mathfrak{m} - \mathfrak{l}|.$$

Proposition 1.7.3 *The mapping $f \in \mathfrak{W}(X) \mapsto T_l(f) \in L(\mathfrak{W}(X)/\mathfrak{J}_l)$, the space of linear mappings of $\mathfrak{W}(X)/\mathfrak{J}_l$ into itself, is a faithful homomorphism. Similarly, the mapping $f \in \mathfrak{W}(X) \mapsto T_r(f) \in L(\mathfrak{W}(X)/\mathfrak{J}_r)$ is a faithful anti-homomorphism.*

Proof We have to show, that $T_l(f) = 0 \Rightarrow f = 0$. Assume $f = \sum c(\mathfrak{m}, \mathfrak{m}')(a^+)^{\mathfrak{m}} a^{\mathfrak{m}'} \neq 0$, and choose $m' = \max\{|\mathfrak{m}'| : c(\mathfrak{m}, \mathfrak{m}') \neq 0\}$. This number exists, as the sum is finite. Choose $\mathfrak{m}_0'$ with $|\mathfrak{m}_0'| = m'$ and $c(\mathfrak{m}, \mathfrak{m}_0') \neq 0$ for some $\mathfrak{m}$. If $c(\mathfrak{m}, \mathfrak{m}') \neq 0$, then $|\mathfrak{m}'| \leq m'$ and $a^{\mathfrak{m}'}|\mathfrak{m}_0'\rangle = \mathfrak{m}_0'!\delta_{\mathfrak{m}',\mathfrak{m}_0'}|0\rangle$ and, furthermore,

$$0 = T_l(f)\big|\mathfrak{m}_0'\big\rangle = \sum_{\mathfrak{m}} \mathfrak{m}_0'!c\big(\mathfrak{m}, \mathfrak{m}_0'\big)\big(a^+\big)^{\mathfrak{m}}|0\rangle = \sum_{\mathfrak{m}} \mathfrak{m}_0'!c\big(\mathfrak{m}, \mathfrak{m}_0'\big)|\mathfrak{m}\rangle.$$

As the $|\mathfrak{m}\rangle$ are linear independent, all $c(\mathfrak{m}, \mathfrak{m}_0') = 0$. This is a contradiction. That T_r is faithful can be proven in an analogous way. □

We consider the vector space

$$(\mathfrak{W}/\mathfrak{J}_l)/\mathfrak{J}_r = (\mathfrak{W}/\mathfrak{J}_r)/\mathfrak{J}_l = \mathfrak{W}/(\mathfrak{J}_l + \mathfrak{J}_r).$$

It is one-dimensional and has the basis

$$1 + \mathfrak{J}_l + \mathfrak{J}_r.$$

Denote by $\langle f \rangle$ the coefficient of 1 when f is expressed in the basis of normal ordered monomials. Then

$$f + \mathfrak{J}_l + \mathfrak{J}_r = \langle f \rangle + \mathfrak{J}_l + \mathfrak{J}_r = \Psi f \Phi.$$

We make the identification

$$\langle f \rangle = \Psi f \Phi = \langle 0|f|0\rangle.$$

If

$$M = a_{x_n}^{\vartheta_n} \cdots a_{x_1}^{\vartheta_1}$$

is a monomial, then

$$\langle M \rangle = \langle 0|M|0\rangle = \sum_{\mathfrak{p} \in \mathfrak{P}_2} \lfloor M \rfloor_{\mathfrak{p}}.$$

Here $\mathfrak{P}_2$ is the set of pair partitions of $[1, n]$; if (r, s), $r > s$, is such a pair, then

$$C(r, s) = \begin{cases} 1 & \text{for } x_r = x_s, \vartheta_r = -1, \vartheta_s = +1, \\ 0 & \text{otherwise.} \end{cases}$$

So

$$C(r, s) = \big\langle a_{x_r}^{\vartheta_r} a_{x_s}^{\vartheta_s} \big\rangle.$$

If n is odd, there exists no pair partition, and $\langle M\rangle = 0$. If $\mathfrak{p} = \{(r_1, s_1), \dots (r_{n/2}, s_{n/2})\}$ with $r_i > s_i$ is pair partition, then

$$\lfloor M \rfloor_{\mathfrak{p}} = \prod_i C(r_i, s_i).$$

Using the anti-isomorphism $M \mapsto M^{\mathsf{T}}$ we obtain

$$\langle M\rangle = \langle M^{\mathsf{T}}\rangle.$$

Define the matrix

$$Q\big((x,\vartheta),(x',\vartheta')\big) = \langle a_x^{\vartheta} a_{x'}^{\vartheta'}\rangle$$

for $x, x' \in X$ and $\vartheta, \vartheta' = \pm 1$; then

$$\langle M\rangle = \gamma_Q(M),$$

where $\gamma_Q(M)$ is the Gaussian functional defined in Sect. 1.5.

We may write Wick's theorem in the form

$$M = \sum_{I \subset [1,n]} :\prod_{i \in I} a_{x_i}^{\vartheta_i}: \Big\langle \prod_{i \in [1,n]\setminus I} a_{x_i}^{\vartheta_i} \Big\rangle.$$

Using the anti-isomorphism $M \mapsto M^{\mathsf{T}}$, we obtain

$$\Phi^{\mathsf{T}} = \Psi,$$

$$\big((a^+)^{\mathfrak{m}}\Phi\big)^{\mathsf{T}} = \big(|\mathfrak{m}\rangle\big)^{\mathsf{T}} = \Psi a^{\mathfrak{m}} = \langle \mathfrak{m}|.$$

The states $|\mathfrak{m}\rangle$ are orthogonal in the sense that

$$\Psi a^{\mathfrak{m}} \big(a^+\big)^{\mathfrak{m}'} \Phi = \langle \mathfrak{m}|\mathfrak{m}'\rangle = \mathfrak{m}!\delta_{\mathfrak{m},\mathfrak{m}'}.$$

In physics, one classically uses instead of $|\mathfrak{m}\rangle$ the states

$$\eta(\mathfrak{m}) = \frac{1}{\sqrt{\mathfrak{m}!}} (a^+)^{\mathfrak{m}} \Phi.$$

They are orthonormal in that

$$\langle \eta(\mathfrak{m})|\eta(\mathfrak{m}')\rangle = \delta_{\mathfrak{m},\mathfrak{m}'}.$$

Define the space $\mathscr{K}(\mathfrak{M}(X))$ of all functions, $\mathfrak{m} \in \mathfrak{M}(X) \mapsto f(\mathfrak{m}) \in \mathbb{C}$ which vanish for $|\mathfrak{m}|$ sufficiently large. Extend the form $\langle \mathfrak{m}|\mathfrak{m}'\rangle$ to a sesquilinear form on $\mathscr{K}(\mathfrak{M}(X))$. Consider the elements of the form

$$|f\rangle = \sum_{\mathfrak{m}} \frac{1}{\mathfrak{m}!} f(\mathfrak{m}) |\mathfrak{m}\rangle$$

$$\langle f| = \sum_{\mathfrak{m}} \frac{1}{\mathfrak{m}!}\overline{f}(\mathfrak{m})\langle \mathfrak{m}|;$$

then

$$\langle f|g\rangle = \sum_{\mathfrak{m}} \frac{1}{\mathfrak{m}!}\overline{f}(\mathfrak{m})g(\mathfrak{m}).$$

Recall that $\mathfrak{X}$ is the set of all finite sequences of elements of X

$$\mathfrak{X} = \{\emptyset\} + X + X^2 + \cdots.$$

If $\xi = (x_1, \dots, x_n) \in X^n$, then the multiset $\xi^\bullet = \kappa\xi = \sum_{x=1}^{n} \mathbf{1}_{x_i}$. We set

$$a_\xi = a_{x_1} \cdots a_{x_n} = a_{\kappa\xi}, \qquad a_\xi^+ = a_{x_1}^+ \cdots a_{x_n}^+ = a_{\kappa\xi}^+,$$
$$|\xi\rangle = |\kappa\xi\rangle, \qquad \langle\xi| = \langle\kappa\xi|.$$

We have

$$\begin{aligned}
(a_y^+)|x_1, \dots, x_n\rangle &= |y, x_1, \dots, x_n\rangle \\
a_y|x_1, \dots, x_n\rangle &= \delta_{y,x_1}|x_2, \dots, x_n\rangle + \delta_{y,x_2}|x_1, x_3, \dots, x_n\rangle \\
&\quad + \cdots + \delta_{y,x_n}|x_1, \dots, x_{n-1}\rangle.
\end{aligned}$$

We denote by $\mathscr{K}_s(\mathfrak{X})$ the space of all *symmetric* functions $\mathfrak{X} \to \mathbb{C}$, which vanish on X^n for n sufficiently big. If $f \in \mathscr{K}_s(X)$, then there exists a unique function $\tilde{f} \in \mathscr{K}(\mathfrak{M}(X))$ with $f = \tilde{f} \circ \kappa$. We obtain

$$|f\rangle = |\tilde{f}\rangle = \sum_{n=0}^{\infty} \frac{1}{n!} \sum_{\xi \in X^n} f(\xi)|\xi\rangle,$$

$$\langle f|g\rangle = \langle\tilde{f}|\tilde{g}\rangle = \sum_{n=0}^{\infty} \frac{1}{n!} \sum_{\xi \in X^n} \overline{f}(\xi)g(\xi).$$

Proposition 1.7.4 *For $x \in X$ define the mappings $a_x, a_x^+ : \mathscr{K}_s(\mathfrak{X}) \to \mathscr{K}_s(\mathfrak{X})$ by*

$$\begin{aligned}
(a_x f)(x_1, \dots, x_n) &= f(x, x_1, \dots, x_n) \\
(a_x^+ f)(x_1, \dots, x_n) &= \delta_{x,x_1} f(x_2, \dots, x_n) + \delta_{x,x_2} f(x_1, x_3, \dots, x_n) \\
&\quad + \cdots + \delta_{x,x_n} f(x_1, \dots, x_{n-1}).
\end{aligned}$$

Then

$$a_x|f\rangle = \sum_{n=0}^{\infty} \frac{1}{n!} \sum_{\xi \in X^n} f(\xi)a_x|\xi\rangle = |a_x f\rangle,$$

$$a_x^+|f\rangle = \sum_{n=0}^{\infty} \frac{1}{n!} \sum_{\xi \in X^n} f(\xi) a^+|\xi\rangle = |a_x^+ f\rangle.$$

Proof We have

$$\begin{aligned}
a_x|f\rangle &= \sum_n \frac{1}{n!} \sum_{x_1,\dots,x_n} f(x_1,\dots,x_n) a_x|x_1,\dots,x_n\rangle \\
&= \sum_n \frac{1}{n!} \sum_{x_1,\dots,x_n} f(x_1,\dots,x_n)\big(\delta_{x,x_1}|x_2,\dots,x_n\rangle + \dots + \delta_{x,x_n}|x_1,\dots,x_{n-1}\rangle\big) \\
&= \sum_n \frac{n}{n!} \sum_{x_2,\dots,x_n} f(x,x_2,\dots,x_n)|x_2,\dots,x_n\rangle = |a_x f\rangle.
\end{aligned}$$

For a_x there is a similar calculation. □

We use the notation of Sect. 1.5. If α is a finite set and $x_\alpha \in X^\alpha$, then

$$a_{x_\alpha} = \prod_{c \in \alpha} a_{x_c}; \qquad a_{x_\alpha}^+ = \prod_{c \in \alpha} a_{x_c}^+; \qquad |x_\alpha\rangle = a_{x_\alpha}^+ \Phi.$$

For $c \notin \alpha$ we have

$$a_{x_c}^+|x_\alpha\rangle = |x_{\alpha+c}\rangle,$$

where we have used the shorthand $\alpha + c = \alpha + \{c\}$. We obtain for $x_c \in X$

$$a_{x_c}|x_\alpha\rangle = \sum_{b \in \alpha} \delta_{x_b,x_c} |x_{\alpha \setminus b}\rangle$$

upon writing $\alpha \setminus b$ for $\alpha \setminus \{b\}$. If $\alpha = (\alpha_0, \alpha_1, \alpha_2, \dots)$ is a sequence of sets with $\#\alpha_n = n$, then, recalling $\Delta\alpha = 1/(\#\alpha)!$, we have

$$|f\rangle = \sum_\alpha (\Delta\alpha) f(x_\alpha)|x_\alpha\rangle,$$

$$\langle f|g\rangle = \sum_\alpha (\Delta\alpha) \overline{f}(x_\alpha) g(x_\alpha).$$

One obtains for an additional index c

$$(a_{x_c} f)(x_\alpha) = f(x_{\alpha+c})$$

and for $x_c \in X$

$$(a_{x_c}^+ f)(x_\alpha) = \sum_{b \in \alpha} \delta_{x_c,x_b} f(x_{\alpha \setminus c}).$$

If $g : X \to \mathbb{C}$ is a function, then define

$$a(g) = \sum_{x \in X} \overline{g}(x) a_x; \qquad a^+(g) = \sum_{x \in X} g(x) a_x^+.$$

We obtain for $f \in \mathscr{K}_s(\mathfrak{X})$

$$\big(a(g)f\big)(x_\alpha) = \sum_{x_c \in X} \overline{g}(x_c) f(x_{\alpha+c}),$$

$$\big(a^+(g)f\big)(x_\alpha) = \sum_{c \in \alpha} g(x_c) f(x_{\alpha \setminus c}).$$

One has also for the commutator

$$\big[a(g), a^+(h)\big] = \langle g | h \rangle.$$

Chapter 2
Continuous Sets of Creation and Annihilation Operators

Abstract We define first the operators $a(\varphi)$ and $a^+(\varphi)$ on the usual Fock space. Then we exhibit a generalization of the sum-integral lemma to measures. We introduce creation and annihilation operators on locally compact spaces, and use these notions to define creation and annihilation operators localized at points.

2.1 Creation and Annihilation Operators on Fock Space

There are many ways to generalize function spaces on finite sets to function spaces on infinite sets. The usual way to generalize creation and annihilation operators employs Hilbert and Fock spaces. Assume we have a measurable space X and a measure λ on X. We consider the Hilbert space $L^2(X, \lambda)$ and a sequence of Hilbert spaces, for $n = 1, 2, \ldots,$

$$L(n) = L^2_s\left(X^n, \lambda^{\otimes n}\right)$$

of *symmetric* square-integrable functions on X^n, with $L(0) = \mathbb{C}$. The Fock space for X is defined as

$$\Gamma(X, \lambda) = \bigoplus_{n=0}^{\infty} L(n).$$

It is provided with the scalar product

$$\langle f|g\rangle_\lambda = \overline{f}_0 g_0 + \sum_{n=1}^{\infty} \frac{1}{n!} \int \lambda(dx_1)\cdots\lambda(dx_n)\overline{f}_n(x_1,\ldots,x_n)g_n(x_1,\ldots,x_n)$$

and the norm

$$\|f\|^2_\Gamma = |f_0|^2 + \sum_{n=1}^{\infty} \frac{1}{n!} \int \lambda(dx_1)\cdots\lambda(dx_n)\left|f_n(x_1,\ldots,x_n)\right|^2$$

for $f = f_0 \oplus f_1 \oplus f_2 \oplus \cdots$ with $f_n \in L(n)$, and g accordingly. So f is in Γ, if only and if $\|f\|_\Gamma < \infty$. We define the subspace $\Gamma_{\text{fin}} \subset \Gamma$ of those f such that $f_n = 0$ for n sufficiently large.

W. von Waldenfels, *A Measure Theoretical Approach to Quantum Stochastic Processes*, Lecture Notes in Physics 878, DOI 10.1007/978-3-642-45082-2_2,

Recall the definition of

$$\mathfrak{X} = \{\emptyset\} + X + X^2 + \cdots$$

and provide $\mathfrak{X}$ with the measure

$$\hat{e}(\lambda)(f) = f(\emptyset) + \sum_{n=1}^{\infty} \frac{1}{n!} \int \lambda(\mathrm{d}x_1) \cdots \lambda(\mathrm{d}x_n) f_n(x_1, \ldots, x_n).$$

We can make the identification

$$L^2_s(\mathfrak{X}, \hat{e}(\lambda)) = \Gamma(X, \lambda).$$

As the values of a function at a given point are generally not defined, we cannot define a_x and a_x^+ for a given $x \in X$. But the definitions at the end of Sect. 1.7 can be generalized. Define for $f \in L(n+1)$ and $g \in L(1)$

$$\big(a(g)f\big)(x_1, \ldots x_n) = \int \lambda(\mathrm{d}x_0) \overline{g}(x_0) f(x_0, x_1, \ldots, x_n)$$

and for $f \in L(n-1)$

$$\big(a^+(g)f\big)(x_1, \ldots, x_n) = \sum_{c \in [1,n]} g(x_c) f(x_{[1,n] \setminus \{c\}}).$$

One obtains in the usual way

$$\big\|a(g)\big\|_\Gamma \leq \sqrt{n+1} \|g\|_\Gamma \|f\|_\Gamma,$$
$$\big\|a^+(g)\big\|_\Gamma \leq \sqrt{n} \|g\|_\Gamma \|f\|_\Gamma$$

with, of course,

$$\|g\|_\Gamma^2 = \int \lambda(\mathrm{d}x) \big|g(x)\big|^2.$$

The mappings $a(g)$ and $a^+(g)$ can be extended to operators $\Gamma_{\mathrm{fin}}(X, \lambda) \to \Gamma_{\mathrm{fin}}(X, \lambda)$, and one has

$$\langle f | a(g) h \rangle = \langle a^+(g) f | h \rangle$$

and the commutator

$$\big[a(f), a^+(g)\big] = \int \lambda(\mathrm{d}x) \overline{f}(x) g(x).$$

2.2 The Sum-Integral Lemma for Measures

In this work we will mainly use another way of generalizing the creation and annihilation operators on finite sets. Instead of $L^2(X, \lambda)$ we will deal with the pairs of spaces of measures and spaces of continuous functions on X. Contrary to the situation described in the last section, we can easily define white noise operators. We have at our disposal the powerful tools of classical measure theory, and may use the positivity of the commutation relations.

This paper is related to the theory of kernels, first used in quantum probability by Maassen [31] and Meyer [34]. The theory of kernels, however, is well known in quantum field theory. Quantum stochastic processes form, to some extent, a quantum field theory in one space coordinate and one time coordinate. Our approach is dual to that of Maassen and Meyer. We introduce the field operators directly and work with them.

The sum-integral lemma is the basic tool of our analysis. It has been well known for diffuse measures for a long time, i.e., for measures where the points have measure 0 [33]. Our lemma is much more general; it holds for all measures.

We shall employ Bourbaki's measure theory. It is a theory of measures on locally compact spaces. If S is a locally compact space, denote by $\mathscr{K}(S)$ the space of complex-valued continuous functions on S with compact support, and by $\mathscr{M}(S)$ the space of complex measures on S. A complex measure is a linear functional $\mu : \mathscr{K}(S) \to \mathbb{C}$, such that for any compact $K \subset S$, there exists a constant C_K such that $|\mu(f)| \leq C_K \max_{x \in S} |f(x)|$ for all $f \in \mathscr{K}(S)$ with support in K. As in other measure theories the set of integrable functions can be extended from functions in $\mathscr{K}(S)$ to much more general functions. All the usual theorems, like the theorem of Lebesgue, are valid. We shall use the *vague* convergence of measures, which is the weak convergence over $\mathscr{K}(S)$, i.e. $\mu_\iota \to \mu$ if $\mu_\iota(f) \to \mu(f)$ for all $f \in \mathscr{K}(S)$.

In order to avoid unnecessary complications, we shall only consider locally compact spaces which are countable at infinity, i.e., which are a union of countably many compact subsets. Assume now that X is a locally compact space, provide X^n with the product topology, and the set

$$\mathfrak{X} = \{\emptyset\} + X + X^2 + \cdots$$

with that topology where the X^n are both open and closed, and where the restrictions to X^n coincide with the natural topology of X^n. Then $\mathfrak{X}$ is locally compact as well, any compact set is contained in a finite union of the X^n, and its intersections with the X^n are compact.

In our case, the space X mostly will be $\mathbb{R}$. But we shall encounter $\mathbb{R} \times \mathbb{S}^2$ and generalizations of $\mathbb{R}$.

If μ is a complex measure on $\mathfrak{X}$, we write

$$\mu = \mu_0 + \mu_1 + \mu_2 + \cdots$$

where μ_n is the restriction of μ to X^n. We denote by Ψ the measure given by

$$\Psi(f) = f(\emptyset).$$

Then μ_0 is a multiple of Ψ. If $A = (A(1), \ldots, A(n))$ is a totally ordered set, we use the notation

$$\mu(\mathrm{d}x_A) = \mu_n(\mathrm{d}x_{A(1)}, \ldots, \mathrm{d}x_{A(n)}).$$

A function f on $\mathfrak{X}$ is called symmetric, if $f(w) = f(\sigma w)$ for all permutations of w. If α is a set without prescribed order and f is symmetric, then $f(x_\alpha)$ is well defined. A measure on X^n is *symmetric*, if for all $f \in \mathscr{K}(X^n)$ and all permutations σ of $[1,n]$, one has $\mu(f) = \mu(\sigma f)$ with $(\sigma f)(w) = f(\sigma w)$ for all $w \in X^n$. A measure on $\mathfrak{X}$ is symmetric if all its restrictions to X^n are symmetric. We then use the notation $\mu(\mathrm{d}x_\alpha)$.

Like a function, a measure μ has an absolute value $|\mu|$. A measure μ is bounded, if the measure of the total space with respect to $|\mu|$ is finite.

If $w \in \mathfrak{X}$, $w = (x_1, \ldots, x_n)$, then we set

$$\Delta w = \frac{1}{\#w!} = \frac{1}{n!}.$$

Theorem 2.2.1 (Sum-integral lemma for measures) *Let there be given a measure*

$$\mu(\mathrm{d}w_1, \ldots, \mathrm{d}w_k)$$

on

$$\mathfrak{X}^k = \sum_{n_1,\ldots,n_k} X^{n_1} \times \cdots X^{n_k},$$

symmetric in each of the variables w_i. Then

$$\mu = \sum_{n_1,\ldots,n_k} \mu_{n_1,\ldots,n_k}$$

where $\mu_{n_1,\ldots,n_k}$ is the restriction of μ to $X^{n_1} \times \cdots \times X^{n_k}$. Assume that

$$\Delta w_1 \cdots \Delta w_k \, \mu(\mathrm{d}w_1, \ldots, \mathrm{d}w_k) = \sum \frac{1}{n_1! \cdots n_k!} \mu_{n_1,\ldots,n_k}(\mathrm{d}w_1, \ldots, \mathrm{d}w_k)$$

is a bounded measure on $\mathfrak{X}^k$. Then

$$\int \cdots \int_{\mathfrak{X}^k} \Delta w_1 \cdots \Delta w_k \, \mu(\mathrm{d}w_1, \ldots, \mathrm{d}w_k) = \int_{\mathfrak{X}} \Delta w \, \nu(\mathrm{d}w)$$

where ν is a measure on $\mathfrak{X}$, and $\sum(1/n!)\nu_n$ is a bounded measure, in which ν_n is the restriction of ν to X^n and

$$\nu_n(\mathrm{d}x_1, \ldots, \mathrm{d}x_n) = \sum_{\beta_1 + \cdots + \beta_k = [1,n]} \mu_{\#\beta_1,\ldots,\#\beta_k}(\mathrm{d}x_{\beta_1}, \ldots, \mathrm{d}x_{\beta_k}),$$

where $\beta_1, \ldots, \beta_k$ are disjoint sets.

Proof

$$\int \cdots \int_{\mathfrak{X}^k} \Delta w_1 \cdots \Delta w_k \, \mu(\mathrm{d}w_1, \ldots, \mathrm{d}w_k)$$
$$= \sum_{n_1,\ldots,n_k} \int_{X^{n_1}} \cdots \int_{X^{n_k}} \frac{1}{n_1! \cdots n_k!} \mu_{n_1,\ldots,n_k}(\mathrm{d}x_{\alpha_1}, \ldots, \mathrm{d}x_{\alpha_n})$$

where the α_i are the intervals

$$\alpha_1 = [1, n_1],$$
$$\alpha_2 = [n_1+1, n_1+n_2], \quad \dots, \quad \alpha_k = [n_1+\cdots n_{k-1}+1, n_1+\cdots+n_k].$$

Fix $n_1, \dots, n_k$ and put $n = n_1 + \cdots n_k$. Then for the summand in the above formula we have

$$\int_{X^{n_1}} \cdots \int_{X^{n_k}} \mu_{n_1,\dots,n_k}(\mathrm{d}x_{\alpha_1}, \dots, \mathrm{d}x_{\alpha_k})$$
$$= \frac{1}{n!} \sum_{\sigma} \int_{X^{n_1}} \cdots \int_{X^{n_k}} \mu_{n_1,\dots,n_k}(\mathrm{d}x_{\sigma(\alpha_1)}, \dots, \mathrm{d}x_{\sigma(\alpha_k)})$$

where the sum runs over all permutations of n elements. The subsets $\sigma(\alpha_i) = \beta_i$ have the property

$$\beta_1 + \cdots + \beta_k = [1, n], \quad \#\beta_i = n_i. \tag{$*$}$$

Fix $\beta_1, \dots, \beta_k$ with property ($*$). There are exactly $n_1! \cdots n_k!$ permutations σ such that

$$\sigma(\alpha_i) = \beta_i \quad \text{for } i = 1, \dots, k.$$

Hence the last integral expression equals

$$\frac{n_1! \cdots n_k!}{n!} \sum_{\beta_1,\dots,\beta_k} \int \cdots \int \mu_{n_1,\dots,n_k}(\mathrm{d}x_{\beta_1}, \dots, \mathrm{d}x_{\beta_k}),$$

for the β_i with ($*$). Hence

$$\sum_{n_1,\dots,n_k} \int_{X^{n_1}} \cdots \int_{X^{n_k}} \frac{1}{n_1! \cdots n_k!} \mu_{n_1,\dots,n_k}(\mathrm{d}x_{\alpha_1}, \dots, \mathrm{d}x_{\alpha_n})$$
$$= \sum_{n} \frac{1}{n!} \sum_{\beta_1,\dots,\beta_k} \int \cdots \int \mu_{n_1,\dots,n_k}(\mathrm{d}x_{\beta_1}, \dots, \mathrm{d}x_{\beta_k}).$$

□

Remark 2.2.1 The proof is purely combinatorial. So analogous assertions hold in similar situations.

We want to use the notation of Sect. 1.7. If $\alpha = \{a_1, \dots, a_n\}$ is a set without a prescribed order and μ is a symmetric measure then

$$\mu(\mathrm{d}x_\alpha) = \mu(\mathrm{d}x_{a_1}, \dots, \mathrm{d}x_{a_n})$$

is well defined. We have

$$\int_{X^n} \mu(\mathrm{d}w)\Delta w = \int_{X^\alpha} \mu(\mathrm{d}x_\alpha)\Delta\alpha = \frac{1}{n!} \int_{X^\alpha} \mu(\mathrm{d}x_\alpha).$$

For a sequence $\alpha = (\alpha_0, \alpha_1, \ldots)$, with $\#\alpha_n = n$, of sets without prescribed ordering we define for a symmetric measure μ on $\mathfrak{X}$

$$\int_{\mathfrak{X}} \Delta w \mu(\mathrm{d}w) = \sum_n \frac{1}{n!} \int_{X^{\alpha_n}} \mu(\mathrm{d}x_{\alpha_n})$$

and write it, for short, as

$$\int_{\mathfrak{X}} \Delta w \mu(\mathrm{d}w) = \int_{X^\alpha} \mu(\mathrm{d}x_\alpha) \Delta\alpha = \int_\alpha \mu(\mathrm{d}x_\alpha) \Delta\alpha.$$

With this notation we want to reformulate the sum-integral lemma.

Theorem 2.2.2 (Variant of sum-integral lemma) *Let $\alpha_i = (\alpha_{i,0}, \alpha_{i,1}, \ldots)$ be sequences of finite sets, with $\#\alpha_{i,n} = n$ and $\alpha_{n,i} \cap \alpha_{n',j} = \emptyset$ for $i \neq j$, and $\beta = (\beta_0, \beta_1, \ldots)$, with $\#\beta_n = n$ and the β_j disjoint from the α_i, then define*

$$\mu(\mathrm{d}x_{\alpha_1}, \ldots, \mathrm{d}x_{\alpha_k}) = \mu_{\#\alpha_1, \ldots, \#\alpha_k}(\mathrm{d}x_{\alpha_1}, \ldots, \mathrm{d}x_{\alpha_k}).$$

We have

$$\int_{\alpha_1} \cdots \int_{\alpha_k} \Delta\alpha_1 \cdots \Delta\alpha_k \mu(\mathrm{d}x_{\alpha_1}, \ldots, \mathrm{d}x_{\alpha_k}) = \int_\beta \Delta\beta \nu(\mathrm{d}x_\beta)$$

with

$$\nu(\mathrm{d}x_\beta) = \sum_{\beta_1 + \cdots + \beta_k = \beta} \mu(\mathrm{d}x_{\beta_1}, \ldots, \mathrm{d}x_{\beta_k}),$$

$$\mu(\mathrm{d}x_{\beta_1}, \ldots, \mathrm{d}x_{\beta_k}) = \mu_{\#\beta_1, \ldots, \#\beta_k}(\mathrm{d}x_{\beta_1}, \ldots, \mathrm{d}x_{\beta_k}).$$

Remark 2.2.2 We introduced the notation

$$\int_{\mathfrak{X}} \Delta w \mu(\mathrm{d}w) = \int_\alpha \Delta(\alpha) \mu(\mathrm{d}x_\alpha).$$

Later we will often skip the $\Delta\alpha$ completely and write for the last expression simply

$$\int_\alpha \mu(\mathrm{d}x_\alpha)$$

and skipping the $\mathrm{d}x$ as well only

$$\int_\alpha \mu(\alpha).$$

With this simplified notation the sum-integral lemma reads

$$\int_{\alpha_1} \cdots \int_{\alpha_k} \mu(\mathrm{d}x_{\alpha_1}, \ldots, \mathrm{d}x_{\alpha_k}) = \int_\alpha \sum_{\alpha_1 + \cdots + \alpha_n = \alpha} \mu(\mathrm{d}x_{\alpha_1}, \ldots, \mathrm{d}x_{\alpha_k})$$

or by neglecting the $\mathrm{d}x$

$$\int_{\alpha_1} \cdots \int_{\alpha_k} \mu(\alpha_1, \ldots, \alpha_k) = \int_{\alpha} \sum_{\alpha_1 + \cdots + \alpha_k = \alpha} \mu(\alpha_1, \ldots, \alpha_k).$$

If $X = \mathbb{R}$ and

$$\mathfrak{R} = \{\emptyset\} + \mathbb{R} + \mathbb{R}^2 + \cdots$$

and if λ is the Lebesgue measure

$$\int_{\alpha} \mathrm{e}(\lambda)(\mathrm{d}x_\alpha) f(x_\alpha) \Delta\alpha = \sum_n \int_{x_1 < \cdots < x_n} \mathrm{d}x_1 \cdots \mathrm{d}x_n \, f(x_1, \ldots, x_n).$$

In the theory of Maassen kernels [34] one defines

$$\int \mathrm{d}\omega f(\omega) = \sum_n \int_{x_1 < \cdots < x_n} \mathrm{d}x_1 \cdots \mathrm{d}x_n \, f(x_1, \ldots, x_n),$$

where ω runs through all finite subsets of $\mathbb{R}$. The mapping

$$(\omega_1, \ldots, \omega_n) \mapsto \omega_1 + \cdots + \omega_n$$

is defined where the ω_i are pairwise disjoint, i.e. Lebesgue almost everywhere. The usual sum-integral lemma is

$$\int \cdots \int \mathrm{d}\omega_1 \cdots \mathrm{d}\omega_k \, f(\omega_1, \ldots, \omega_k) = \int \mathrm{d}\omega \sum_{\omega_1 + \cdots + \omega_k = \omega} f(\omega).$$

It can be easily derived from the sum-integral lemma for measures, as multisets with multiple points have Lebesgue measure 0.

2.3 Creation and Annihilation Operators on Locally Compact Spaces

We use the duality between measures and continuous functions of compact support. We define creation and annihilation operators for symmetric functions and measures on $\mathfrak{X}$. Assume given a function $\varphi \in \mathscr{K}(X)$, a function $f \in \mathscr{K}_s(\mathfrak{X})$, the space of symmetric continuous functions on $\mathfrak{X}$ of compact support, a measure $\nu \in \mathscr{M}(X)$, and a measure $\mu \in \mathscr{M}_s(\mathfrak{X})$, the space of symmetric measures on $\mathfrak{X}$. We define

$$\big(a(\nu) f\big)(x_1, \ldots, x_n) = \int \overline{\nu}(\mathrm{d}x_0) f(x_0, x_1, \ldots, x_n)$$

or in another notation, where $\alpha + c = \alpha + \{c\}$ means that the point c is added to the set α, and similarly using $\alpha \setminus c = \alpha \setminus \{c\}$, we can continue with

$$\big(a(\nu)f\big)(x_\alpha) = \int \overline{\nu}(\mathrm{d}x_c) f(x_{\alpha+c}),$$

$$\big(a^+(\varphi)f\big)(x_\alpha) = \sum_{c\in\alpha} \varphi(x_c) f(x_{\alpha\setminus c}),$$

$$\big(a^+(\nu)\mu\big)(\mathrm{d}x_\alpha) = \sum_{c\in\alpha} \nu(\mathrm{d}x_c)\mu(\mathrm{d}x_{\alpha\setminus c}),$$

$$\big(a(\varphi)\mu\big)(\mathrm{d}x_\alpha) = \int \overline{\varphi(x_c)}\mu(\mathrm{d}x_{\alpha+c}).$$

If Φ is the function defined by

$$\Phi(\emptyset) = 1; \qquad \Phi(x_\alpha) = 0 \quad \text{for } \alpha \neq \emptyset$$

then

$$a(\nu)\Phi = 0.$$

Similarly if Ψ is the measure defined by

$$\Psi(f) = \langle\Psi|f\rangle = f(\emptyset),$$

then

$$a(\varphi)\Psi = 0.$$

We have therefore

$$\langle\Psi|\Phi\rangle = 1.$$

We define the mapping

$$\mu \in \mathscr{M}(\mathfrak{X}) \mapsto \mu(\Phi)$$

and use the notation for it

$$\mu(\Phi) = \Phi(\mu) = \langle\Phi|\mu\rangle.$$

One obtains

$$\big\langle\Psi|a(\nu)a^+(\varphi)\Phi\big\rangle = \int_X \overline{\nu}(\mathrm{d}x)\varphi(x) = \langle\nu|\varphi\rangle$$

$$\big\langle\Phi|a(\varphi)a^+(\nu)\Psi\big\rangle = \int_X \nu(\mathrm{d}x)\overline{\varphi(x)} = \langle\varphi|\nu\rangle$$

and the commutation relations

$$\big[a(\nu), a^+(\varphi)\big] = \int \overline{\nu}(\mathrm{d}x)\varphi(x) = \langle\nu|\varphi\rangle,$$

$$\left[a(\varphi), a^+(\nu)\right] = \int \nu(\mathrm{d}x)\overline{\varphi}(x) = \langle\varphi|\nu\rangle.$$

We define

$$\langle\mu|f\rangle = \int_{\mathfrak{X}} \Delta w\, \overline{\mu}(\mathrm{d}w) f(w) = \int_\alpha \Delta\alpha\, \overline{\mu}(\mathrm{d}x_\alpha) f(x_\alpha),$$
$$\langle f|\mu\rangle = \overline{\langle\mu|f\rangle}.$$

Proposition 2.3.1 *We have*

$$\left\langle a^+(\nu)\mu|f\right\rangle = \left\langle\mu|a(\nu)f\right\rangle,$$
$$\left\langle a(\varphi)\mu|f\right\rangle = \left\langle\mu|a(\varphi)^+f\right\rangle$$

or

$$\int \Delta w\, \overline{\left(a^+(\nu)\mu\right)}(\mathrm{d}w) f(w) = \int \Delta w\, \overline{\mu}(\mathrm{d}w)\left(a(\nu)f\right)(w),$$
$$\int \Delta w\, \overline{\left(a(\varphi)\mu\right)}(\mathrm{d}w) f(w) = \int \Delta w\, \overline{\mu}(\mathrm{d}w)\left(a^+(\varphi)f\right)(w).$$

Proof We prove only one of the equations by using the sum-integral lemma

$$\int_\beta \Delta\beta \overline{\left(a^+(\nu)\mu\right)}(\mathrm{d}x_\beta) f(x_\beta) = \int_\beta \Delta\beta \sum_{c\in\beta} \overline{\nu(\mathrm{d}x_c)\mu(\mathrm{d}x_{\beta\setminus c})} f(x_\beta).$$

Introduce the sequence consisting of $\{c\}$ alone, and the sequence $\alpha = (\alpha_0, \alpha_1, \ldots)$, by putting $\alpha_{n-1} = \beta_n \setminus c$. In this way the integral becomes

$$\int_\alpha \int_c \Delta\alpha\, \overline{\nu(\mathrm{d}x_c)\mu(\mathrm{d}x_\alpha)} f(x_{\alpha+c}) = \left\langle\mu|a(\nu)f\right\rangle.$$ □

We define the exponential measures and functions

$$\mathrm{e}(\varphi) = \Phi + \varphi + \varphi^{\otimes 2} + \cdots = e^{a^+(\varphi)}\Phi,$$
$$\mathrm{e}(\nu) = \Psi + \nu + \nu^{\otimes 2} + \cdots = e^{a^+(\nu)}\Psi.$$

So, for $\alpha = \{a_1, \ldots, a_n\}$,

$$\mathrm{e}(\varphi)(x_\alpha) = \varphi(x_{a_1})\cdots\varphi(x_{a_n}),$$
$$\mathrm{e}(\nu)(\mathrm{d}x_\alpha) = \nu(\mathrm{d}x_{a_1})\cdots\nu(\mathrm{d}x_{a_n}).$$

2.4 Introduction of Point Measures

We consider the function

$$\varepsilon : x \in X \mapsto \varepsilon_x \in \mathscr{M}(X), \qquad \int \varepsilon_x(\mathrm{d}y)\varphi(y) = \varphi(x).$$

So ε_x is the *point measure* at the point $x \in X$.

Lemma 2.4.1 *If μ is a measure on X^n, then*

$$\int_{x_1} \varepsilon_{x_1}(\mathrm{d}y)\mu(\mathrm{d}x_1, \mathrm{d}x_2, \ldots, \mathrm{d}x_n) = \mu(\mathrm{d}y, dx_2, \ldots, \mathrm{d}x_n),$$

where the subscript variable x_1 on the integral indicates integration over the range X of that variable.

Proof If $\varphi \in \mathscr{K}(X)$ then

$$\begin{aligned}\int_y \int_{x_1} \varphi(y)\varepsilon_{x_1}(\mathrm{d}y)\mu(\mathrm{d}x_1, dx_2, \ldots, \mathrm{d}x_n) &= \int_{x_1} \varphi(x_1)\mu(\mathrm{d}x_1, \mathrm{d}x_2, \ldots, \mathrm{d}x_n) \\ &= \int_y \varphi(y)\mu(\mathrm{d}y, \mathrm{d}x_2, \ldots, \mathrm{d}x_n).\end{aligned}$$
□

We can easily define the mapping

$$a(x) = a(\varepsilon_x) : \mathscr{K}_s(\mathfrak{X}) \to \mathscr{K}_s(\mathfrak{X}),$$

$$\big(a(x)f\big)(x_1, \ldots, x_n) = \int_{x_0} \varepsilon_x(\mathrm{d}x_0) f(x_0, x_1, \ldots, x_n) = f(x, x_1, \ldots, x_n).$$

If $\mu \in \mathscr{M}_s(\mathfrak{X})$ then

$$a^+(\varepsilon_x)\mu(\mathrm{d}x_\alpha) = \sum_{c \in \alpha} \varepsilon_x(\mathrm{d}x_c)\mu(\mathrm{d}x_{\alpha \setminus c}).$$

If ν is a measure on X, then

$$a(\nu) = \int \overline{\nu}(\mathrm{d}x) a(x).$$

We will mostly use the symbol $a^+(\mathrm{d}x)$ for a mapping from $\mathscr{K}_s(\mathfrak{X})$ into the measures on X, which we will now introduce and explain.

If S is a locally compact space, μ a measure on S, and f a Borel function, we define the product $f\mu$ by the formula

$$\int (f\mu)(ds)\varphi(s) = \int \mu(ds) f(s)\varphi(s)$$

for $\varphi \in \mathscr{K}(S)$, and write

$$(f\mu)(ds) = (\mu f)(ds) = f(s)\mu(ds).$$

Let S and T be locally compact spaces. We consider a function $f : S \to \mathscr{M}(T)$, with target the space of measures on T. It can be considered as a function

$$f : S \times \mathscr{K}(T) \to \mathbb{C}$$

and we write it

$$f = f(s, \mathrm{d}t).$$

We extend the notion of the creation operator to functions $f = f(x, \mathrm{d}y) : X \to \mathscr{M}(X)$, where using x indicates the variable and the $\mathrm{d}y$ reminds us that the value is a measure, and define for $g \in \mathscr{K}_s(\mathfrak{X})$

$$\big(a^+(f)g\big)(x_\alpha, \mathrm{d}y) = \sum_{c\in\alpha} f(x_c, \mathrm{d}y)g(x_{\alpha\setminus c}).$$

We apply this notion to the function $\varepsilon : x \mapsto \varepsilon_x$ and write

$$\big(a^+(\mathrm{d}y)g\big)(x_\alpha) = \big(a^+\big(\varepsilon(\mathrm{d}y)\big)g\big)(x_\alpha) = \sum_{c\in\alpha} \varepsilon_{x_c}(\mathrm{d}y)g(x_{\alpha\setminus c}).$$

We may consider $a^+(\varepsilon)$ as an operator-valued measure and write

$$a^+(\varepsilon) = a^+(\varepsilon)(\mathrm{d}y).$$

If $\varphi \in \mathscr{K}(X)$, i.e., φ has one variable, then

$$a^+(\varphi)f = \int a^+(\mathrm{d}x)\varphi(x).$$

We obtain the commutation relations

$$\begin{aligned}\big[a(\varepsilon_x), a(\varepsilon_y)\big] &= 0,\\ \big[a^+(\varepsilon)(\mathrm{d}x), a^+(\varepsilon)(\mathrm{d}y)\big] &= 0,\\ \big[a(\varepsilon_x), a^+(\varepsilon)(\mathrm{d}y)\big] &= \varepsilon_x(\mathrm{d}y).\end{aligned}$$

We extend this notion to any Borel function $g : y \in X \mapsto g_y \in \mathscr{K}_s(\mathfrak{X})$ and write

$$\big(a^+(\varepsilon)g_y\big)(x_\alpha, \mathrm{d}y) = \sum_{c\in\alpha} \varepsilon_{x_c}(\mathrm{d}y)g_y(x_{\alpha\setminus c}).$$

In this equation the product of the measure $\varepsilon_{x_c}(\mathrm{d}y)$ with the function g_y appears.
A special case arises if $g_y = a(\varepsilon_y)f$.

Proposition 2.4.1

$$\big(a^+(\varepsilon)a(\varepsilon_y)f\big)(x_\alpha, \mathrm{d}y) = \sum_{c\in\alpha} \varepsilon_{x_c}(\mathrm{d}y) f(x_\alpha).$$

Proof

$$\begin{aligned}\big(a^+(\varepsilon)a(\varepsilon_y)f\big)(x_\alpha, \mathrm{d}y) &= \sum_{c\in\alpha} \varepsilon_{x_c}(\mathrm{d}y)\big(a(\varepsilon_y)f\big)(x_{\alpha\backslash c})\\ &= \sum_{c\in\alpha} \varepsilon_{x_c}(\mathrm{d}y) f\big(x_{\alpha\backslash c} + \{y\}\big) = \sum_{c\in\alpha} \varepsilon_{x_c}(\mathrm{d}y) f\big(x_{\alpha\backslash c} + \{x_c\}\big)\\ &= \sum_{c\in\alpha} \varepsilon_{x_c}(\mathrm{d}y) f(x_\alpha)\end{aligned}$$

as

$$\varepsilon(x, \mathrm{d}y)g(y) = \varepsilon(x, \mathrm{d}y)g(x).$$ □

So

$$\mathrm{n}(\mathrm{d}y) = a^+(\varepsilon)(\mathrm{d}y)a(\varepsilon_y)$$

is the operator analogous to the number operator $a_x^+ a_x$ in the case of finitely many x considered in Sect. 1.7.

We single out a positive measure λ on X, and introduce in $\mathscr{K}_s(\mathfrak{X})$ the positive sesquilinear form considered already in Sect. 1.7,

$$\langle f|g\rangle_\lambda = \int_\alpha \Delta\alpha\ \mathrm{e}(\lambda)(\mathrm{d}x_\alpha)\overline{f}(x_\alpha)g(x_\alpha) = \big\langle f\ \mathrm{e}(\lambda)|g\big\rangle = \big\langle f|g\ \mathrm{e}(\lambda)\big\rangle,$$

using the product of a function with the measure

$$\mathrm{e}(\lambda) = \Psi + \lambda + \lambda^{\otimes 2} + \cdots .$$

More generally, if ν is a measure on X, we have

$$\big\langle f|a(\nu)g\big\rangle_\lambda = \big\langle a^+(\nu)\mathrm{e}(\lambda)f|g\big\rangle.$$

We introduced in Sect. 2.1 the operator $a^+(\varphi)$. One obtains now

$$\big\langle a^+(\varphi)f|g\big\rangle_\lambda = \big\langle f|a(\varphi\lambda)g\big\rangle_\lambda.$$

So $a(\varphi\lambda)$ corresponds to the operator $a(\varphi)$ introduced in Sect. 2.1.

If μ is a symmetric measure on $\mathfrak{X}$, one has

$$\big(a(\varepsilon)(\mathrm{d}x_c)\mu\big)(\mathrm{d}x_\alpha) = \mu(\mathrm{d}x_{\alpha+c})$$

as

$$\big(a(\varepsilon)(\mathrm{d}y)\mu\big)(\mathrm{d}x_\alpha) = \int_{x_c} \varepsilon_{x_c}(\mathrm{d}y)\mu(\mathrm{d}x_{\alpha+c}) = \mu(\mathrm{d}x_\alpha, \mathrm{d}y).$$

We can calculate

$$\langle \mu | a^+(\varepsilon(\mathrm{d}y)) f \rangle = \langle a(\varepsilon(\mathrm{d}y))\mu | f \rangle.$$

Proposition 2.4.2 *For $f, g \in \mathscr{K}_s(\mathfrak{X})$*

$$\langle f | a^+(\varepsilon(\mathrm{d}y)) g \rangle_\lambda = \lambda(\mathrm{d}y) \langle a(\varepsilon(y)) f | g \rangle_\lambda,$$

$$\int_y \langle f | \mathfrak{n}(dy) g \rangle_\lambda = \langle f | N g \rangle_\lambda$$

where N is the operator on the space of functions on $\mathfrak{X}$ given by

$$(Nf)(x_1, \ldots, x_n) = n f(x_1, \ldots, x_n).$$

Proof

$$\begin{aligned}\langle f | a^+(\varepsilon(\mathrm{d}x_c)) g \rangle_\lambda &= \langle a(\varepsilon(\mathrm{d}x_c)) f \mathrm{e}(\lambda) | g \rangle = \int (\mathrm{e}(\lambda)\overline{f})(\mathrm{d}x_{\alpha+c}) g(x_\alpha) \Delta(\alpha) \\ &= \lambda(\mathrm{d}x_c) \int \overline{f}(x_{\alpha+c})(\mathrm{e}(\lambda))(\mathrm{d}x_\alpha) g(x_\alpha) \Delta(\alpha) \\ &= \lambda(\mathrm{d}x_c) \langle a(x_c) f | g \rangle_\lambda.\end{aligned}$$

Hence

$$\langle f | a^+(\varepsilon(\mathrm{d}y)) g \rangle_\lambda = \lambda(\mathrm{d}y) \langle a(\varepsilon(y)) f | g \rangle_\lambda.$$

One obtains, from the definition of $\mathfrak{n}$

$$\langle f | \mathfrak{n}(\mathrm{d}y) g \rangle_\lambda = \langle f | a^+(\varepsilon(\mathrm{d}y)) a(\varepsilon_y) g \rangle_\lambda = \lambda(\mathrm{d}y) \langle a(\varepsilon_y) f | a(\varepsilon_y) g \rangle_\lambda$$

and

$$\begin{aligned}&\int_y \lambda(\mathrm{d}y) \langle a(\varepsilon_y) f | a(\varepsilon_y) g \rangle_\lambda \\ &\quad = \sum_{n=0}^{\infty} (1/n!) \int \lambda(\mathrm{d}y) \int \lambda(\mathrm{d}x_1) \cdots \lambda(\mathrm{d}x_n) f(y, x_1, \ldots, x_n) g(y, x_1, \ldots, x_n) \\ &\quad = \langle f | N g \rangle_\lambda.\end{aligned}$$

□

If V is a complex vector space with the scalar product $\langle \cdot | \cdot \rangle$, we may write $|f\rangle$ for $f \in V$, and $\langle f|$ for the semilinear functional $g = |g\rangle \mapsto \langle f | g \rangle$. If $c \in \mathbb{C}$, then $\langle cf| = \overline{c}\langle f|$. Given an operator $A : V \to V$, we define the operator $A^\dagger$ operating on $\langle f|$ to the left by

$$\langle f | A^\dagger = \langle Af |.$$

There might be, or there might not be, an operator A^+ acting on $|g\rangle$ to the right with $A^\dagger = A^+$ or $\langle Af|g\rangle = \langle f|A^\dagger g\rangle = \langle f|A^+ g\rangle$.

We apply this definition to $\mathscr{K}_s(\mathfrak{X})$ provided with the scalar product $\langle\cdot|\cdot\rangle_\lambda$ and, as a corollary of Proposition 2.4.2, we have

$$a^+\big(\varepsilon(\mathrm{d}y)\big) = a^\dagger(\varepsilon_y)\lambda(\mathrm{d}y).$$

We use Bourbaki's terminology in denoting by ε_x the point measure at the point $x \in X$. We compare it to the δ-function on $\mathbb{R}$, as used in physical literature. The δ-function has three different meanings, depending on the differentials with which it is multiplied:

$$\delta(x-y)\mathrm{d}y = \varepsilon_x(\mathrm{d}y),$$
$$\delta(x-y)\mathrm{d}x = \varepsilon_y(\mathrm{d}x),$$
$$\delta(x-y)\mathrm{d}x\,\mathrm{d}y = \Lambda(\mathrm{d}x,\mathrm{d}y),$$

where

$$\int \Lambda(\mathrm{d}x,\mathrm{d}y)f(x,y) = \int \mathrm{d}x f(x,x).$$

Recall

$$\mathfrak{R} = \{\emptyset\} + \mathbb{R} + \mathbb{R}^2 + \cdots,$$

use for λ the Lebesgue measure, treat the δ-function formally as an ordinary function, and put $\delta_x(y) = \delta(x-y)$; then

$$\big(a^+(\delta_x)f\big)(x_\alpha) = \sum_{c\in\alpha} \delta(x-x_c)f(x_{\alpha\setminus c}),$$

$$\big(a(\delta_{x_c})f\big)(x_\alpha) = \int \mathrm{d}x_b\,\delta(x_c-x_b)f(x_{\alpha+b}) = f(x_{\alpha+c}).$$

We have, with this notation, the nice duality relation

$$\big\langle f|a^+(\delta_x)g\big\rangle_\lambda = \big\langle a(\delta_x)f|g\big\rangle_\lambda.$$

For many calculations it is advantageous to work with the δ-function. In doing so there is no difference between a^+ and $a^\dagger$. But the author hopes that the mathematics has become clearer through the use of the ε-measures.

In some calculations we use the terminology of Laurent Schwartz and write

$$\varepsilon_0(\mathrm{d}x) = \delta(x)\mathrm{d}x.$$

Chapter 3
One-Parameter Groups

Abstract In the first section, starting from the resolvent equation we study strongly continuous one-parameter groups, their resolvents and their generators. In the second section, we introduce the spectral Schwartz distribution.

3.1 Resolvent and Generator

We follow, for quite a while, the book of Hille and Phillips [24]. Assume we have a Banach space V. Denote by $L(V)$ the space of all bounded linear operators from V to V provided with the usual operator norm. If $a \in L(V)$ the *resolvent set* of a is

$$\rho(z) = \left\{ z \in \mathbb{C} : (z-a)^{-1} \text{ exists} \right\},$$

where $z - a$ stands for $z1 - a$, as usual. The set $\rho(z)$ is open. The function

$$R(z) : z \in \rho(z) \mapsto (z-a)^{-1}$$

is called the *resolvent* of a. The resolvent satisfies the *resolvent equation*

$$R(z_1) - R(z_2) = (z_2 - z_1) R(z_1) R(z_2).$$

Approaching matters the other way round, assume we have an open set $G \subset \mathbb{C}$ and a function $R(z) : G \to L(V)$ satisfying the resolvent equation. Such a function is called a *pseudoresolvent*; the resolvent equation implies that the $R(z)$, $z \in G$, commute. From

$$\big(1 + (z_2 - z_1) R(z_1)\big) R(z_2) = R(z_1)$$

one concludes, that for $|z_2 - z_1| \, \|R(z_1)\| < 1$ the inverse of $(1 + (z_2 - z_1)R(z_1))$ exists and

$$R(z_2) = \big(1 + (z_2 - z_1) R(z_1)\big)^{-1} R(z_1).$$

Hence $R(z)$ is *holomorphic* in G.

Proposition 3.1.1 *If $R(z) : G \to L(V)$ is a pseudoresolvent, then*

$$D = R(z)V$$

W. von Waldenfels, *A Measure Theoretical Approach to Quantum Stochastic Processes*, Lecture Notes in Physics 878, DOI 10.1007/978-3-642-45082-2_3,

is a subset independent of $z \in G$. *If* $R(z_0)$ *is injective for one* $z_0 \in G$, *then* $R(z)$ *is injective for all* $z \in G$, *and there exists a mapping* $a : D \to V$ *such that*

$$(z-a)R(z)f = f \quad \text{for } f \in V,$$
$$R(z)(z-a)f = f \quad \text{for } f \in D,$$

or

$$R(z) = (z-a)^{-1};$$

furthermore,

$$aR(z) = -1 + zR(z) \quad \text{and} \quad R(z)a = -1 + zR(z).$$

The operator a *is closed. If* $V_0 \subset V$ *is a dense subspace, then* a *is the closure of its restriction to* $R(z)V_0$, *where* z *is an element of the resolvent set.*

Proof If $f \in R(z_0)V$, then there is a $g \in V$ such that

$$f = R(z_0)g = \big(1 + (z - z_0)R(z_0)\big)R(z)g,$$

so $f \in R(z)V$. Assume $R(z_0)$ to be injective and denote by $R(z_0)^{-1} : D \to V$ its inverse. Define $a = z_0 - R(z_0)^{-1}$, then, for $f \in D$,

$$\begin{aligned} R(z)(z-a)f &= R(z)\big(z - z_0 + R(z_0)^{-1}\big)f \\ &= R(z)\big(1 + (z - z_0)R(z_0)\big)R(z_0)^{-1}f \\ &= \big(R(z) + (z - z_0)R(z)R(z_0)\big)R(z_0)^{-1}f = f. \end{aligned}$$

The other equality is proven in the same way.

The graph of a is the subset

$$G = \{(f, af) : f \in D\}.$$

We have to show, that G is closed. Assume we have a sequence (f_n, af_n) converging in $V \times V$ to (f, h). Then we may take g_n so that $f_n = R(z)g_n$, and

$$(z-a)f_n = (z-a)R(z)g_n = g_n \to zf - h = g$$

defines g, for which $f_n = R(z)g_n \to f = R(z)g$. So $f \in D$ and

$$af = aR(z)g = -g + zR(z)g = -g + zf = h.$$

If $(f, af) \in G$ then $f = R(z)g$, and there exists a sequence $g_n \in V_0$, such that $g_n \to g$. Hence $R(z)g_n \to R(z)g$ and

$$aR(z)g_n = -g_n + zR(z)g_n \to -g + zR(z)g = af. \qquad \square$$

Proposition 3.1.2 *Assume we have an operator a defined on a subset D of V, and a function $R(z)$ defined on an open set $G \subset \mathbb{C}$ such that $R(z) : V \to D$, for $z \in G$, and*

$$\begin{aligned} (z-a)R(z)f &= f \quad \text{for } f \in V, \\ R(z)(z-a)f &= f \quad \text{for } f \in D. \end{aligned}$$

Then $R(z)$ fulfills the resolvent equation.

Proof We have

$$\begin{aligned} R(z_1)(z_1 - z_2)R(z_2) &= R(z_1)(z_2 - a)R(z_1) - R(z_1)(z_1 - a)R(z_2) \\ &= R(z_1) - R(z_2). \end{aligned}$$ □

If the assumptions of the last proposition are fulfilled, we call $R(z)$ the *resolvent of the operator a*.

A *strongly continuous one-parameter group* in $L(V)$ is a family $T(t)$, $t \in \mathbb{R}$, of operators in $L(V)$ such that

$$\begin{aligned} &T(0) = 1, \\ &T(s+t) = T(s)T(t) \quad \text{for } s, t \in \mathbb{R}, \end{aligned}$$

and for $f \in V$ the function

$$t \mapsto T(t)f$$

is norm continuous in V. Furthermore, we assume that there exists a constant $r \geq 0$ such that, for $t \in \mathbb{R}$,

$$\|T(t)\| \leq \text{const}\, \mathrm{e}^{r|t|}.$$

From now on, all one-parameter groups $T(t)$ will be assumed to be strongly continuous and to satisfy the bound on growth given just above.

Define, as we now are sure we can,

$$R(z) = \begin{cases} -\mathrm{i}\int_0^\infty \mathrm{e}^{\mathrm{i}zt}T(t)\mathrm{d}t & \text{for } \operatorname{Im} z > r, \\ \mathrm{i}\int_{-\infty}^0 \mathrm{e}^{\mathrm{i}zt}T(t)\mathrm{d}t & \text{for } \operatorname{Im} z < r. \end{cases}$$

Proposition 3.1.3 *Consider a family of operators $T(t)$, $t \in \mathbb{R}$, with $T(0) = 1$ and $t \mapsto T(t)f$ norm continuous for $f \in V$, and $\|T(t)\| \leq \text{const}\, \mathrm{e}^{rt}$; then $T(t)$ is a one-parameter group if and only if $R(z)$ satisfies the resolvent equation for $|\operatorname{Im} z| > r$.*

Proof Assume to begin with that $\operatorname{Im} z_1 > r$ and $\operatorname{Im} z_2 > r$; then

$$-(z_2 - z_1)\int_0^\infty \int_0^\infty \mathrm{e}^{\mathrm{i}z_1 t_1 + \mathrm{i}z_2 t_2}\big(T(t_1 + t_2) - T(t_1)T(t_2)\big)\mathrm{d}t_1\mathrm{d}t_2$$

$$= -(z_2 - z_1)\int_0^\infty \mathrm{d}t \int_0^t \mathrm{d}t_1 \mathrm{e}^{\mathrm{i}z_1 t_1 + \mathrm{i}z_2(t-t_1)} T(t) - (z_2 - z_1) R(z_1) R(z_2)$$

$$= R(z_1) - R(z_2) - (z_2 - z_1) R(z_1) R(z_2).$$

If $\mathrm{Im}\, z_1 > r$ and $\mathrm{Im}\, z_2 < r$, then

$$(z_2 - z_1)\int_0^\infty \int_{-\infty}^0 \mathrm{e}^{\mathrm{i}z_1 t_t + \mathrm{i}z_2 t_2} \big(T(t_1 + t_2) - T(t_1) T(t_2)\big) \mathrm{d}t_1 \mathrm{d}t_2$$

$$= (z_2 - z_1)\int_0^\infty \mathrm{d}t \int_t^\infty \mathrm{d}t_1 \mathrm{e}^{\mathrm{i}z_1 t_1 + \mathrm{i}z_2(t-t_1)} T(t)$$

$$+ (z_2 - z_1)\int_{-\infty}^0 \mathrm{d}t \int_{-\infty}^t \mathrm{d}t_2 \mathrm{e}^{\mathrm{i}z_1(t-t_2) + \mathrm{i}z_2 t_2} T(t) - (z_2 - z_1) R(z_1) R(z_2)$$

$$= R(z_1) - R(z_2) - (z_2 - z_1) R(z_1) R(z_2).$$

The proposition follows from the uniqueness of Laplace transform. □

We call $R(z)$ the *resolvent of the one-parameter group* $T(t),\ t \in \mathbb{R}$.
We have, for $y > r$,

$$\mathrm{i}y\, R(\mathrm{i}y) = y \int_0^\infty \mathrm{d}t\, \mathrm{e}^{-yt} T(t) = \int \mathrm{d}t\, Y(t) \mathrm{e}^{-yt} T(t),$$

with $Y(t) = \mathbf{1}_{t>0}$. Using the convergence for $y \uparrow \infty$

$$yY(t)\mathrm{e}^{-yt} \to \delta(t)$$

one obtains the lemma:

Lemma 3.1.1 *If $R(z)$ is the resolvent of a one-parameter group, then for $y \uparrow \infty$ and $f \in V$*

$$\mathrm{i}y\, R(\mathrm{i}y) f \to f$$

in norm.

From there one obtains

Proposition 3.1.4 *If $R(z)$ is the resolvent of a one-parameter group, then the set $D = R(z)V$ is dense in V, and, furthermore, the mapping $R(z) : V \to D$ is injective.*

Proof That the set $D = R(z)V$ is dense in V follows directly from the preceding lemma. For the second assertion we have to prove

$$R(z) f = 0 \Rightarrow f = 0.$$

But $R(z)f - R(\mathrm{i}y)f = (z - \mathrm{i}y)R(\mathrm{i}y)R(z)f = 0$, so $R(\mathrm{i}y)f = 0$ and

$$f = \lim_{y\uparrow\infty} \mathrm{i}y\, R(\mathrm{i}y)f = 0. \qquad \square$$

The *generator* S of the group $T(t)$ has the domain

$$D_S = \left\{ f \in V : \lim_{t\to 0} \frac{T(t) - 1}{t} f \text{ exists} \right\}$$

and, for $f \in D_S$,

$$Sf = \lim_{t\to 0} \frac{T(t) - 1}{t} f.$$

Proposition 3.1.5 *Define the operator a as in Proposition* 3.1.1, *and $R(z)$ as in Proposition* 3.1.2. *We have $D_S = R(z)V = D$ and*

$$S = (-\mathrm{i})\big(1 - R(z)^{-1}\big) = -\mathrm{i}a.$$

Proof Calculate, for $\operatorname{Im} z > r$,

$$\begin{aligned}(1/s)\big(T(s) - 1\big)R(z) &= 1/(\mathrm{i}s)\left(\int_0^\infty \mathrm{e}^{\mathrm{i}zt}T(t+s)\mathrm{d}t - \int_0^\infty \mathrm{e}^{\mathrm{i}zt}T(t)\mathrm{d}t\right)\\ &= 1/(\mathrm{i}s)\left(\int_s^\infty \big(\mathrm{e}^{-\mathrm{i}zs} - 1\big)\mathrm{e}^{\mathrm{i}zt}T(t)\mathrm{d}t - \int_0^s \mathrm{e}^{\mathrm{i}zt}T(t)\mathrm{d}t\right).\end{aligned}$$

For $f \in V$

$$(1/s)\big(T(s) - 1\big)R(z)f \to -\mathrm{i}zR(z)f + \mathrm{i}f.$$

Hence $R(z)f \in D_S$, and

$$Sf = -\mathrm{i}zR(z)f + \mathrm{i}f = -\mathrm{i}af.$$

So $D \subset D_S$. On the other hand, if $f \in D_S$, then

$$(1/s)\big(T(s) - 1\big)R(z)f = R(z)(1/s)\big(T(s) - 1\big)f \to -\mathrm{i}R(z)f + \mathrm{i}f = R(z)Sf$$

and also $f \in R(z)V = D$ and $D_S \subset D$. $\square$

Assume now that V is Hilbert space with scalar product $(f|g)$. Denote the adjoint of a bounded operator K by K^*.

Proposition 3.1.6 *With the current definition of $R(z)$, the one-parameter group $T(t)$ is unitary if and only if*

$$R(z)^* = R(\overline{z}).$$

In this case is $\|T(t)\| = 1$.

Proof Calculate, for $\operatorname{Im} z > r$,

$$R(z)^* = \mathrm{i}\int_0^\infty \mathrm{e}^{-\mathrm{i}\overline{z}t} T(t)^* \mathrm{d}t,$$

$$R(\overline{z}) = \mathrm{i}\int_{-\infty}^0 \mathrm{e}^{\mathrm{i}\overline{z}t} T(t)\mathrm{d}t = \mathrm{i}\int_0^\infty \mathrm{e}^{-\mathrm{i}\overline{z}t} T(-t)\mathrm{d}t.$$

By the uniqueness of the Laplace transform, we have

$$T(t)^* = T(-t) = T(t)^{-1}$$

for $t > 0$. For $t < 0$ a similar argument holds. □

Definition 3.1.1 If $-ia$ is the generator of a unitary strongly continuous one-parameter group $U(t)$, we call a the *Hamiltonian* of the group and denote it by H.

Proposition 3.1.7 *Assume given a pseudoresolvent* $z \in G \mapsto R(z)$ *with values in a Hilbert space* V; *assume* $z, \overline{z} \in G$, $\operatorname{Im} z \neq 0$, *that* $R(z)^* = R(\overline{z})$ *and* $R(z)$ *is injective, and that* $D = R(z)V$ *is dense in* V. *Then*

$$a = H = z - R(z)^{-1},$$

$$H : D = R(z)V \to V \quad \textit{is selfadjoint},$$

and $(H - \lambda)^{-1}$ *exists for* $\operatorname{Im}\lambda \neq 0$. *The Hamiltonian* $H : D \to V$ *of a unitary group is selfadjoint.*

Proof We show first that H is symmetric, i.e., that

$$(f|Hg) = (Hf|g)$$

for $f, g \in D$, or

$$\big(R(z)h|HR(z)k\big) = \big(HR(z)h|R(z)k\big),$$

for $h, k \in V$. This can be done by a straightforward calculation, as

$$\big(h|R(\overline{z})HR(z)k\big) = \big(h|\big(-1 + \overline{z}R(\overline{z})\big)R(z)k\big) = \big(\big(-1 + zR(z)\big)h|k\big).$$

We still have to prove that the domain of the adjoint is D. The domain D_{H^*} of the adjoint H^*, which is usually unbounded, is the set of all $f \in V$ such that there exists a $g \in V$ with

$$(Hh|f) = (h|g)$$

for all $h \in D$. So

$$\big(HR(z)k|f\big) = \big(R(z)k|g\big)$$

for all $k \in V$, or

$$\big((-1+zR(z))k|f\big) = \big(k|(-1+\overline{z}R(\overline{z}))f\big) = \big(k|R(\overline{z})g\big).$$

Therefore

$$-f+\overline{z}R(\overline{z})f = R(\overline{z})g$$

and $f \in D$, and thus $D_{H^*} \subset D$. The symmetry of H, and that D is dense in V, implies that $D_{H^*} \supset D$.

Define

$$U = 1+(\overline{z}-z)R(z).$$

Then

$$UU^* = U^*U = 1,$$

U is unitary and $\|U\| = 1$. So

$$(U-\zeta)^{-1}$$

exists for $\zeta \in \mathbb{C}$, $|\zeta| \neq 1$, as the corresponding power series converge. We have for $\lambda \neq z$

$$\begin{aligned} U - \frac{\overline{z}-\lambda}{z-\lambda} &= 1+(\overline{z}-z)R(z) - \left(1+\frac{\overline{z}-z}{z-\lambda}\right) \\ &= \frac{\overline{z}-z}{z-\lambda}\big((z-\lambda)R(z)-1\big) = \frac{\overline{z}-z}{z-\lambda}(H-\lambda)R(z) \end{aligned}$$

and

$$\left|\frac{\overline{z}-\lambda}{z-\lambda}\right| \neq 1 \Longleftrightarrow \operatorname{Im}\lambda \neq 0.$$

So $(H-\lambda)R(z)$ is bijective, and since $R(z)$ is bijective, $H-\lambda$ is bijective. □

As a corollary of the two last propositions we have

Proposition 3.1.8 *If $U(t)$, $t \in \mathbb{R}$, is a unitary strongly continuous one-parameter group, then its generator is $S = -\mathrm{i}H$ and the Hamiltonian H is selfadjoint.*

We will have to study, in Chaps. 8 and 9, the following situation. Let there be a unitary group $U(t)$ and a dense subspace $V_0 \subset V$. Assume given a subspace $D_0 \subset V$ and $z, \overline{z}$ in the resolvent set of the Hamiltonian, and furthermore that $R(z)V_0$ and $R(\overline{z})V_0$ are contained in D_0. Let there be a *symmetric* operator H_0: $D_0 \to V$, i.e., for $f, g \in D_0$,

$$(f|H_0g) = (H_0g|f)$$

and assume that, for $\xi \in V_0$,

$$H_0 R(z)\xi = -\xi + zR(z)\xi,$$
$$H_0 R(\overline{z})\xi = -\xi + \overline{z}R(\overline{z})\xi.$$

Proposition 3.1.9 *With the definitions in the previous paragraph, the subspace D_0 is dense in V, $D_0 \subset D$, and*

$$H_0 = H \upharpoonright D_0,$$

and H is the closure of H_0.

Proof We know already by Propositions 3.1.1 and 3.1.4, that $R(z)V_0$, and hence D_0, is dense in V, and also that H is the closure of its restriction to $R(z)V_0$. Consider the matrix elements, for $\xi \in V_0$ and $f \in D_0$,

$$\big(\xi \,|\, R(z)H_0 f\big) = \big(R(\overline{z})\xi \,|\, H_0 f\big).$$

Now $R(\overline{z})\xi$ is in D_0, and using the symmetry of H_0 the last expression equals

$$\big(H_0 R(\overline{z})\xi \,|\, f\big) = \big(-\xi + \overline{z}R(\overline{z})\xi \,|\, f\big) = \big(\xi \,|\, -f + zR(z)f\big).$$

As V_0 is dense in V, we obtain

$$R(z)H_0 f = -f + zR(z)f.$$

So

$$f = zR(z)f - R(z)H_0 f \in R(z)V = D$$

and

$$(z - H)f = zf - H_0 f$$

and

$$Hf = H_0 f. \qquad \square$$

3.2 The Spectral Schwartz Distribution

If $G \subset \mathbb{C}$ is open, and the function $f : G \to \mathbb{C},\ f(z) = f(x + \mathrm{i}y)$ has a continuous derivative, set

$$\partial f = \frac{\mathrm{d}f}{\mathrm{d}z} = \frac{1}{2}\left(\frac{\partial f}{\partial x} - \mathrm{i}\frac{\partial f}{\partial y}\right), \qquad \overline{\partial} f = \frac{\mathrm{d}f}{\mathrm{d}\overline{z}} = \frac{1}{2}\left(\frac{\partial f}{\partial x} + \mathrm{i}\frac{\partial f}{\partial y}\right).$$

The function f is holomorphic if and only if $\overline{\partial} f = 0$. In an analogous way one defines these derivatives for Schwartz distributions [37]. The function $z \mapsto 1/z$ is locally integrable, and one obtains

$$\overline{\partial}(1/z) = \pi\delta(z),$$

where $\delta(z)$ is the δ-function in the complex plane. Assume we are given an open set $G \subset \mathbb{C}$ and a function $f : G \setminus \mathbb{R} \to \mathbb{C}$, which is holomorphic and is the restriction for $z = x + \mathrm{i}y \in G,\, y > 0$ of a continuous function on $z \in G,\, y \geq 0$, and for $z \in G,\, y < 0$ it is the restriction of a continuous function on $z \in G,\, y \leq 0$. This is equivalent to the statement, that the limit

$$\lim_{\varepsilon \downarrow 0} f(x \pm \mathrm{i}\varepsilon) = f(x \pm \mathrm{i}0)$$

exists locally uniformly. Hence $f(x \pm \mathrm{i}0)$ exists and is continuous. We have

$$\overline{\partial} f(x + \mathrm{i}y) = (\mathrm{i}/2)\big(f(x + \mathrm{i}0) - f(x - \mathrm{i}0)\big)\delta(y). \qquad (*)$$

In the following we call a *test function* an infinitely differentiable function with compact support, and the space of these is usually denoted C_c^∞, so we say we have a C_c^∞-function.

In the symmetrical form of the Dirac notation for spaces in duality, one uses two verticals in the notation, so that for instance below we write $(f|R(z)|g)$ where we could have just written as before $(f|R(z)g)$. This emphasizes the duality and clarifies the calculations we make.

Proposition 3.2.1 *Assume given a function $R(z) : G \to L(V)$, defined and obeying the resolvent equation almost everywhere, and a subspace $V_0 \subset V$ such that $z \mapsto (f|R(z)|g)$ is locally integrable for all $f, g \in V_0$; then*

$$z_1, z_2 \mapsto (f|R(z_1)R(z_2)|g)$$

is also locally integrable, and for the Schwartz derivatives one has the formula

$$\overline{\partial}_1\overline{\partial}_2(f|R(z_1)R(z_2)|g) = \pi\delta(z_1 - z_2)\overline{\partial}(f|R(z_1)|g).$$

Proof The resolvent equation has as a consequence that $z_1, z_2 \mapsto (f|R(z_1)R(z_2)|g)$ is locally integrable, as e.g.,

$$z_1, z_2 \mapsto \frac{1}{z_2 - z_1}(f|R(z_1)|g)$$

is locally integrable.

Given two test functions φ_1, φ_2, then

$$\iint \mathrm{d}z_1\mathrm{d}z_2\big(\overline{\partial}_1\overline{\partial}_2(f|R(z_1)R(z_2)|g)\big)\varphi_1(z_1)\varphi_2(z_2)$$

$$= \iint dz_1 dz_2 \frac{1}{z_2 - z_1}(f|\big(R(z_1) - R(z_2)\big)|g)\bar{\partial}_1\varphi_1(z_1)\bar{\partial}_2\varphi_2(z_2),$$

and, looking at the first summand on the right-hand side and integrating by parts over z_2, we have

$$\iint dz_1 dz_2 \frac{1}{z_2 - z_1}(f|R(z_1)|g)\bar{\partial}_1\varphi_1(z_1)\bar{\partial}_2\varphi_2(z_2)$$
$$= -\pi \int dz(f|R(z)|g)\bar{\partial}\varphi_1(z)\varphi_2(z).$$

For the second summand we have a similar calculation with a result differing in overall sign, and we obtain

$$\iint dz_1 dz_2 \big(\bar{\partial}_1\bar{\partial}_2(f|R(z_1)R(z_2)|g)\big)\varphi_1(z_1)\varphi_2(z_2)$$
$$= -\pi \int dz(f|R(z)|g)\bar{\partial}\big(\varphi_1(z)\varphi_2(z)\big)$$
$$= \iint dz_1 dz_2 \pi\delta(z_1 - z_2)\bar{\partial}(f|R(z_1)|g)\varphi_1(z_1)\varphi_2(z_2). \qquad \square$$

Definition 3.2.1 Under the assumptions of the last proposition we call

$$M = (1/\pi)\bar{\partial}R$$

defined scalarly for $f, g \in V_0$ by

$$(f|M(z)|g) = (1/\pi)\bar{\partial}(f|R(z)|g)$$

the *spectral Schwartz distribution* of R.

Corollary 3.2.1 *As corollary of the last proposition we have*

$$\iint dz_1 dz_2 (f|M(z_1)M(z_2)|g)\varphi_1(z_1)\varphi_2(z_2) = \int dz(f|M(z)|g)\varphi_1(z)\varphi_2(z).$$

This can be written

$$M(z_1)M(z_2) = \delta(z_1 - z_2)M(z_1)$$

or

$$M(\varphi_1)M(\varphi_2) = M(\varphi_1\varphi_2).$$

Proposition 3.2.2 *Under the assumptions of the last proposition and under the additional assumption, that $R(z)$ is injective, denote again by a the operator defined by the resolvent. Then we have*

$$aM(z) = zM(z)$$

or more precisely

$$(f|aM(z)|g) = z(f|M(z)|g).$$

Proof We have

$$-\int dz(f|aR(z)|g)\bar{\partial}(\varphi(z) = -\int dz(f|(-1+zR(z)|g)\bar{\partial}(\varphi(z)$$
$$= \int dz z\varphi(z)\bar{\partial}(f|R(z)|g)$$

as

$$\bar{\partial}z = 0.$$ □

Remark 3.2.1 The spectral distribution seems to be an interesting object. Suppose we have a matrix A with the resolvent

$$R(z) = \frac{1}{z-A} = \sum_i \frac{1}{z-\lambda_i} p_i$$

where p_i are the eigenprojectors, so that $p_i p_j = p_i \delta_{ij}$. Then

$$M(z) = (1/\pi)\bar{\partial}R(z) = \sum_i \delta(z-\lambda_i)p_i.$$

We have, for test functions φ_1, φ_2,

$$M(\varphi_1)M(\varphi_2) = \iint dz_1 dz_2 M(z_1)M(z_2)\varphi(z_1)\varphi(z_2) = M(\varphi_1\varphi_2).$$

The last equation also holds if A is nilpotent, e.g., $A^2 = 0$. Then one has to take

$$R(z) = \frac{1}{z} + A\mathscr{P}\frac{1}{z^2},$$

where $\mathscr{P}$ denotes the principal value. Then

$$M(z) = \delta(z) - A\partial\delta(z).$$

We consider again, as in Proposition 3.1.7, a pseudoresolvent $z \in G \mapsto R(z)$ with values in a Hilbert space V, and assume $z, \bar{z} \in G$, $\operatorname{Im} z \neq 0$, $R(z)^* = R(\bar{z})$, and $R(z)$ injective, and that $D = R(z)V$ is dense in V. Then $R(z)$ can be extended to the set of all z with $\operatorname{Im} z \neq 0$.

Proposition 3.2.3 *If* $z \mapsto (f|R(z)|g)$, *for* $f, g \in V_0$, *is locally integrable, then*

$$(f|\mu(x)|g) = \lim_{\varepsilon\downarrow 0} \frac{1}{2\pi i}(f|(R(x-i0) - R(x+i0)|g)$$

exists in the sense of Schwartz distributions. Furthermore

$$(f|\mu(x)|f)$$

is a measure ≥ 0*, and*

$$(f|M(x+\mathrm{i}y)|g) = (f|\mu(x)|g)\delta(y).$$

If $(f|R(x \pm 0)|f)$ *exists locally uniformly and is therefore continuous, then* $(f|\mu(x)|f)$ *has continuous density* ≥ 0 *with respect to Lebesgue measure, and is given by*

$$(f|\mu(x)|f) = \frac{1}{2\pi \mathrm{i}}(f|\big(R(x-\mathrm{i}0) - R(x+\mathrm{i}0)|f\big).$$

Proof We write $z = x + \mathrm{i}y = (x, y)$ and use both notations. We use the abbreviations $(f|R(z)|f) = F(z)$ and $(f|M(z)|f) = G(z)$. The other matrix elements can be obtained by polarization. The distribution

$$G(z) = \frac{1}{\pi}\overline{\partial}F(z)$$

has as support the real line, the function $F(z)$ is holomorphic in $\operatorname{Im} z \neq 0$ and locally integrable. If φ is test function, we define

$$\|\varphi\|_1 = \sup\{|\varphi(z)| + |\partial_x\varphi(z)| + |\partial_y\varphi(z)|\}.$$

As

$$\int G(z)\varphi(z)\,\mathrm{d}z = -\frac{1}{\pi}\int F(z)\overline{\partial}\varphi(z)\,\mathrm{d}z$$

we have that for any compact subset K of $\mathbb{C}$ there exists a constant C_K such that

$$\left|\int G(z)\varphi(z)\mathrm{d}z\right| \leq C_K\|\varphi\|_1.$$

Define a test function ρ on $\mathbb{R}$

$$0 \leq \rho(y) \leq 1,$$

$$\rho(y) = \begin{cases} 1 & \text{for } 0 < |y| < 1/2, \\ 0 & \text{for } |y| > 1. \end{cases}$$

If ψ is test function, then

$$\left\|y^2\psi(z)\rho(y/\varepsilon)\right\|_1 = O(\varepsilon)$$

for $\varepsilon \downarrow 0$. Hence

$$\int \mathrm{d}z\, G(z) y^2\psi(y) = 0, \qquad \text{(i)}$$

because

$$y^2\psi(z)\bigl(1-\rho(y/\varepsilon)\bigr)$$

has its support in $\mathbb{C}\setminus\mathbb{R}$, and we have

$$\left|\int \mathrm{d}z\, G(z)y^2\psi(y)\right| = \left|\int \mathrm{d}z\, G(z)y^2\psi(y)\rho(y/\varepsilon)\right| \le C_K \left\|y^2\psi(z)\rho(y/\varepsilon)\right\|_1 = O(\varepsilon),$$

if ψ has its support in the compact set K.

Choose a test function φ and r so large that the support of φ is in the strip $\{|y|<r\}$. Then

$$\begin{aligned}\int \mathrm{d}z\,\varphi(z)G(z) &= \int \mathrm{d}z\varphi(z)\rho(y/r)G(z)\\ &= \int \mathrm{d}z\bigl(\varphi(x,0)+\partial_y\varphi(x,0)y+y^2\psi(z)\bigr)\rho(y/r)G(z).\end{aligned}$$

Taking into account (i) we obtain

$$G(z) = G_0(x)\delta(y) - G_1(x)\delta'(y) \tag{ii}$$

with, for any test function $\chi(x)$,

$$\begin{aligned}\int \mathrm{d}x\, G_0(x)\chi(x) &= \int \mathrm{d}z\, G(z)\chi(x)\rho(y/r),\\ \int \mathrm{d}x\, G_1(x)\chi(x) &= \int \mathrm{d}z\, G(z)\chi(x)y\rho(y/r),\end{aligned}$$

where these expressions are independent of r, provided that the support of φ is in the strip $\{|y|<r\}$. Equation (ii) is a special case of a theorem due to L. Schwartz [37].

We calculate

$$\overline{\int \mathrm{d}z (f|R(z)|f)\bar\partial\varphi(z)} = \overline{\iint \mathrm{d}x\mathrm{d}y (f|R(x+\mathrm{i}y)|f)(1/2)(\partial_x+\mathrm{i}\partial_y)\varphi(x,y)}.$$

Now

$$\overline{(f|R(x+\mathrm{i}y)|f)} = (f|R(x+\mathrm{i}y)^*|f) = (f|R(x-\mathrm{i}y)|f)$$

and we obtain

$$\begin{aligned}&\iint \mathrm{d}x\mathrm{d}y (f|R(x-\mathrm{i}y)|f)(1/2)(\partial_x-\mathrm{i}\partial_y)\overline{\varphi}(x,y)\\ &= \iint \mathrm{d}x\,\mathrm{d}y (f|R(x+\mathrm{i}y)|f)(1/2)(\partial_x+\mathrm{i}\partial_y)\overline{\varphi}(x,-y)\\ &= \int \mathrm{d}z (f|R(z)|f)\bar\partial\tilde\varphi(z)\end{aligned}$$

with

$$\tilde{\varphi}(z) = \overline{\varphi}(\overline{z}) = \overline{\varphi}(x, -y).$$

We continue with the estimate

$$\begin{aligned}
0 &\le (f|\Big(\int dz_1 R(z_1)\bar{\partial}\varphi(z_1)\Big)^{+}\Big(\int dz_2 R(z_2)\bar{\partial}\varphi(z_2)\Big)|f) \\
&= \iint dz_1 dz_2 (f|R(z_1)R(z_2)|f)\bar{\partial}\tilde{\varphi}(z_1)\bar{\partial}\varphi(z_2) \\
&= -\pi \int dz (f|R(z)|f)\bar{\partial}\big(\tilde{\varphi}(z)\varphi(z)\big).
\end{aligned}$$

Hence

$$\int dz (f|M(z)|f)\tilde{\varphi}(z)\varphi(z) = \int dz G(z)\tilde{\varphi}(z)\varphi(z) \ge 0. \tag{iii}$$

Use Eq. (ii) and obtain

$$0 \le \int dx\, G_0(x)\big|\varphi(x,0)\big|^2 + \int dx\, G_1(x)\big(-\partial_y\overline{\varphi}(x,0)\varphi(x,0) + \overline{\varphi}(x,0)\partial_y\varphi(x,0)\big).$$

As $\partial_y(\varphi(x,0))$ can be chosen arbitrarily, we conclude that $G_1 = 0$, and, again using L. Schwartz [37], that $G_0 = (f|\mu|f)$ is a measure ≥ 0.

We have

$$\begin{aligned}
\int dx\, G_0(x)\chi(x) &= \int dz G(z)\chi(x)\rho(y/r) = -\frac{1}{\pi}\int dz\, F(z)\bar{\partial}\chi(x)\rho(y/r) \\
&= -\frac{1}{\pi}\lim_{\varepsilon\downarrow 0}\int_{|y|>\varepsilon} dz\, F(z)\bar{\partial}\chi(x)\rho(y/r) \\
&= \frac{i}{2\pi}\lim_{\varepsilon\downarrow 0}\int dx\big(F(x+i\varepsilon) - F(x-i\varepsilon)\big)\chi(x).
\end{aligned}$$

This is the equation for μ in the proposition. If $F(x \pm i0)$ exists in the usual sense locally uniformly, then it is continous and we have by Eq. ($*$) at the beginning of the section, that

$$(f|\mu(x)|f) = \frac{1}{2\pi i}(f|\big(R(x-i0) - R(x+i0)|f\big).$$

Hence $(f|\mu(x)|f)$ is a continuous function ≥ 0, identified with the measure whose density it is. □

Proposition 3.2.4 *Assume furthermore that μ is a bounded measure. If $\varphi \in C_c^\infty(\mathbb{C})$ is a test function, then*

$$\psi = \int d\zeta\, \varphi(\zeta)/(z-\zeta)$$

is a C^∞ function vanishing at ∞. We have in the sense of distributions,

$$\int dx(f|\mu(x)|f)/(z-x) = (f|R(z)|f).$$

Proof We have

$$\int dx(f|\mu(x)|f)\psi(x) = \int d\zeta\varphi(\zeta)\int dx\big(f|\mu(x)f\big)/(\zeta - x)$$

and

$$= \int (f|M(\zeta)|f)\psi(\zeta)d\zeta = -\frac{1}{\pi}\int d\zeta(f|R(\zeta)|f)\bar{\partial}_\zeta\psi(\zeta)$$
$$= \int d\zeta(f|R(\zeta)|f)\varphi(\zeta).$$

□

Remark 3.2.2 Compare this result with the formula of the spectral theorem

$$(f|1/(z-H)|f) = (f|R(z)|f) = \int (f|dE_x|f)1/(z-x)$$

for $f \in V$. Then for $f \in V_0$ one concludes

$$(f|dE_x|f) = (f|\mu(x)|f)dx,$$

where $(E_x, x \in \mathbb{R})$ is the spectral family of the self-adjoint operator H.

Example Consider the multiplication operator Ω in $L^2(\mathbb{R})$, given by $(\Omega f)(\omega) = \omega f(\omega)$. The resolvent

$$R_\Omega(z) = (z - \Omega)^{-1}$$

is holomorphic off the real line. The domain of Ω is the space $D = R_\Omega(z)L^2$, the space of all L^2 functions f such that Ωf is square integrable. Here we have defined Ωf for all functions in a natural way. The corresponding strongly continuous one-parameter group is

$$U(t) = e^{-i\Omega t}, \qquad \big(U(t)f\big)(\omega) = e^{-i\omega t}f(\omega).$$

The group is clearly unitary, as is confirmed by the equation $R_\Omega(z)^* = R_\Omega(\overline{z})$.

For $f, g \in C_c^1$

$$(f|R_\Omega(x \pm i0)|g) = \int d\omega\overline{f}(\omega)g(\omega)\mathscr{P}/(x-\omega) \mp i\pi\overline{f}(x)g(x).$$

So

$$(f|\mu(x)|g) = \overline{f}(x)g(x).$$

The generalized eigenfunctions are

$$\delta_x(\omega) = \delta(x - \omega).$$

Using the Dirac formalism of bra and ket vectors we obtain

$$\mu(x) = |\delta_x)(\delta_x|.$$

We have

$$R_\Omega(z) = \int \mathrm{d}x \frac{1}{z - x} \mu(x) = \int \mathrm{d}x \frac{1}{z - x} |\delta_x)(\delta_x|,$$

and

$$\Omega|\delta_x) = x|\delta_x).$$

The eigenvectors δ_x form a *generalized orthonormal basis*, i.e.,

$$(\delta_x|\delta_y) = \delta(x - y), \qquad \int \mathrm{d}x |\delta_x)(\delta_x| = 1.$$

The first equation can be checked directly and follows from Proposition 3.2.1. The second equation says that $\int \mathrm{d}x \, \mu(x) = 1$.

Chapter 4
Four Explicitly Calculable One-Excitation Processes

Abstract We consider in this chapter four examples which can be treated without much apparatus. Three of them are of physical interest. We do not need the full Fock space but only its one-particle and zero-particle subspaces. We calculate the time development explicitly and give the Hamiltonian. We obtain its spectral decomposition with the help of generalized eigenvectors.using a method, which has been applied in the study of radiative transfer by Gariy V. Efimov and the author in J. Spectrosc. Radiat. Transf. 53, 59–74 (1953).

4.1 Krein's Formula

The formula in the theorem below is an important tool in our discussions. I know it as Krein's formula, and we'll call it that. If M is a quadratic matrix, its resolvent

$$R(z) = (z - M)^{-1} = \frac{1}{z - M}$$

is a meromorphic matrix-valued function for $z \in \mathbb{C}$. Its poles are the eigenvalues of M, its residues at the poles are the projectors onto the eigenspaces, and the Laurent expansions at the poles give the principal eigenvectors often called generalized eigenvectors. We shall use the term generalized eigenvectors in another sense below. We allow fractions for non-commutative quantities, if the numerator and denominator commute.

Theorem 4.1.1 (Krein's formula) *Given a matrix of the form*

$$H = \begin{pmatrix} 0 & L \\ G & K \end{pmatrix},$$

where $0, K, G, L$ *are block matrices, then the resolvent can be written*

$$R(z) = \frac{1}{z - H} = \begin{pmatrix} 0 & 0 \\ 0 & R_K \end{pmatrix} + \begin{pmatrix} 1 \\ R_K G \end{pmatrix} C^{-1}(1, LR_K)$$

W. von Waldenfels, *A Measure Theoretical Approach to Quantum Stochastic Processes*, Lecture Notes in Physics 878, DOI 10.1007/978-3-642-45082-2_4,

with

$$R_K = R_K(z) = \frac{1}{z-K}$$

and

$$C = C(z) = z - LR_K(z)G.$$

Proof We have to check, that

$$HR(z) = -1 + zR(z).$$

Now

$$HR = \begin{pmatrix} 0 & LR_K \\ 0 & KR_K \end{pmatrix} + \begin{pmatrix} LR_KG \\ G + KR_KG \end{pmatrix} C^{-1}(1, LR_K).$$

Use

$$KR_K = -1 + zR_K \quad \text{and} \quad LR_KGC^{-1} = -1 + zC^{-1}$$

and a short calculation provides the proof. □

For $\operatorname{Im} z$ sufficiently large, positive or negative,

$$R(z) = \begin{cases} -\mathrm{i}\int_0^\infty \mathrm{d}t\, \mathrm{e}^{-\mathrm{i}Ht+\mathrm{i}zt} & \text{for } \operatorname{Im} z > 0, \\ \mathrm{i}\int_{-\infty}^0 \mathrm{d}t\, \mathrm{e}^{-\mathrm{i}Ht+\mathrm{i}zt} & \text{for } \operatorname{Im} z < 0. \end{cases}$$

Set

$$U(t) = \mathrm{e}^{-\mathrm{i}Ht}, \qquad U_K(t) = \mathrm{e}^{-\mathrm{i}Kt}$$

and write for the Heaviside function

$$Y(t) = \mathbf{1}_{t>0}, \qquad \check{Y}(t) = Y(-t) = \mathbf{1}_{t<0}.$$

Define the Laplace transform for a function of t

$$(\mathscr{L}f)(z) = \int \mathrm{d}t\, \mathrm{e}^{\mathrm{i}zt} f(t).$$

Then

$$R(z) = \begin{cases} -\mathrm{i}(\mathscr{L}UY)(z) & \text{for } \operatorname{Im} z > 0, \\ \mathrm{i}(\mathscr{L}U\check{Y})(z) & \text{for } \operatorname{Im} z < 0. \end{cases}$$

Using the Schwartz distributions δ and δ' we have

$$1 = \mathscr{L}\delta, \qquad z = \mathrm{i}\mathscr{L}\delta'.$$

Define the function $Z(t)$ such that

$$C^{-1}(z) = \begin{cases} -\mathrm{i}(\mathscr{L}ZY)(z) & \text{for } \operatorname{Im} z > 0, \\ \mathrm{i}(\mathscr{L}Z\check{Y})(z) & \text{for } \operatorname{Im} z < 0. \end{cases}$$

The equation

$$(z - LR_K G)C^{-1} = 1$$

becomes, since convolution is mapped into ordinary multiplication by the Laplace transform, a pair of formulas, one for positive t and one for negative t, namely

$$(\delta' + LU_K YG) * ZY = \delta,$$

$$Z' = -\int_0^t dt_1 LU_K(t - t_1)GZ(t_1), \qquad Z(0) = 1, \qquad Z(t) = 0 \quad \text{for } t < 0$$

and

$$(-\delta' + LU_K \check{Y}G) * Z\check{Y} = \delta,$$

$$Z' = \int_t^0 dt_1 LU_K(t - t_1)GZ(t_1), \qquad Z(0) = 1, \qquad Z(t) = 0 \quad \text{for } t > 0.$$

Krein's formula becomes under the Laplace transform, for $t > 0$,

$$UY = \begin{pmatrix} 0 & 0 \\ 0 & U_K Y \end{pmatrix} + \begin{pmatrix} \delta \\ -\mathrm{i}U_K YG \end{pmatrix} * ZY * (\delta, -\mathrm{i}LU_K Y)$$

upon canceling one of the factors of $-\mathrm{i}$ that occurs throughout, and for $t < 0$ similarly becomes

$$U\check{Y} = \begin{pmatrix} 0 & 0 \\ 0 & U_K \check{Y} \end{pmatrix} + \begin{pmatrix} \delta \\ \mathrm{i}U_K \check{Y}G \end{pmatrix} * Z\check{Y} * (\delta, \mathrm{i}LU_K \check{Y}).$$

4.2 A Two-Level Atom Coupled to a Heat Bath of Oscillators

4.2.1 Discussion of the Model

The two-level atom in a heat bath of oscillators is equivalent to the harmonic oscillator in a heat bath of oscillators. The heat bath causes transitions from the upper to the lower level. The oscillator is being damped [40].

In quantum mechanics a harmonic oscillator with frequency ω is described by two operators a and a^+ with the commutation relation $[a, a^+] = 1$. So they generate a Weyl algebra. Their representation has been described in Sect. 1.7. We have a vacuum $|0\rangle$ and the vectors $|n\rangle = (a^+)^n|0\rangle$, $n \geq 1$. They span a pre-Hilbert space with the elements

$$f = \sum_{n=0}^{\infty} (1/n!) f_n |n\rangle$$

and the scalar product given by

$$\langle n|n'\rangle = n!\,\delta_{n,n'}$$

hence

$$\langle f|g\rangle = \sum_{n=0}^{\infty} (1/n!) \overline{f_n} g_n .$$

Our notation differs from the one common in quantum mechanics. The vectors $|n\rangle$ are usually normalized with the factor $(n!)^{-1/2}$. The Hamiltonian is

$$H = \omega a^+ a.$$

So $H|n\rangle = \omega n|n\rangle$ and $\exp(-\mathrm{i}Ht)$ can be defined so that

$$\mathrm{e}^{-\mathrm{i}Ht}|n\rangle = \mathrm{e}^{-\mathrm{i}n\omega t}|n\rangle.$$

One obtains

$$\mathrm{e}^{\mathrm{i}Ht} a \mathrm{e}^{-\mathrm{i}Ht} = \mathrm{e}^{-\mathrm{i}\omega t} a, \qquad \mathrm{e}^{\mathrm{i}Ht} a^+ \mathrm{e}^{-\mathrm{i}Ht} = \mathrm{e}^{\mathrm{i}\omega t} a^+.$$

Consider now a finite system of oscillators, with frequencies ω_λ, given by the creation and annihilation operators $a_\lambda, a_\lambda^+, \lambda \in \Lambda$. The representation space is a pre-Hilbert space spanned by the vectors $|\mathfrak{m}\rangle = (a^+)^{\mathfrak{m}}|0\rangle$, where $\mathfrak{m}$ runs through all multisets of Λ. The Hamiltonian is

$$H_0 = \sum_{\lambda \in \Lambda} \omega_\lambda a_\lambda^+ a_\lambda$$

and

$$\mathrm{e}^{-\mathrm{i}H_0 t}|\mathfrak{m}\rangle = \exp\left(-\mathrm{i}\sum_{\lambda\in\Lambda} m_\lambda \omega_\lambda t\right)|\mathfrak{m}\rangle$$

for $\mathfrak{m} = \sum_{\lambda\in\Lambda} m_\lambda \mathbf{1}_\lambda$. We have

$$\mathrm{e}^{\mathrm{i}H_0 t} a_\lambda \mathrm{e}^{-\mathrm{i}H_0 t} = \mathrm{e}^{-\mathrm{i}\omega_\lambda t} a_\lambda, \qquad \mathrm{e}^{\mathrm{i}H_0 t} a_\lambda^+ \mathrm{e}^{-\mathrm{i}H_0 t} = \mathrm{e}^{\mathrm{i}\omega_\lambda t} a_\lambda^+.$$

A non-degenerate two-level atom is described by a two-dimensional Hilbert space spanned by $|+\rangle$ and $|-\rangle$. The Hamiltonian is given by

$$H_{\text{atom}}|+\rangle = \omega_0|+\rangle, \qquad H_{\text{atom}}|-\rangle = 0$$

or, upon defining $|\pm\rangle\langle\pm| = E_{\pm,\pm}$,

$$H_{\text{atom}} = \omega_0 E_{++}.$$

We are interested in a two-level atom coupled to a system of oscillators. The total Hamiltonian is, with coupling constants g_λ and h_λ,

$$\begin{aligned}H_{\text{tot}} &= H_0 + H_{\text{atom}} + H_{\text{int}}\\ &= \sum_{\lambda\in\Lambda} \omega_\lambda a_\lambda^+ a_\lambda + \omega_0 E_{++}\\ &\quad + \sum_{\lambda\in\Lambda} \left(g_\lambda a_\lambda E_{+-} + \overline{g}_\lambda a_\lambda^+ E_{-+} + h_\lambda a_\lambda E_{-+} + \overline{h}_\lambda a_\lambda^+ E_{+-}\right).\end{aligned}$$

We calculate the interaction Hamiltonian in the so-called interaction representation

$$\begin{aligned}H'_{\text{int}}(t) &= \exp\bigl(\mathrm{i}(H_0 + H_{\text{atom}})t\bigr) H_{\text{int}} \exp\bigl(-\mathrm{i}(H_0 + H_{\text{atom}})t\bigr)\\ &= \sum_{\lambda\in\Lambda} \bigl(g_\lambda a_\lambda E_{+-} \mathrm{e}^{\mathrm{i}(-\omega_\lambda+\omega_0)t} + \overline{g}_\lambda a_\lambda^+ E_{-+} \mathrm{e}^{-\mathrm{i}(-\omega_\lambda+\omega_0)t}\\ &\quad + h_\lambda a_\lambda E_{-+} \mathrm{e}^{\mathrm{i}(-\omega_\lambda-\omega_0)t} + \overline{h}_\lambda a_\lambda^+ E_{+-} \mathrm{e}^{\mathrm{i}(+\omega_\lambda+\omega_0)t}\bigr).\end{aligned}$$

Assume now $|\omega_\lambda - \omega_0| \ll \omega_0$, then the terms including h_λ vary rapidly and can be neglected. This is the so-called *rotating wave approximation*. Define $\omega'_\lambda = \omega_\lambda - \omega_0$, then

$$H_{\text{tot}} = \omega_0 \left(\sum_{\lambda\in\Lambda} a_\lambda^+ a_\lambda + E_{++}\right) + \sum_{\lambda\in\Lambda} \omega'_\lambda a_\lambda^+ a_\lambda + \sum_{\lambda\in\Lambda} \left(g_\lambda a_\lambda E_{+-} + \overline{g}_\lambda a_\lambda^+ E_{-+}\right).$$

The expression $\sum_{\lambda\in\Lambda} a_\lambda^+ a_\lambda + E_{++}$ is the operator corresponding to the number of excitations in the system. It commutes with H_{tot} and gives a background contribution, which is neglected in the dynamics that are being calculated. So we take a simplified total Hamiltonian

$$H_{\text{tot}} = \sum_{\lambda\in\Lambda} \omega'_\lambda a_\lambda^+ a_\lambda + \sum_{\lambda\in\Lambda} \left(g_\lambda a_\lambda E_{+-} + \overline{g}_\lambda a_\lambda^+ E_{-+}\right).$$

The interaction Hamiltonian in the interaction representation now becomes

$$H'_{\text{int}}(t) = \sum_{\lambda\in\Lambda} \left(g_\lambda a_\lambda \mathrm{e}^{-\mathrm{i}\omega'_\lambda t} E_{+-} + \overline{g}_\lambda a_\lambda^+ \mathrm{e}^{\mathrm{i}\omega'_\lambda t} E_{-+}\right).$$

Define

$$F(t) = \sum_{\lambda\in\Lambda} g_\lambda a_\lambda \mathrm{e}^{-\mathrm{i}\omega'_\lambda t};$$

we interpret it as *coloured quantum noise* [23]. It has the commutator

$$\left[F(t), F^{+}(t')\right] = \sum_{\lambda\in\Lambda} |g_\lambda|^2 \mathrm{e}^{-\mathrm{i}\omega'_\lambda(t-t')}.$$

We obtain

$$H'_{\text{int}}(t) = F(t)E_{+-} + F^{+}(t)E_{-+}.$$

Assume, that the number of excitations is 1, then we have only to consider the states

$$|+\rangle \otimes |0\rangle, |-\rangle \otimes |\mathbf{1}_\lambda\rangle, \quad \text{for } \lambda \in \Lambda,$$

and we can represent H_{tot} by the matrix H in the space $\mathbb{C} \oplus \mathbb{C}^{\Lambda}$

$$H = \begin{pmatrix} 0 & \langle g| \\ |g\rangle & \Omega \end{pmatrix},$$

where $|g\rangle$ is the column vector in $\mathbb{C}^{\Lambda}$ with the elements g_λ and $\langle g|$ is the row vector with the entries $\overline{g}_\lambda$; also Ω is the $\Lambda \times \Lambda$-matrix with entries $\omega'_\lambda \delta_{\lambda,\lambda'}$.

If we assume continuous sets of frequencies, and make the rotating wave approximation, we arrive by analogy at

$$\begin{aligned} H_{\text{tot}} &= H_0 + H_{\text{atom}} + H_{\text{int}} \\ &= \omega_0\left(\int a^{+}(\mathrm{d}\omega)a(\omega) + E_{++}\right) \\ &\quad + \int \omega a^{+}(\mathrm{d}\omega)a(\omega) + \int g(\omega)a(\omega)\mathrm{d}\omega E_{+-} + \int \omega\overline{g}(\omega)a^{+}(\mathrm{d}\omega)E_{-+}. \end{aligned}$$

The term

$$\int a^{+}(\mathrm{d}\omega)a(\omega) + E_{++}$$

is the number of excitations. We assume it to be 1 and disregard it. In the interaction representation we obtain

$$\begin{aligned} H'_{\text{int}}(t) &= \int \mathrm{d}\omega\ g(\omega)a(\omega)\mathrm{e}^{-\mathrm{i}\omega t}E_{+-} + \int \overline{g}(\omega)a^{+}(\mathrm{d}\omega)\mathrm{e}^{\mathrm{i}\omega t}E_{-+} \\ &= F(t)E_{+-} + F^{+}(t)E_{-+} \end{aligned}$$

where

$$F(t) = \int \mathrm{d}\omega\ g(\omega)a_\lambda \mathrm{e}^{-\mathrm{i}\omega t}$$

is the quantum coloured noise with the commutator

$$\left[F(t), F^{+}(t')\right] = \langle 0|F(t)F^{+}(t')|0\rangle = \int \mathrm{d}\omega |g(\omega)|^2 \mathrm{e}^{-\mathrm{i}\omega(t-t')}.$$

Under these assumption, we may write H_{tot} in the form of a matrix over $\mathbb{C} \oplus D_\Omega$, where Ω is the multiplication operator acting on functions on $\mathbb{R}$, and D_Ω is its domain, so that

$$H_g = \begin{pmatrix} 0 & \langle g| \\ |g\rangle & \Omega \end{pmatrix} = \Omega E_{--} + E_{+-}\langle g| + E_{-+}|g\rangle$$

with

$$|g\rangle = \big(g(\omega)\big)_{\omega \in \mathbb{R}} \in L^2(\mathbb{R}).$$

We want to change g in such a way, that

$$\left[F(t), F^+(t')\right] = \langle 0|F(t)F^+(t')|0\rangle = 2\pi\delta(t-t');$$

this means that g approaches 1. This is the so-called *singular coupling limit.*

There are other physical situations, which yield the same mathematical problem. Consider a harmonic oscillator with frequency ω_0 in a heat bath of oscillators. Describe the oscillators by the creation and annihilation operators b^+, b. Then Hamiltonian of the *damped oscillator* in the rotating wave approximation is

$$H_{\text{tot}} = H_0 + H_{\text{osc}} + H_{\text{int}} = \sum_{\lambda \in \Lambda} \omega_\lambda a_\lambda^+ a_\lambda + \omega_0 b^+ b + \sum_{\lambda \in \Lambda} \big(g_\lambda a_\lambda b^+ + \overline{g}_\lambda a_\lambda^+ b\big).$$

The number of excitations is

$$\sum_{\lambda \in \Lambda} a_\lambda^+ a_\lambda + b^+ b.$$

It commutes with the Hamiltonian and is set to 1. Then we arrive, as before, at

$$H_{\text{tot}} = \sum_{\lambda \in \Lambda} \omega_\lambda' a_\lambda^+ a_\lambda + \sum_{\lambda \in \Lambda} \big(g_\lambda a_\lambda b^+ + \overline{g}_\lambda a_\lambda^+ b\big).$$

A third possibility is the *Heisenberg equation for the damped oscillator.* If A is an operator,

$$\eta_t(A) = \exp(-\mathrm{i}H_{\text{tot}})A\exp(+\mathrm{i}H_{\text{tot}}).$$

Then

$$\frac{\mathrm{d}}{\mathrm{d}t}\begin{pmatrix} \eta_t(b) \\ \eta_t(a_\lambda) \end{pmatrix} = \sum_{\lambda'} \begin{pmatrix} 0 & \langle g| \\ |g\rangle & \Omega \end{pmatrix}_{\lambda,\lambda'} \begin{pmatrix} \eta_t(b) \\ \eta_t(a_{\lambda'}) \end{pmatrix}.$$

Continue as before.

4.2.2 Singular Coupling Limit

We define the Hilbert space

$$\mathfrak{H} = \mathbb{C} \oplus L^2(\mathbb{R})$$

with the scalar product

$$\big\langle (c, f) \big| (c', g) \big\rangle = \overline{c} c' + \int \mathrm{d}x \overline{f}(x) g(x).$$

As explained in the last subsection, we discuss the operator on $\mathbb{C} \oplus D_\Omega \subset \mathfrak{H}$ given by the matrix

$$H_g = \begin{pmatrix} 0 & \langle g| \\ |g\rangle & \Omega \end{pmatrix},$$

where $g \in L^2$ and Ω is the multiplication operator considered already at the end of Sect. 3.2.

Proposition 4.2.1 *The resolvent $R_g(z)$ of H_g is given by*

$$\frac{1}{z - H_g} = R_g(z) = \begin{pmatrix} 0 & 0 \\ 0 & R_\Omega(z) \end{pmatrix} + \begin{pmatrix} 1 \\ R_\Omega(z)|g\rangle \end{pmatrix} \frac{1}{C_g(z)} \big(1, \langle g| R_\Omega(z)\big)$$

with

$$C_g(z) = z - \langle g| R_\Omega(z) |g\rangle = z - \int \frac{|g(\omega)|^2}{z - \omega} \mathrm{d}\omega.$$

The resolvent is defined for $\operatorname{Im} z \neq 0$ *and we have the equation*

$$R_g(z)^+ = R_g(\overline{z}).$$

Proof One checks immediately that $R(z)$ is defined for $\operatorname{Im} z \neq 0$, that $R(z)^+ = R(\overline{z})$, and that $R(z)$ maps $\mathfrak{H}$ into the domain $\mathbb{C} \oplus D_\Omega$ of H_g. By the same calculations as the ones we were using for Krein's formula in the matrix case, we establish that

$$(z - H_g) R_g(z) = 1,$$
$$R_g(z)(z - H_g) = 1.$$

By Proposition 3.1.1, we see that $R_g(z)$ is the resolvent of H_g. □

We want to replace g by the constant 1. We denote by E the constant function 1 and by the bra-vector $\langle E|$ the linear functional

$$f \in L^1(\mathbb{R}) \mapsto \langle E|f\rangle = \int \mathrm{d}x\, f(x) \in \mathbb{C},$$

and by the ket-vector $|E\rangle$ the semilinear functional

$$f \in L^1(\mathbb{R}) \mapsto \langle f|E\rangle = \int \mathrm{d}x\, \overline{f(x)} = \overline{\langle E|f\rangle} \in \mathbb{C}.$$

We perform the so-called singular coupling limit. We consider a sequence g_n of square-integrable functions, converging to E pointwise, uniformly bounded by some constant function, with the property

$$g_n(\omega) = \overline{g_n(-\omega)}.$$

Then, for fixed z with $\operatorname{Im} z \neq 0$, the resolvents $R_{g_n}(z)$ converge in operator norm to

$$R(z) = \begin{pmatrix} 0 & 0 \\ 0 & R_\Omega(z) \end{pmatrix} + \begin{pmatrix} 1 \\ R_\Omega(z)|E\rangle \end{pmatrix} \frac{1}{C(z)} \big(1, \langle E|R_\Omega(z)\rangle.$$

The function

$$C(z) = z + \mathrm{i}\pi\sigma(z)$$

with

$$\sigma(z) = \begin{cases} 1 & \text{for } \operatorname{Im} z > 0, \\ -1 & \text{for } \operatorname{Im} z < 0 \end{cases}$$

is holomorphic in the upper and lower half-planes and continuous at the boundaries. We extend the operator $R_\Omega(z) = (z - \Omega)^{-1}$ to all functions on the real line. So for $f \in L^2$

$$\langle f|R_\Omega(z)|E\rangle = \overline{\langle E|R_\Omega(z)|f\rangle} = \int \overline{f}(\omega)/(z-\omega)\,\mathrm{d}\omega.$$

The function $R(z)$ is defined for $\operatorname{Im} z \neq 0$, and, as a limit of resolvents in operator norm, the function $R(z)$ fulfills the resolvent equation. Furthermore $R(z)^+ = R(\overline{z})$, which could be seen immediately directly.

We want now to discuss existence and the shape of the Hamiltonian of $R(z)$. We fix a number $z \in \mathbb{C}$, $\operatorname{Im} z \neq 0$. The domain of the Hamiltonian can be directly determined by the resolvent with the help of the formula

$$D = R(z)\mathfrak{H}.$$

Hence

$$D = \left\{ f = \begin{pmatrix} 0 \\ R_\Omega(z)\tilde{f} \end{pmatrix} + c\begin{pmatrix} 1 \\ R_\Omega E \end{pmatrix} : \tilde{f} \in L^2, c \in \mathbb{C} \right\}.$$

One concludes at first, that the obvious guess for H is wrong:

$$H \neq \begin{pmatrix} 0 & \langle E| \\ |E\rangle & \Omega \end{pmatrix}$$

as, e.g.,

$$\langle E|R_\Omega E\rangle = \int \frac{\mathrm{d}\omega}{z-\omega}$$

is not defined.

We propose a more refined construction. Define the space of functions on $\mathbb{R}$

$$\mathfrak{E} = \{f : f = \tilde{f} + cE \text{ with } \tilde{f} \in L^2(\mathbb{R}), c \in \mathbb{C}\}.$$

We then define the subspace $\mathfrak{L} \subset L^2$

$$\mathfrak{L} = R_\Omega(z)\mathfrak{E} = \{f = R_\Omega(z)(cE + \tilde{f}) : \tilde{f} \in L^2\}$$

or more explicitly $f \in \mathfrak{L}$ if and only if, for some $c \in \mathbb{C}$ and $\tilde{f} \in L^2$,

$$f(\omega) = \frac{1}{z - \omega}\big(c + \tilde{f}(\omega)\big).$$

The space $\mathfrak{L}$ is independent of the chosen z as

$$R_\Omega(z_0)(cE + \tilde{f}) = R_\Omega(z)\big(1 + (z - z_0)R_\Omega(z_0)\big)(cE + \tilde{f})$$

and $\tilde{f} + R_\Omega(z_0)(cE + \tilde{f}) \in L^2$.

Denote by $\mathfrak{L}^*$ the algebraic dual of $\mathfrak{L}$, i.e. the set of all linear functionals $\mathfrak{L} \to \mathbb{C}$, and by $\mathfrak{L}^\dagger$ the set of all semilinear functionals $\mathfrak{L} \to \mathbb{C}$. A semilinear functional φ is additive and $\varphi(cf) = \overline{c}\varphi(f)$ for $f \in \mathbb{C}$. By the scalar product $\langle g|f\rangle = \int d\omega\, \overline{g}(\omega) f(\omega)$ we associate to any $f \in L^2$ a semilinear functional φ on $\mathfrak{L}$,

$$\varphi(\xi) = \langle \xi | f\rangle.$$

As $\mathfrak{L}$ is dense in L^2, the functional determines f. So we may imbed L^2 into $\mathfrak{L}^\dagger$ and

$$\mathfrak{L} \subset L^2 \subset \mathfrak{L}^\dagger.$$

Define the functionals $\langle \hat{E}| \in \mathfrak{L}^*$, and also $|\hat{E}\rangle \in \mathfrak{L}^\dagger$, for f of the form given above, by

$$\langle \hat{E}|f\rangle = \lim_{r\to\infty} \int_{-r}^{r} f(\omega)d\omega = -i\pi c\sigma(z) + \int \frac{1}{z-\omega}\tilde{f}(\omega)d\omega$$

and

$$\langle f|\hat{E}\rangle = \overline{\langle \hat{E}|f\rangle}.$$

Define the operator

$$\hat{\Omega} : \mathfrak{L} \to \mathfrak{L}^\dagger,$$

$$\langle g|\hat{\Omega} f\rangle = \lim_{r\to\infty} \int_{-r}^{r} d\omega\, \overline{g}(\omega)\omega f(\omega),$$

$$\hat{\Omega} f = -c|\hat{E}\rangle - \tilde{f} + zf.$$

Compare it to the equation holding pointwise

$$\Omega f = -cE - \tilde{f} + zf.$$

We have, in particular,

$$\langle \hat{E}|R_\Omega(z)|E\rangle = -\mathrm{i}\pi\sigma(z),$$
$$\hat{\Omega}_\Omega R(z)|E\rangle = -|\hat{E}\rangle + zR(z)|E\rangle.$$

Define the operator

$$\hat{H} : \mathbb{C} \oplus \mathfrak{L} \to \mathbb{C} \oplus \mathfrak{L}^\dagger,$$
$$\hat{H} = \begin{pmatrix} 0 & \langle \hat{E}| \\ |\hat{E}\rangle & \hat{\Omega} \end{pmatrix}.$$

Theorem 4.2.1 *The operator $\hat{H}$ maps $\xi \in \mathbb{C} \oplus \mathfrak{L} \to \hat{H}\xi \in \mathbb{C} \oplus L^2 = \mathfrak{H}$ if and only if*

$$\xi \in D$$

i.e.

$$\xi = \begin{pmatrix} c \\ R_\Omega(z)(cE + \tilde{f}) \end{pmatrix}$$

with $c \in \mathbb{C}$, $\tilde{f} \in L^2$. We have

$$\hat{H}R(z)f = -f + zR(z)f.$$

So the Hamiltonian H exists and is the restriction of $\hat{H}$ to D.

Proof Assume

$$\xi = \begin{pmatrix} c' \\ f \end{pmatrix} = \begin{pmatrix} c' \\ R_\Omega(z)(cE + \tilde{f}) \end{pmatrix}.$$

We obtain

$$\hat{H}\begin{pmatrix} c' \\ f \end{pmatrix} = \begin{pmatrix} \langle \hat{E}|f\rangle \\ c'\langle \hat{E}| - c\langle \hat{E}| - \tilde{f} + zf \end{pmatrix}.$$

Hence

$$\hat{H}\xi \in \mathfrak{H} \Leftrightarrow \xi \in D.$$

Using the same calculations as in the matrix case in Sect. 4.1, namely Krein's formula, one obtains

$$\hat{H}R(z)f = -f + zR(z)f.$$

From there one concludes, that $R(z)$ is injective. By Proposition 3.1.1, it gives rise to a Hamiltonian H, which is selfadjoint. It is the restriction of $\hat{H}$ to D. □

Remark 4.2.1 By direct calculation one establishes, that $\hat{H}$ is symmetric, i.e. that

$$\langle f|\hat{H}g\rangle = \overline{\langle g|\hat{H}f\rangle} = \langle \hat{H}f|g\rangle$$

for $f, g \in \mathfrak{L}$.

4.2.3 Time Evolution

The function $R(z)$ is determined by a function $U(t)$ for $t \in \mathbb{R}$ whose values are operators on $\mathfrak{H}$:

$$R(z) = \begin{cases} -\mathrm{i}(\mathscr{L}UY)(z) & \text{for } \operatorname{Im} z > 0, \\ \mathrm{i}(\mathscr{L}U\check{Y})(z) & \text{for } \operatorname{Im} z < 0. \end{cases}$$

Hence, for $t > 0$,

$$UY = \begin{pmatrix} 0 & 0 \\ 0 & U_\Omega Y \end{pmatrix} + \begin{pmatrix} \delta \\ -\mathrm{i}U_\Omega Y|E\rangle \end{pmatrix} * ZY * \left(\delta, -\mathrm{i}\langle E|U_\Omega Y\right)$$

and, for $t < 0$,

$$U\check{Y} = \begin{pmatrix} 0 & 0 \\ 0 & U_\Omega \check{Y} \end{pmatrix} + \begin{pmatrix} \delta \\ \mathrm{i}U_\Omega \check{Y}|E\rangle \end{pmatrix} * Z\check{Y} * \left(\delta, \mathrm{i}\langle E|U_\Omega \check{Y}\right)$$

with

$$Z = \mathrm{e}^{-\pi|t|}.$$

Writing the convolutions in an explicit way we have, for $t > 0$,

$$U(t) = \begin{pmatrix} U_{00} & U_{01} \\ U_{10} & U_{11} \end{pmatrix}$$

with

$$U_{00} = \mathrm{e}^{-\pi t},$$

$$U_{01} = -\mathrm{i}\int_0^t \mathrm{d}t_1 \mathrm{e}^{-\pi(t-t_1)}\langle E|\mathrm{e}^{-\mathrm{i}\Omega t_1},$$

$$U_{10} = -\mathrm{i}\int_0^t \mathrm{d}t_1 \mathrm{e}^{-\mathrm{i}\Omega(t-t_1)}|E\rangle \mathrm{e}^{-\pi t_1},$$

$$U_{11} = \mathrm{e}^{-\mathrm{i}\Omega t} - \iint_{0<t_1<t_2<t} \mathrm{d}t_1\mathrm{d}t_2 \mathrm{e}^{-\mathrm{i}\Omega(t-t_2)}|E\rangle \mathrm{e}^{-\pi(t_2-t_1)}\langle E|\mathrm{e}^{-\mathrm{i}\Omega t_1}.$$

Lemma 4.2.1 *The operator $U(t)$ depends continuously on t, and $\|U(t)\| = O(\sqrt{t})$ for $t \to \infty$.*

Proof We prove the lemma for $t > 0$. The proof for $t < 0$ is similar. It suffices to show the assertion for the U_{ik}. It is clear for U_{00}.

$$U_{10}(t)(\omega) = -\mathrm{i}\int_0^t \mathrm{d}t_1 \mathrm{e}^{-\mathrm{i}\omega(t-t_1)}\mathrm{e}^{-\pi t_1} = -\mathrm{i}\frac{1}{\mathrm{i}\omega - \pi}\left(\mathrm{e}^{-\pi t} - \mathrm{e}^{-\mathrm{i}\omega t}\right).$$

Then

$$\left\|U_{10}(t)\right\|^2 = \int \mathrm{d}\omega \left|U_{10}(t)(\omega)\right|^2 \le \int \mathrm{d}\omega \frac{4}{\pi^2 + \omega^2}$$

is bounded. The function $U_{10}(t)(\omega)$ is continuous in L^2-norm by the theorem of Lebesgue, as it is a continuous function bounded by a fixed L^2-function. We have

$$\langle U_{01}(t)|f\rangle = \int \mathrm{d}\omega\, U_{10}(t)(\omega) f(\omega)$$

and one obtains the desired result from that for U_{10}. The continuity and norm bound are trivial for $\mathrm{e}^{-\mathrm{i}\Omega t}$. For the second term of U_{11} we have to consider

$$\begin{aligned} F(t)(\omega) &= -\iint_{0<t_1<t_2<t} \mathrm{d}t_1 \mathrm{d}t_2 \mathrm{e}^{-\mathrm{i}\omega(t-t_2)}\mathrm{e}^{-\pi(t_2-t_1)} \int \mathrm{d}\omega_1 \mathrm{e}^{-\mathrm{i}\omega_1 t_1} f(\omega_1) \\ &= -\mathrm{i}\int_0^t \mathrm{d}t_1 U_{10}(t-t_1)(\omega)\tilde{f}(t_1) \end{aligned}$$

with

$$\tilde{f}(t_1) = \int \mathrm{d}\omega_1 \mathrm{e}^{-\mathrm{i}\omega_1 t_1} f(\omega_1).$$

We calculate

$$\begin{aligned} \left\|F(t)\right\|^2 &= \int \mathrm{d}\omega \left|F(t)(\omega)\right|^2 \le \int \mathrm{d}\omega \int_0^t \mathrm{d}t_1 \left|U_{10}(t-t_1)(\omega)\right|^2 \int_0^t \mathrm{d}t_1 \left|\tilde{f}(t_1)\right|^2 \\ &\le \int \mathrm{d}\omega \frac{4t}{\pi^2+\omega^2} \int_{-\infty}^{\infty} \mathrm{d}t_1 \left|\tilde{f}(t_1)\right|^2 = \int \mathrm{d}\omega \frac{8\pi t}{\pi^2+\omega^2} \|f\|^2. \end{aligned}$$ □

With the results of Sect. 3.1 we obtain the theorem

Theorem 4.2.2 *The $U(t)$ form a one-parameter unitary strongly continuous group generated by $-\mathrm{i}H$, where $H : D \to \mathfrak{H}$ is defined by Theorem* 4.2.1.

Physical Interpretation The term $U_{00}(t)$ is the probability amplitude that the atom started at $t = 0$ in the upper state and stayed there until the time t. So the probability that the atom is at time t in the upper state is $\mathrm{e}^{-2\pi t}$. Then $U_{10}(t)(\omega)$ gives the probability amplitude that the atom is at time $t = 0$ in the upper state, and jumps at time t to the lower state, emitting a photon of frequency ω. The asymptotic frequency distribution for $t \to \infty$ is

$$\lim_{t\to\infty} \left|U_{10}(t)(\omega)\right|^2 = \frac{1}{\pi^2+\omega^2},$$

the well-known Lorentz or Cauchy distribution. $U_{01}(t)(\omega)$ is the probability amplitude that at time 0 the atom is in the lower state, and a photon of frequency ω is absorbed between the times 0 and t. The matrix element

$$(\omega'|U_{11}(t)|\omega) = \mathrm{e}^{-\mathrm{i}\omega t}\delta(\omega'-\omega) - \iint_{0<t_1<t_2<t} \mathrm{d}t_1\mathrm{d}t_2 \mathrm{e}^{-\mathrm{i}\omega'(t-t_2)}\mathrm{e}^{-\pi(t_2-t_1)}\mathrm{e}^{-\mathrm{i}\omega t_1}$$

corresponds to the case that an incoming photon of frequency ω either passes by unperturbed, or is absorbed and reemitted with frequency ω'.

4.2.4 Replacing Frequencies by Formal Times

By the use of the Fourier transform we replace frequencies labelled ω by formal times labelled τ. This is used in quantum stochastic differential equations and makes them similar to classical stochastic differential equations. In addition, it gives some insight into the physical situation. Introduce

$$\psi_\omega(\omega') = \delta(\omega-\omega'), \qquad \varphi_\tau(\omega) = (2\pi)^{-1/2}\mathrm{e}^{\mathrm{i}\omega\tau}.$$

Then

$$\langle\psi_\omega|\varphi_\tau\rangle = (2\pi)^{-1/2}\mathrm{e}^{\mathrm{i}\omega\tau}, \qquad \langle\varphi_\tau|\psi_\omega\rangle = (2\pi)^{-1/2}\mathrm{e}^{-\mathrm{i}\omega\tau}.$$

Define

$$\mathscr{F}f(\tau) = (2\pi)^{-1/2}\int \mathrm{d}\omega\, \mathrm{e}^{-\mathrm{i}\omega\tau} f(\omega) = (\varphi_\tau|f).$$

Calculate

$$\mathscr{F}\left(\mathrm{e}^{-\mathrm{i}\Omega t}f\right) = \mathscr{F}f(t+\tau) = \left(\Theta(t)\mathscr{F}f\right)(\tau),$$

where

$$\left(\Theta(t)g\right)(\tau) = g(t+\tau)$$

is the right shift. So

$$\mathscr{F}\mathrm{e}^{-\mathrm{i}\Omega t} = \Theta(t)\mathscr{F}.$$

One finds

$$\mathscr{F}E(\tau) = (2\pi)^{-1/2}\delta(\tau).$$

Define

$$R_\Theta(z) = \mathscr{F} R_\Omega(z) \mathscr{F}^{-1} = \begin{cases} -\mathrm{i}\int_0^\infty \mathrm{d}t\, \mathrm{e}^{\mathrm{i}zt}\Theta(t), & \text{for } \operatorname{Im} z > 0, \\ \mathrm{i}\int_{-\infty}^0 \mathrm{d}t\, \mathrm{e}^{\mathrm{i}zt}\Theta(t), & \text{for } \operatorname{Im} z < 0. \end{cases}$$

Then $R_\Theta(z)L^2$ is the Sobolev space of those functions on $\mathbb{R}$ which are L^2 and the Schwartz derivatives of which are L^2 as well. For $\operatorname{Im} z > 0$, one has

$$\big(\mathscr{F} R_\Omega(z)E\big)(\tau) = \big(R_\Theta(z)(2\pi)^{-1/2}\delta\big)(\tau) = -\mathrm{i}(2\pi)^{-1/2}\mathbf{1}\{\tau < 0\}\mathrm{e}^{-\mathrm{i}z\tau}.$$

The space

$$\mathscr{F}\mathfrak{L} = \big\{R_\Theta(z)(f + c\delta) : f \in L^2, c \in \mathbb{C}\big\}$$

consists of functions which are L^2, the derivatives of which are L^2 on $\mathbb{R}\setminus\{0\}$, and which have a jump at 0, and where the left and right limits exist. Define

$$\langle\hat{\delta}, f\rangle = (1/2)\big(f(0+) - f(0-)\big),$$
$$\hat{\partial} f = \partial_c f + \big(f(0+) - f(0-)\big)\hat{\delta}$$

where $\partial_c f$ is the restriction of ∂f to $\mathbb{R}\setminus\{0\}$. One obtains

$$\mathscr{F}\hat{H}\mathscr{F}^{-1} = \begin{pmatrix} 0 & \sqrt{2\pi}\,\langle\hat{\delta}| \\ \sqrt{2\pi}\,|\hat{\delta}\rangle & \mathrm{i}\hat{\partial} \end{pmatrix}.$$

Recall

$$U_{00} = \mathrm{e}^{-\pi t},$$
$$U_{01} = -\mathrm{i}\int_0^t \mathrm{d}t_1 \mathrm{e}^{-\pi(t-t_1)}\langle E|\mathrm{e}^{-\mathrm{i}\Omega t_1},$$
$$U_{10} = -\mathrm{i}\int_0^t \mathrm{d}t_1 \mathrm{e}^{-\mathrm{i}\Omega(t-t_1)}|E\rangle\mathrm{e}^{-\pi t_1},$$
$$U_{11} = \mathrm{e}^{-\mathrm{i}\Omega t} - \iint_{0<t_1<t_2<t} \mathrm{d}t_1\mathrm{d}t_2 \mathrm{e}^{-\mathrm{i}\Omega(t-t_2)}|E\rangle\mathrm{e}^{-\pi(t_2-t_1)}\langle E|\mathrm{e}^{-\mathrm{i}\Omega t_1}.$$

Factorize

$$U(t) = \begin{pmatrix} 1 & 0 \\ 0 & U_\Omega(t) \end{pmatrix} V(t).$$

So $V(t)$ is an interaction representation of $U(t)$. We have

$$V_{00}(t) = \mathrm{e}^{-\pi t},$$
$$V_{01}(t) = -\mathrm{i}\int_0^t \mathrm{d}t_1 \mathrm{e}^{-\pi(t-t_1)}\langle E|\mathrm{e}^{-\Omega t_1},$$

$$V_{10}(t) = -\mathrm{i}\int_0^t \mathrm{d}t_1 \mathrm{e}^{\mathrm{i}\Omega t_1}|E\rangle \mathrm{e}^{-\pi t_1},$$

$$V_{11}(t) = 1 - \iint_{0<t_1<t_2<t} \mathrm{d}t_1\mathrm{d}t_2 \mathrm{e}^{\mathrm{i}\Omega t_2}|E\rangle \mathrm{e}^{-\pi(t_2-t_1)}\langle E|\mathrm{e}^{-\mathrm{i}\Omega t_1}.$$

We obtain in τ-representation

$$V_{00}(t) = \mathrm{e}^{-\pi t},$$

$$\big(V_{01}(t)|\tau\big) = -\mathrm{i}(2\pi)^{1/2}\int_0^t \mathrm{d}t_1 \mathrm{e}^{-\pi(t-t_1)}\delta(\tau - t_1),$$

$$\big(\tau|V_{10}(t)\big) = -\mathrm{i}(2\pi)^{1/2}\int_0^t \mathrm{d}t_1 \delta(t_1 - \tau)\mathrm{e}^{-\pi t_1},$$

$$(\tau_2|V_{11}(t)|\tau_1) = \delta(\tau_1 - \tau_2) - 2\pi \iint_{0<t_1<t_2<t} \mathrm{d}t_1\mathrm{d}t_2 \delta(\tau_2 - t_2)\mathrm{e}^{-\pi(t_2-t_1)}\delta(t_1 - \tau_1).$$

So $V_{11}(t)$ corresponds to the case that at time 0 the atom stays in the upper level and no emission occurs. Then $(V_{01}(t)|\tau)$ is the probability amplitude for the case that between 0 and t at time τ a photon, with the label τ, is absorbed, and $(\tau|V_{10}(t))$ is the probability amplitude that at time τ between 0 and t a photon, with label τ, is emitted. Finally $(\tau_2|V_{11}(t)|\tau_1)$ is the probability amplitude that at time τ_1 a photon with label τ_1 is absorbed, and at time $\tau_2 > \tau_1$ a photon with label τ_2 is emitted, all with $0 < \tau_1 < \tau_2 < t$, or that the photon passes undisturbed.

Remark that $V(t)$ is related to the solution of the quantum stochastic differential equation

$$(\mathrm{d}/\mathrm{d}t)U(t) = -\mathrm{i}\sqrt{2\pi}a^\dagger(t)E_{-+}U(t) - \mathrm{i}\sqrt{2\pi}E_{+-}U(t)a(t) - \pi E_{++}U(t).$$

Here, as in Sect. 4.2.1, $E_{\pm\pm}$ are the matrix units of two-dimensional matrices. The differential equation leaves the number of excitations

$$\int a^+(\mathrm{d}\omega)a(\omega) + E_{++}$$

invariant, and $V(t)$ is the restriction of $U(t)$ to the subspace of one excitation. Quantum stochastic differential equations will be discussed below in Chap. 8.

4.2.5 The Eigenvalue Problem

We start with the well-known formula

$$\frac{1}{x \pm \mathrm{i}0} = \frac{\mathscr{P}}{x} \mp \mathrm{i}\pi\delta(x).$$

Here $\mathscr{P}/x$ denotes the principal value. If f is a function differentiable at 0, then

$$\int \mathrm{d}x \frac{\mathscr{P}}{x} f(x) = \lim_{\varepsilon \to 0} \int_{-\varepsilon}^{\varepsilon} \mathrm{d}x \frac{f(x)}{x} = \int \mathrm{d}x \frac{f(x) - f(0)\mathbf{1}\{|x| \leq 1\}}{x}.$$

The equation means that, for $f \in C_c^1$, the space of once continuously differentiable functions with compact support,

$$\lim_{\varepsilon \downarrow 0} \int \mathrm{d}x \frac{f(x)}{x \pm \mathrm{i}\varepsilon} = \int \mathrm{d}x f(x) \left(\frac{\mathscr{P}}{x} \mp \mathrm{i}\pi \delta(x) \right).$$

We continue with the observations

$$\frac{1}{x \pm \mathrm{i}0 - \omega} = \frac{\mathscr{P}}{x - \omega} \mp \mathrm{i}\pi \delta(x - \omega)$$

and

$$R_\Omega(x \pm \mathrm{i}0) = \frac{1}{x \pm \mathrm{i}0 - \Omega} = \frac{\mathscr{P}}{x - \Omega} \mp \mathrm{i}\pi \delta(x - \Omega) = \frac{\mathscr{P}}{x - \Omega} \mp \mathrm{i}\pi |\delta_x\rangle\langle\delta_x|$$

as, for $f, g \in C_c^1$,

$$\langle f | R_\Omega(x \pm \mathrm{i}0) | g \rangle = \int \mathrm{d}\omega \overline{f}(\omega) \frac{\mathscr{P}}{x - \omega} g(\omega) \mp \mathrm{i}\pi \overline{f}(x) g(x).$$

For $f \in C_c^1$, we have the limits

$$\big\langle E | R(x \pm \mathrm{i}0) f \big\rangle = \int \mathrm{d}\omega \frac{\mathscr{P}}{x - \omega} f(\omega) \mp \mathrm{i}\pi f(x),$$

$$\big\langle f | R(x \pm \mathrm{i}0) E \big\rangle = \int \mathrm{d}\omega \frac{\mathscr{P}}{x - \omega} \overline{f}(\omega) \mp \mathrm{i}\pi \overline{f}(x).$$

We define the subspace $\mathfrak{H}_0 \subset \mathfrak{H} = \mathbb{C} \otimes L^2(\mathbb{R})$

$$\mathfrak{H}_0 = \left\{ \begin{pmatrix} c \\ f \end{pmatrix} : c \in \mathbb{C},\, f \in C_c^1(\mathbb{R}) \right\}.$$

Recall the spectral Schwartz distribution and the formulae of Sect. 3.2:

$$\overline{\partial} R(z) = \pi M(z),$$

$$M(x + \mathrm{i}y) = \mu(x)\delta(y),$$

$$\mu(x) = \frac{1}{2\pi \mathrm{i}} \big(R(x - \mathrm{i}0) - R(x + \mathrm{i}0) \big).$$

Proposition 4.2.2 *For $\xi_1, \xi_2 \in \mathfrak{H}_0$ we have*

$$\langle \xi_1 | 1/(2\pi \mathrm{i}) \big(R(x - \mathrm{i}0) - R(x + \mathrm{i}0) | \xi_2 \big\rangle = \langle \xi_1 | \mu(x) | \xi_2 \rangle = \langle \xi_1 | \alpha_x \rangle \langle \alpha_x | \xi_2 \rangle$$

with

$$|\alpha_x\rangle = \frac{1}{\sqrt{x^2+\pi^2}}\begin{pmatrix} 1 \\ x|\delta_x\rangle + \frac{\mathscr{P}}{x-\Omega}|E\rangle \end{pmatrix}.$$

Proof Recall that

$$R(z) = \begin{pmatrix} 0 & 0 \\ 0 & R_\Omega(z) \end{pmatrix} + \begin{pmatrix} 1 \\ R_\Omega(z)|E\rangle \end{pmatrix} \frac{1}{z+\mathrm{i}\pi\sigma(z)}(1, \langle E|R_\Omega(z)).$$

Write

$$\begin{pmatrix} 1 \\ R_\Omega(x\pm \mathrm{i}0)|E\rangle \end{pmatrix} = a \mp \mathrm{i}\pi b$$

with

$$a = \begin{pmatrix} 1 \\ \frac{\mathscr{P}}{x-\Omega}|E\rangle \end{pmatrix}, \qquad b = \begin{pmatrix} 0 \\ |\delta_x\rangle \end{pmatrix}$$

and

$$a^+ = \left(1, \quad \langle E|\frac{\mathscr{P}}{x-\Omega}\right), \qquad b^+ = \left(0, \quad \langle \delta_x|\right).$$

Then

$$\begin{aligned} &\frac{1}{2\pi \mathrm{i}}\big(R(x-\mathrm{i}0) - R(x+\mathrm{i}0)\big) \\ &\quad = bb^+ + \frac{1}{2\pi\mathrm{i}}\left((a+\mathrm{i}\pi b)\frac{1}{x-\mathrm{i}\pi}(a^+ + \mathrm{i}\pi b^+) - (a-\mathrm{i}\pi b)\frac{1}{x+\mathrm{i}\pi}(a^+ - \mathrm{i}\pi b^+)\right) \\ &\quad = \frac{1}{x^2+\pi^2}(a+xb)(a^+ + xb^+) = |\alpha_x\rangle\langle\alpha_x|. \end{aligned}$$

The first term comes directly from the equations for $R_\Omega(x\pm 0)$ given recently above. The rest of the equation requires arithmetic and the definition of $|\alpha_x\rangle$. □

Recall the space $\mathfrak{E}$ of Sect. 4.2.1, and define the subspace

$$\mathfrak{E}_0 = \{cE + f : c \in \mathbb{C}, f \in C^1 \cap L^2\}$$

and the space of distributions

$$\mathfrak{L}'_x = \sum_{\pm} R(x\pm \mathrm{i}0)\mathfrak{E}_0 = \left\{ f = c_1\frac{\mathscr{P}}{x-\Omega}g + c_2\delta(x-\Omega)g : g \in \mathfrak{E}_0 \right\}.$$

We extend the functional $\hat{E}$ to $\mathfrak{L}'_x$ and define

$$\langle \hat{E}|f\rangle = \lim_{r\to\infty}\int_{-r}^{r} \mathrm{d}\omega f(\omega).$$

As

$$\left\langle \hat{E} \middle| \frac{\mathscr{P}}{x-\Omega} E \right\rangle = \lim_{r\to\infty} \int_{-r}^{r} \mathrm{d}\omega \frac{\mathscr{P}}{x-\omega} = 0$$

$$\left\langle \hat{E} \middle| \delta(x-\Omega) E \right\rangle = \lim_{r\to\infty} \int_{-r}^{r} \mathrm{d}\omega\, \delta(x-\omega) = 1$$

we obtain

$$\langle \hat{E} | R_\Omega(x \pm \mathrm{i}0) | E \rangle = \langle E | R_\Omega(x \pm \mathrm{i}0) | \hat{E} \rangle = \mp \mathrm{i}\pi.$$

As $R_\Omega(x \pm \mathrm{i}0)|\hat{E}\rangle = R_\Omega(x \pm \mathrm{i}0)|E\rangle$, we have

$$\langle \hat{E} | R_\Omega(x \pm \mathrm{i}0) | \hat{E} \rangle = \mp \mathrm{i}\pi.$$

Define

$$\mathfrak{L}_0 = R(z)\mathfrak{E}_0.$$

Extend the operator $\hat{\Omega}$ in the same way as in Sect. 4.2.2 and obtain an operator

$$\hat{\Omega} : \mathfrak{L}'_x \to \mathfrak{L}_0^\dagger;$$

$\hat{\Omega}$ acting on semilinear functionals $\mathfrak{L}_0 \to \mathbb{C}$ has the following specific properties:

$$\hat{\Omega}|\delta_x\rangle = x|\delta_x\rangle, \qquad \langle \delta_x | \hat{\Omega} = \langle \delta_x | x,$$

$$\hat{\Omega} \frac{\mathscr{P}}{x-\Omega} |E\rangle = -|\hat{E}\rangle + x \frac{\mathscr{P}}{x-\Omega} |E\rangle, \qquad \langle E | \frac{\mathscr{P}}{x-\Omega} \hat{\Omega} = -\langle \hat{E} | + x \langle E | \frac{\mathscr{P}}{x-\Omega}.$$

Use these equations and obtain

Proposition 4.2.3 $|\alpha_x\rangle$ *is an eigenvector of* $\hat{H}$ *for the eigenvalue* x, *i.e.,*

$$\hat{H}|\alpha_x\rangle = x|\alpha_x\rangle.$$

We cite the definition of a generalized eigenvector due to Gelfand-Vilenkin ([18, p. 105]) "Let A be an operator in a linear topological space Φ. A linear functional F on Φ such that

$$F(A\varphi) = \lambda F(\varphi)$$

for all $\varphi \in \Phi$ is called a *generalized eigenvector* corresponding to λ." We can adapt this definition to our situation.

Proposition 4.2.4 *If* $\xi \in \mathfrak{H}_0$, *then*

$$\left\langle \alpha_x | R(z)\xi \right\rangle = \frac{1}{z-x} \langle \alpha_x | \xi \rangle.$$

So α_x is a generalized eigenvector of $R(z)$ for the eigenvalue $1/(z-x)$ in the sense of Gelfand-Vilenkin.

If ξ in the domain of H is of the form

$$\xi = c\begin{pmatrix} 1 \\ R_\Omega(z)E \end{pmatrix} + \begin{pmatrix} 0 \\ f \end{pmatrix}$$

with $c \in \mathbb{C}$ and $f \in C_c^1$, then

$$\langle \alpha_x | H | \xi \rangle = x \langle H | \xi \rangle.$$

So ξ is a generalized eigenvector of H for the eigenvalue x in the sense of Gelfand-Vilenkin.

Proof The proof is carried out by straightforward calculation using the equation

$$\begin{aligned}\langle E | \frac{\mathscr{P}}{x-\Omega}\frac{1}{z-\Omega}|E\rangle &= \langle \hat{E} | \frac{\mathscr{P}}{x-\Omega}\frac{1}{z-\Omega}|E\rangle \\ &= \langle \hat{E} | \frac{1}{z-x}\left(\frac{\mathscr{P}}{x-\Omega} - \frac{1}{z-\Omega}\right)|E\rangle = \frac{1}{z-x} \mathrm{i}\sigma(z)\pi.\end{aligned}$$

Recollect $\sigma(z)$ is the sign of the imaginary part of z. □

As

$$\frac{\mathscr{P}}{x} = \frac{\mathrm{d}\log|x|}{\mathrm{d}x}$$

in the sense of Schwartz distributions, and since $\log|x|$ is locally integrable, the function

$$x \mapsto \int \mathrm{d}y f(y) \frac{\mathscr{P}}{(x-y)} = \int \mathrm{d}y f'(y) \log|x-y|$$

is continuous for $f \in C_c^1$ and is continuously differentiable for $f \in C_c^2$.

Lemma 4.2.2 *We have the formula*

$$\frac{\mathscr{P}}{x-\omega}\frac{\mathscr{P}}{y-\omega} = \frac{1}{y-x}\left(\frac{\mathscr{P}}{x-\omega} - \frac{\mathscr{P}}{y-\omega}\right) + \pi^2\delta(x-\omega)\delta(y-\omega),$$

which means explicitly, for $f, g, h \in C_c^2$, that

$$\omega \mapsto \int \mathrm{d}x f(x) \frac{\mathscr{P}}{x-\omega}, \qquad \omega \mapsto \int \mathrm{d}y g(y) \frac{\mathscr{P}}{y-\omega}$$

are square integrable, and

$$(x,y) \mapsto \frac{1}{y-x}\left(\int \mathrm{d}\omega\, h(\omega)\frac{\mathscr{P}}{x-\omega} - \int \mathrm{d}\omega\, h(\omega)\frac{\mathscr{P}}{y-\omega}\right)$$

is continuous, and

$$\iiint \mathrm{d}x\,\mathrm{d}y\,\mathrm{d}\omega\, f(x)g(y)h(\omega)\frac{\mathscr{P}}{x-\omega}\frac{\mathscr{P}}{y-\omega}$$
$$= \iint \mathrm{d}x\,\mathrm{d}y\, f(x)g(y)\left(\frac{1}{y-x}\int \mathrm{d}\omega\, h(\omega)\left(\frac{\mathscr{P}}{x-\omega}-\frac{\mathscr{P}}{y-\omega}\right)\right)$$
$$+\pi^2\int \mathrm{d}\omega\, f(\omega)g(\omega)h(\omega).$$

Proof We calculate

$$\iint \mathrm{d}x\,\mathrm{d}y f(x)g(y)\left(\frac{1}{y-x}\int \mathrm{d}\omega\, h(\omega)\left(\frac{\mathscr{P}}{x-\omega}-\frac{\mathscr{P}}{y-\omega}\right)\right)$$
$$= \lim_{\varepsilon\to 0}\iint \mathrm{d}x\,\mathrm{d}y\, f(x)g(y)$$
$$\times \frac{1}{y-x}\left(\int \mathrm{d}\omega\, h(\omega)\left(\frac{x-\omega}{(x-\omega)^2+\varepsilon^2}-\frac{y-\omega}{(y-\omega)^2+\varepsilon^2}\right)\right)$$
$$= \lim_{\varepsilon\to 0}\iiint \mathrm{d}x\,\mathrm{d}y\,\mathrm{d}\omega\, f(x)g(y)h(\omega)\frac{(x-\omega)(y-\omega)-\varepsilon^2}{((x-\omega)^2+\varepsilon^2)((y-\omega)^2+\varepsilon^2)}$$
$$= \iiint \mathrm{d}x\,\mathrm{d}y\,\mathrm{d}\omega f(x)g(y)h(\omega)\frac{\mathscr{P}}{x-\omega}\frac{\mathscr{P}}{y-\omega}-\pi^2\int \mathrm{d}\omega\, f(\omega)g(\omega)h(\omega).$$

Here, at the end, we have employed the well-known limits

$$\lim_{\varepsilon\to 0}\frac{x-\omega}{(x-\omega)^2+\varepsilon^2}=\frac{\mathscr{P}}{x-\omega},$$
$$\lim_{\varepsilon\downarrow 0}\frac{\varepsilon}{(x-\omega)^2+\varepsilon^2}=\pi\delta(x-\omega). \qquad \square$$

Lemma 4.2.3 *We have*

$$\int \mathrm{d}\omega\frac{\mathscr{P}}{x-\omega}\frac{\mathscr{P}}{y-\omega}=\pi^2\delta(x-y)$$

or explicitly, for $f, g\in C_c^2$,

$$\iiint \mathrm{d}x\,\mathrm{d}y\,\mathrm{d}\omega f(x)g(y)\frac{\mathscr{P}}{x-\omega}\frac{\mathscr{P}}{y-\omega}=\pi^2\int \mathrm{d}\omega f(\omega)g(\omega).$$

Proof We show the first expression on the right-hand side in the preceding lemma goes to 0. Replace the function h in the lemma by h_ε, with $h_\varepsilon(\omega)=h(\varepsilon\omega)$ and $h(\omega)=1/(1+\omega^2)$. Then

$$\frac{1}{y-x}\int \mathrm{d}\omega\, h_\varepsilon(\omega)\left(\frac{\mathscr{P}}{x-\omega}-\frac{\mathscr{P}}{y-\omega}\right)$$

$$= \frac{1}{y-x}\int \mathrm{d}\omega\, \varepsilon h'(\varepsilon\omega) \log \frac{|x-\omega|}{|y-\omega|}$$

$$= \frac{1}{y-x}\int \mathrm{d}\omega\, h'(\omega) \log \frac{|x-\omega/\varepsilon|}{|y-\omega/\varepsilon|} = \frac{1}{y-x}\int \mathrm{d}\omega\, h'(\omega) \log \frac{|1-\varepsilon x/\omega|}{|1-\varepsilon y/\omega|}$$

$$\sim \varepsilon \int \mathrm{d}\omega\, h'(\omega)/\omega$$

for $\varepsilon \to 0$ and

$$\int \mathrm{d}\omega\, h'(\omega)/\omega < \infty.$$

The variables x and y can supposed to be bounded, as f and g are of compact support. From there one obtains the result. □

Remark 4.2.2 The equation of the last lemma is well known. The equation is the basis of the *Hilbert transform.*

Proposition 4.2.5 *The α_x are orthonormal in the generalized sense that*

$$\langle \alpha_x | \alpha_y \rangle = \delta(x-y).$$

More precisely, if $f \in C_c^2$, then

$$\int \mathrm{d}x\, f(x)\alpha_x \in \mathfrak{H} = \mathbb{C} \oplus L^2(\mathbb{R})$$

and, if $g \in C_c^2$ as well, then

$$\left\langle \int \mathrm{d}x\, f(x)\alpha_x \middle| \int \mathrm{d}y\, g(y)\alpha_y \right\rangle = \iint \overline{f}(x)g(y)\delta(x-y)$$

$$= \int \mathrm{d}\omega\, \overline{f}(\omega)g(\omega) = \langle f|g\rangle_{L^2}.$$

Proof Calculate

$$\left\langle \int \mathrm{d}x\, f(x)\alpha_x \middle| \int \mathrm{d}y\, g(y)\alpha_y \right\rangle$$

$$= \left\langle \int \mathrm{d}x\, \overline{f}(x) \frac{1}{\sqrt{x^2+\pi^2}} (1, x\delta_x(\Omega) + \mathscr{P}/(x-\Omega)1) \right|$$

$$\left. \int \mathrm{d}y\, g(y) \frac{1}{\sqrt{y^2+\pi^2}} \left(\frac{1}{(1, y\delta_y(\Omega) + \mathscr{P}/(y-\Omega)1} \right) \right\rangle$$

$$= \int \mathrm{d}\omega \iint \mathrm{d}x\mathrm{d}y\, \overline{f}(x)g(y) \frac{1}{\sqrt{x^2+\pi^2}} \frac{1}{\sqrt{y^2+\pi^2}}$$

$$\times\left(1+\left(x\delta(x-\omega)+\frac{\mathscr{P}}{x-\omega}\right)\left(y\delta(y-\omega)+\frac{\mathscr{P}}{y-\omega}\right)\right).$$

Interchange the order of integration, use the properties of the δ-function and the $\mathscr{P}$-function, and the preceding lemma, to get

$$=\iint \mathrm{d}x\mathrm{d}y\,\overline{f}(x)g(y)\delta(x-y)=(f|g).$$ □

Corollary 4.2.1 *For the spectral Schwartz distribution we have the formula*

$$M(z_1)M(z_2)=\delta(z_1-z_2)M(z_1)$$

or, as $M(x+\mathrm{i}y)=\mu(x)\delta(y)$,

$$\mu(x_1)\mu(x_2)=\delta(x_1-x_2)\mu(x_1).$$

More precisely, if $\xi=\binom{c}{f}$, $c\in\mathbb{C}$, $f\in C_c^2$, $g\in C_c^2$, *then*

$$\int \mathrm{d}x\, g(x)\mu(x)|\xi\rangle=\int \mathrm{d}x\, g(x)|\alpha_x\rangle\langle\alpha_x|\xi\rangle$$

belongs to L^2, *and*

$$\left\langle\int \mathrm{d}x_1\, g_1(x_1)\mu(x_1)\middle|\int \mathrm{d}x_2\, g_2(x_2)\mu(x_2)\right\rangle=\int \mathrm{d}x\,\overline{g_1(x)}g_2(x)\mu(x).$$

Proof Use the preceding proposition and that

$$x\mapsto\langle\alpha_x|\xi\rangle$$

is a bounded C^2 function. □

Remark 4.2.3 Compare the last formula to the result holding for spectral families $(E_x,\ x\in\mathbb{R})$ namely

$$\int \mathrm{d}E_{x_1}g_1(x_1)\int \mathrm{d}E_{x_2}g_2(x_2)=\int \mathrm{d}E_x\, g_1(x)g_2(x)$$

which holds for bounded Borel functions g_1 and g_2.

Proposition 4.2.6 *The orthonormal system of the* α_x *is complete, so*

$$\int \mathrm{d}x\,|\alpha_x\rangle\langle\alpha_x|=1,$$

or more precisely for $\xi=\binom{c}{f}$, $f\in C_c^1$,

$$x\mapsto\langle\xi|\alpha_x\rangle\in L^2$$

and, for $\xi = \binom{c}{f}$, $\eta = \binom{c'}{g}$, $f, g \in C^1_c$, *one has*

$$\int \mathrm{d}x \langle \xi | \alpha_x \rangle \langle \alpha_x | \eta \rangle = \langle \xi | \eta \rangle.$$

For the resolvent one obtains

$$R(z) = \int \mathrm{d}x \frac{1}{z - x} |\alpha_x\rangle\langle\alpha_x|,$$

or more precisely with ξ, η *as above*

$$\langle \xi | R(z) | \eta \rangle = \int \mathrm{d}x \frac{1}{z - x} \langle \xi | \alpha_x \rangle \langle \alpha_x | \eta \rangle.$$

Proof The resolvent

$$R(z) = \begin{pmatrix} 0 & 0 \\ 0 & R_\Omega(z) \end{pmatrix} + \begin{pmatrix} 1 \\ R_\Omega(z)|E\rangle \end{pmatrix} \frac{1}{z + \mathrm{i}\sigma(z)} \bigl(1, \langle E | R_\Omega(z)\rangle\bigr)$$

is holomorphic for $\operatorname{Im} z \neq 0$; the function $\langle \xi | R(z) | \eta \rangle$ is continuous at the boundary. By deforming the boundary one obtains that

$$\int_{-r}^{r} \mathrm{d}x \, \langle \xi | R(x \pm \mathrm{i}0) | \eta \rangle = \mp \int_{\Gamma_\pm} \mathrm{d}z \, \langle \xi | R(z) | \eta \rangle,$$

where $\Gamma_\pm$ is the semicircle of radius r joining $-r$ and r in the upper, resp. lower, half-plane. Then

$$\begin{aligned} \int_{-r}^{r} \mathrm{d}x \, \langle \xi | \alpha_x \rangle \langle \alpha_x | \eta \rangle &= \frac{1}{2\pi \mathrm{i}} \int_{-r}^{r} \mathrm{d}x \, \langle \xi | \bigl(R(x - \mathrm{i}0) - R(x + \mathrm{i}0) \bigr) | \eta \rangle \\ &= \frac{1}{2\pi \mathrm{i}} \int_{\Gamma} \mathrm{d}z \langle \xi | R(z) | \eta \rangle, \end{aligned}$$

where Γ is the circle of radius r. As f and g are of compact support

$$\langle \xi | R(z) | \eta \rangle = \frac{1}{z} \langle f | g \rangle + O\bigl(z^{-2}\bigr)$$

and one obtains the first assertion by taking $r \to \infty$.

Assume, e.g., $\operatorname{Im} z > 0$ and put $F(z) = \langle \xi | R(z) | \eta \rangle$, then

$$F(z) = \frac{1}{2\pi \mathrm{i}} \int_{\gamma} \mathrm{d}\zeta \frac{F(\zeta)}{\zeta - z},$$

where γ is a small circle in the upper half-plane encircling z. By blowing γ up so it consists of the interval $[-r, r]$ and the semi-circle Γ_+ one arrives at

$$F(z) = \frac{1}{2\pi \mathrm{i}} \left(\int_{\Gamma_+} \mathrm{d}\zeta \frac{F(\zeta)}{\zeta - z} + \int_{-r}^{r} \mathrm{d}x \frac{F(x + \mathrm{i}0)}{x - z} \right).$$

In the lower half-plane one obtains similarly

$$0 = \frac{1}{2\pi \mathrm{i}} \left(\int_{\Gamma_-} \mathrm{d}\zeta \frac{F(\zeta)}{\zeta - z} - \int_{-r}^{r} \mathrm{d}x \frac{F(x - \mathrm{i}0)}{x - z} \right).$$

So

$$F(z) = \frac{1}{2\pi \mathrm{i}} \left(\int_{\Gamma} \mathrm{d}\zeta \frac{F(\zeta)}{\zeta - z} + \int_{-r}^{r} \mathrm{d}x \frac{F(x + \mathrm{i}0) - F(x - \mathrm{i}0)}{x - z} \right).$$

As $F(\zeta) = O(\zeta^{-2})$, we take $r \to \infty$ and obtain the second assertion. □

Corollary 4.2.2 *If E_x is the spectral family of H, then*

$$\langle \xi | \mathrm{d}E_x | \eta \rangle = \langle \xi | \alpha_x \rangle \langle \alpha_x | \eta \rangle \mathrm{d}x.$$

4.3 A Two-Level Atom Interacting with Polarized Radiation

4.3.1 Physical Considerations

We discuss a two-level atom with transition frequency ω_0. The levels are supposed not degenerate, and have the wave functions $\psi_1(\mathbf{x})$ for the upper level and $\psi_0(\mathbf{x})$ for the lower level. We shall use relativistic units with $\hbar = 1$ and the velocity of light $c = 1$. In these units the square of charge of the electron is $e^2 = 1/137$.

The radiation field is a system of independent oscillators labelled by $\lambda \in \Lambda$

$$\Lambda = \left\{ \mathbf{m} = (m_1, m_2, m_3) \in \mathbb{Z}^3, \sum |m_i| \le M \right\} \times \{1, 2\}.$$

Associate to $\mathbf{k} \in \mathbb{R}^3$ two unit vectors $\mathbf{e}_1(\mathbf{k}), \mathbf{e}_2(\mathbf{k})$ such that the three vectors $\mathbf{k}/|\mathbf{k}|, \mathbf{e}_1(\mathbf{k}), \mathbf{e}_2(\mathbf{k})$ form a right-handed coordinate system (a trihedron) in $\mathbb{R}^3$. Choose a large number $L > 0$, and define

$$\mathbf{k}_\lambda = \mathbf{k}_{\mathbf{m},j} = \frac{2\pi}{L} \mathbf{m}, \qquad \omega_\lambda = |\mathbf{k}_\lambda| = \frac{2\pi}{L} |\mathbf{m}|, \qquad \mathbf{e}_\lambda = \mathbf{e}_j(\mathbf{k}_\lambda).$$

We have to consider the finite system of oscillators, labelled by $\lambda \in \Lambda$ with frequencies ω_λ, given by the creation and annihilation operators $a_\lambda, a_\lambda^+, \lambda \in \Lambda$. The representation space is a pre-Hilbert space spanned by the vectors $|\mathfrak{m}\rangle = (a^+)^{\mathfrak{m}} |0\rangle$, where $\mathfrak{m}$ runs through all multisets of Λ. The Hamiltonian is

$$H_{\text{rad}} = \sum_{\lambda \in \Lambda} H_\lambda = \sum_{\lambda \in \Lambda} \omega_\lambda a_\lambda^+ a_\lambda.$$

Use the notation $E_{10} = |\psi_1\rangle \langle \psi_0|$ etc., then in rotating wave approximation

$$H_{\text{tot}} = H_{\text{rad}} + H_{\text{atom}} + H_{\text{int}} = \sum_{\lambda \in \Lambda} \omega_\lambda a_\lambda^+ a_\lambda + \omega_0 E_{11} + \sum_\lambda \left(g_\lambda a_\lambda E_{10} + \overline{g}_\lambda a_\lambda^+ E_{01} \right).$$

One has

$$g_\lambda = -\frac{e}{m_e}\sqrt{\frac{2\pi}{\omega_\lambda}}L^{-3/2}\langle\psi_1|\mathbf{p}.\mathbf{e}_\lambda \exp(\mathrm{i}\mathbf{k}_\lambda.\mathbf{x})|\psi_0\rangle,$$

where e is the electron charge and m_e the electron mass. $\mathbf{p}$ is the momentum operator $\mathbf{p} = (p_1, p_2, p_3)$; $p_i = -\mathrm{i}d/dx_i$. If a is an estimate of the atomic radius. Then frequency $\omega_0 \approx e^2/a$, so

$$a\omega_0 \approx e^2 = 1/137.$$

The function $\exp(\mathrm{i}\mathbf{k}_\lambda.\mathbf{x})$ is approximately constantly 1 until frequencies of the order $1/a$. We mutilate g_λ

$$g_\lambda = -\frac{e}{m_e}\sqrt{2\pi/\omega_\lambda}L^{-3/2}\langle\psi_1|\mathbf{p}.\mathbf{e}_\lambda|\psi_0\rangle c(\omega_\lambda - \omega_0),$$

where $0 \le c(\omega) \le 1$ and $c(\omega) = 1$ for $|\omega_1| \ll \omega_0$ and is 0 for $|\omega| > \omega_1$. To justify this mutilation is outside the scope of this work. Using the relation

$$\langle\psi_{1,i}|\mathbf{p}/m_e|\psi_0\rangle = \mathrm{i}\omega_0\langle\psi_1|\mathbf{x}|\psi_0\rangle$$

and $\omega_\lambda \approx \omega_0$ we arrive at

$$g_\lambda = \mathrm{i}e\sqrt{2\pi\omega_0}L^{-3/2}c(\omega_\lambda - \omega_0)\langle\psi_1|\mathbf{p}.\mathbf{e}_\lambda|\psi_0\rangle.$$

Introduce

$$\Lambda' = \left\{\mathbf{m} = (m_1, m_2, m_3) \in \mathbb{Z}^3, \sum |m_i| \le M\right\} \times \{1, 2, 3\}.$$

We imbed $\mathbb{C}^\Lambda$ into $\mathbb{C}^{\Lambda'}$. If $\mathfrak{e}_\lambda = \mathfrak{e}_{\mathbf{m},j}$, resp. $\mathfrak{e}'_{\mathbf{m},i}$ are the standard basis vectors of $\mathbb{C}^\Lambda$, resp. $\mathbb{C}^{\Lambda'}$, we map

$$\mathfrak{e}_{\mathbf{m},j} \mapsto \sum_{j=1,2}(\mathbf{e}_{\mathbf{m},j})_i\mathfrak{e}'_{\mathbf{m},i}.$$

This means, if we consider the elements of $\mathbb{C}^{\Lambda'}$ as vector fields, we affix to the point $\mathbf{m}$ the vectors $\mathbf{e}_{\mathbf{m},j}$. Similar we define annihilation and creation operators $b_{\mathbf{m},i}$ and $b^+_{\mathbf{m}.i}$ for $(\mathbf{m}, i) \in \Lambda'$. We express the annihilation and creation operators indexed by Λ in terms of those of indexed by Λ',

$$a_{\mathbf{m},j} = \sum_{j=1,2}(\mathbf{e}_{\mathbf{m},j})_i b_{\mathbf{m},i}, \qquad a^+_{\mathbf{m},j} = \sum_{j=1,2}(\mathbf{e}_{\mathbf{m},j})_i b^+_{\mathbf{m},i}.$$

This means physically, that we have introduced a fictitious longitudinal polarization. Denote by $\Pi(\mathbf{m})$ the orthogonal projector onto the plane perpendicular to $\mathbf{m}$, then

$$\Pi(\mathbf{m})_{il} = \sum_{j=1,2}(\mathbf{e}_{\mathbf{m},j})_i(\mathbf{e}_{\mathbf{m},j})_l = \delta_{il} - \mathbf{m}_i\mathbf{m}_l/|\mathbf{m}|^2.$$

We have then, with $\mathbf{k}_\mathbf{m} = (2\pi/L)\mathbf{m}$ and $\omega_\mathbf{m} = (2\pi/L)|\mathbf{m}|$,

$$H_{\mathrm{rad}} = \sum_{\mathbf{m}} \omega_\mathbf{m} \sum_{i,l} \Pi(\mathbf{m})_{il} b^+_{\mathbf{m},i} b_{\mathbf{m},l},$$

$$H_{\mathrm{int}} = \sum_{\mathbf{m},i,l} e\sqrt{2\pi\omega_0}\, L^{-3/2} c(\omega_\mathbf{m} - \omega_0)\Pi(\mathbf{m})_{i,l} \big(\mathrm{i}E_{10}(\psi_1|\mathbf{x}|\psi_0)_i b_{\mathbf{m},l}$$
$$- \mathrm{i}E_{01}(\psi_0|\mathbf{x}|\psi_1)_i b^+_{\mathbf{m},l}\big).$$

Introduce the space

$$X = \mathbb{R}^3 \times \{1, 2, 3\}.$$

Define the cube

$$C_\mathbf{m} = \big\{\mathbf{k} = (k_1, k_2, k_3) : \big|k_i - (\mathbf{k}_\mathbf{m})_i\big| < \pi/L\big\}$$

with the volume $C = (2\pi/L)^3$. Put

$$b_{\mathbf{m},i} = C^{-1/2} \int \mathrm{d}\mathbf{k} 1_{C_\mathbf{m}}(\mathbf{k}) a(\mathbf{k}, i) = C^{-1/2} a(C_{\mathbf{m},i}).$$

For $f, g \in \mathscr{K}_s(\mathfrak{X})$, one obtains, since $a(C_{\mathbf{m},i}) f \approx C a(\mathbf{k}, i) f$,

$$(f|H_{\mathrm{rad}}|f) = \sum_{\mathbf{m},i,l} \omega_\mathbf{m} \Pi(\mathbf{m})_{il} C^{-1} \big\langle a(C_{\mathbf{m},i}) f \big| a(C_{\mathbf{m},l}) | f\big\rangle$$
$$\approx \sum_{\mathbf{m},i,l} \omega_\mathbf{m} \Pi(\mathbf{m})_{il} C \big\langle a(\mathbf{k}_{\mathbf{m},i}) f \big| a(\mathbf{k}_{\mathbf{m},l}) | f\big\rangle$$
$$\approx \int \mathrm{d}\mathbf{k} \sum_{i,l} \Pi(\mathbf{k})_{il} |\mathbf{k}| \big\langle a(\mathbf{k}, i) f \big| a(\mathbf{k}, l) | f\big\rangle$$

and finally

$$H_{\mathrm{rad}} = \int \mathrm{d}\mathbf{k} |\mathbf{k}| \sum_{i,l} \Pi(\mathbf{k})_{i,l} a^\dagger(\mathbf{k}, i) a(\mathbf{k}, l),$$

$$H_{\mathrm{int}} = \int \mathrm{d}\mathbf{k} \sum_{i,l} e\sqrt{\omega_0}/(2\pi) c\big(|\mathbf{k}| - \omega_0\big) \Pi(\mathbf{k})_{i,l}$$
$$\times \big(-\mathrm{i}E_{10}(\psi_1|\mathbf{x}|\psi_0)_i a(\mathbf{k}, l) + \mathrm{i}E_{01}(\psi_0|\mathbf{x}|\psi_1)_i a(\mathbf{k}, l)^\dagger\big).$$

The quantity

$$N = \int \mathrm{d}\mathbf{k} \sum_{i,l} \Pi(\mathbf{k})_{i,l} a^\dagger(\mathbf{k}, i) a(\mathbf{k}, l) + E_{11}$$

is the operator for the total number of excitations and commutes with H_{tot}. As it gives only a trivial contribution we just consider $H_{\mathrm{tot}} - \omega_0 N$ and call it H_{tot} once more. So

$$H_{\mathrm{tot}} = \int \mathrm{d}\mathbf{k}\big(|\mathbf{k}| - \omega_0\big) \sum_{i,l} \Pi(\mathbf{k})_{i,l} a^+(\mathbf{k}, i) a(\mathbf{k}, l) + H_{\mathrm{int}}.$$

We introduce polar coordinates in a slightly modified way

$$\mathbf{k} = (\omega + \omega_0)\mathbf{n}, \qquad \mathrm{d}\mathbf{k} = (\omega + \omega_0)^2 \mathrm{d}\omega\, \mathrm{d}\mathbf{n}.$$

Here $\mathbf{n} \in \mathbb{S}^2$ and $\mathrm{d}\mathbf{n}$ is the surface element on the sphere $\mathbb{S}^2$ normalized such that $\int_{\mathbb{S}^2} \mathrm{d}\mathbf{n} = 4\pi$. As $c(|\mathbf{k}| - \omega_0) = c(\omega)$ vanishes for $|\omega| > \omega_1$ we have only to consider ω for $|\omega| < \omega_1$. As we assumed $\omega_1 \ll \omega_0$,

$$\mathrm{d}\mathbf{k} = \omega_0^2 \mathrm{d}\omega\, \mathrm{d}\mathbf{n},$$

and we may allow ω to go from $-\infty$ to $+\infty$. So for the radiation our basic space X becomes

$$X_{\mathrm{rad}} = \mathbb{R} \times \mathbb{S}^2 \times \{1, 2, 3\},$$

where $\omega \in \mathbb{R}$ is the frequency, $\mathbf{n} \in \mathbb{S}^2$ the direction, and i corresponding to $\mathfrak{e}_i$ in the standard basis of $\mathbb{R}^3$ is the polarization. Remark that we have introduced a superfluous direction of polarization, the direction of $\mathbf{n}$. We have

$$\begin{aligned} H_{\mathrm{tot}} = & \int \mathrm{d}\omega\, \mathrm{d}\mathbf{n}\, \omega_0^2 \omega \sum_{i,l} \Pi(\mathbf{n})_{i,l} a^\dagger(\omega, \mathbf{n}, i) a(\omega, \mathbf{n}, l) \\ & + \int \mathrm{d}\omega\, \mathrm{d}\mathbf{n}\, \omega_0^2 \sum_{i,l} e\sqrt{\omega_0}/(2\pi) c(\omega) \Pi(\mathbf{k})_{i,l} \\ & \times \big(-\mathrm{i}E_{10} \langle \psi_1 | \mathbf{x} | \psi_0 \rangle_i a(\omega, \mathbf{n}, l) + \mathrm{i}E_{01} \langle \psi_0 | \mathbf{x} | \psi_1 \rangle_i a(\omega, \mathbf{n}, l)^\dagger\big). \end{aligned}$$

We restrict ourselves to the case of one excitation. Then we have only to consider the cases, that either we have the photon vacuum Φ and the atom is in the upper level or the atom is in the lower level and a photon $(\omega, \mathbf{n}, i)$ is present. We restrict our state space to the space generated by the states $\psi_1 \otimes \Phi$ or $\psi_0 \otimes a^+(\omega, \mathbf{n}, i)\Phi$. So we may use as Hilbert space

$$\mathfrak{H} = \mathbb{C} \oplus L^2(X_{\mathrm{rad}}, \lambda)$$

where λ is now the measure on X_{rad} given by

$$\langle \lambda | f \rangle = \iint \mathrm{d}\omega \omega_0^2\, \mathrm{d}\mathbf{n} \sum_{i=1,2,3} f(\omega, \mathbf{n}, i).$$

Consider the elements $(c, f) \in \mathfrak{H}$, $c \in \mathbb{C}$, $f \in \mathscr{K}(X_{\text{rad}})$, where $\mathscr{K}(X_{\text{rad}})$ is the space of continuous functions with compact support, and use the notation

$$\Psi(c, f) = c\psi_1 \otimes \Phi + \int \mathrm{d}\lambda f(\omega, \mathbf{n}, i)\big(\psi_0 \otimes a^+(\omega, \mathbf{n}, i)\Phi\big).$$

Then

$$\big\langle \Psi(c, f)\big|H_{\text{tot}}\big|\Psi(c', f')\big\rangle = (c, f)\begin{pmatrix} 0 & \langle g| \\ |g\rangle & K \end{pmatrix}\begin{pmatrix} c' \\ f' \end{pmatrix}$$

with

$$g(\omega, \mathbf{n}, i) = \mathrm{i}c(\omega)\sum_l e\frac{\sqrt{\omega_0}}{2\pi}\Pi(\mathbf{n})_{i,l}\langle\psi_0|\mathbf{x}|\psi_1\rangle_l,$$

$$(Kf)(\omega, \mathbf{n}, i) = \omega\sum_l \Pi(n)_{i,l} f(\omega, \mathbf{n}, l).$$

4.3.2 Singular Coupling

We rewrite the results of the last subsection. We consider the space

$$\mathfrak{H} = \mathbb{C} \oplus L^2\big(\mathbb{R} \times \mathbb{S}^2 \times \{1, 2, 3\}\big)$$

provided with the measure λ given by

$$\langle\lambda|f\rangle = \iint \mathrm{d}\omega\, \omega_0^2\, \mathbf{dn} \sum_{i=1,2,3} f(\omega, \mathbf{n}, i).$$

We consider the elements of $L^2(\mathbb{R} \times \mathbb{S}^2 \times \{1, 2, 3\})$ as vector-valued functions on $\mathbb{R} \times \mathbb{S}^2$. Then $\mathfrak{H}$ becomes

$$\mathfrak{H} = \mathbb{C} \oplus L^2\big(\mathbb{R} \times \mathbb{S}^2, \mathbb{C}^3\big).$$

We will be studying the operator given by the matrix

$$H_c = \begin{pmatrix} 0 & \langle \mathbf{g}| \\ |\mathbf{g}\rangle & K \end{pmatrix} = \begin{pmatrix} 0 & \langle c\mathbf{v}| \\ |\mathbf{v}c\rangle & A\Omega A \end{pmatrix};$$

here

$$\mathbf{g}(\omega, \mathbf{n}) = c(\omega)\mathbf{v}(\mathbf{n}),$$

$$\mathbf{v}(\mathbf{n}) = \mathrm{i}e\frac{\sqrt{\omega_0}}{2\pi}\Pi(\mathbf{n})\langle\psi_0|\mathbf{x}|\psi_1\rangle,$$

$$(\Omega\mathbf{f})(\omega\mathbf{n}) = \omega f(\omega, \mathbf{n}),$$

$$(A\mathbf{f})(\omega, \mathbf{n}) = \Pi(n)\mathbf{f}(\omega, \mathbf{n}),$$
$$K = A\Omega A = A\Omega = \Omega A,$$
$$\Pi(\mathbf{n})_{ij} = \delta_{ij} - \mathbf{n}_i\mathbf{n}_j;$$

the function c is one with the properties $0 \leq c(\omega) \leq 1$ and $c(\omega) = 0$ for $|\omega| \geq \omega_1$, and the operator A is a projector, so $A^2 = A$ and $A\mathbf{v} = \mathbf{v}$.

By Krein's formula we can calculate the resolvent

$$R_c(z) = \frac{1}{z - H_c} = \begin{pmatrix} 0 & 0 \\ 0 & R_K(z) \end{pmatrix} + \begin{pmatrix} 1 \\ R_K|\mathbf{g}\rangle \end{pmatrix} \frac{1}{z - \langle \mathbf{g}|R_K(z)|\mathbf{g}\rangle} \big(1, \langle \mathbf{g}|R_K\big)$$

with

$$R_K(z) = \frac{1}{z - K} = \frac{1}{z - A\Omega A} = A\frac{1}{z - \Omega}A + \frac{1}{z}(1 - A).$$

Since

$$A\mathbf{v} = \mathbf{v}$$

we obtain, with $R_\Omega(z) = 1/(z - \Omega)$,

$$R_c(z) = \begin{pmatrix} 0 & 0 \\ 0 & AR_\Omega(z)A + \frac{1}{z}(1 - A) \end{pmatrix}$$
$$+ \begin{pmatrix} 1 \\ R_\Omega(z)|\mathbf{v}c\rangle \end{pmatrix} \frac{1}{z - \langle \mathbf{g}|R_K(z)|\mathbf{g}\rangle} \big(1, \langle c\mathbf{v}|R_\Omega(z)\big).$$

We now perform the singular coupling limit and make the function c converge to the constant function E: $E(\omega) = 1$, in such a way that c stays bounded by E and $c(\omega) = c(-\omega)$. Then

$$\langle \mathbf{g}|R_K(z)|\mathbf{g}\rangle = \langle c|R_\Omega(z)|c\rangle\langle \mathbf{v}|A|\mathbf{v}\rangle = \int \mathrm{d}\omega\, \frac{c(\omega)^2}{z - \omega}\langle \mathbf{v}|\mathbf{v}\rangle \to -\mathrm{i}\pi\sigma(z)\gamma$$

with

$$\gamma = \langle \mathbf{v}|\mathbf{v}\rangle = \int \mathrm{d}\mathbf{n}\omega_0^2\big\langle \mathbf{v}(\mathbf{n})|\mathbf{v}(\mathbf{n})\big\rangle = e^2\frac{2\omega_0^3}{3\pi}\big|\langle \psi_0|\mathbf{x}|\psi_1\rangle\big|^2.$$

Here $\sigma(z)$ is the sign of $\mathrm{Im}\, z$. The resolvent becomes

$$R(z) = \begin{pmatrix} 0 & 0 \\ 0 & AR_\Omega(z)A \end{pmatrix} + \begin{pmatrix} 0 & 0 \\ 0 & \frac{1}{z}(1 - A) \end{pmatrix}$$
$$+ \begin{pmatrix} 1 \\ R_\Omega(z)|\mathbf{v}E\rangle \end{pmatrix} \frac{1}{z + \mathrm{i}\pi\sigma(z)\gamma} \big(1, \langle E\mathbf{v}|R_\Omega(z)\big).$$

The term

$$\begin{pmatrix} 0 & 0 \\ 0 & \frac{1}{z}(1-A) \end{pmatrix}$$

is the contribution of the fictitious longitudinally polarized photons and need not to be considered further. The expression $\langle E|$ is the linear functional $f \mapsto \langle E|f\rangle = \int d\omega f(\omega)$, and $E = |E\rangle$ is the semilinear functional given by $\langle f|E\rangle = \overline{\langle E|f\rangle}$.

For the time development we obtain, similarly to Sect. 4.2.4,

$$U(t) = \begin{pmatrix} U_{00} & U_{01} \\ U_{10} & U_{11} \end{pmatrix}$$

with

$$\begin{aligned}
U_{00} &= e^{-\pi\gamma t}, \\
U_{01} &= -i\int_0^t dt_1 e^{-\pi\gamma(t-t_1)} \langle E|e^{-i\Omega t_1} \otimes \langle \mathbf{v}|, \\
U_{10} &= -i\int_0^t dt_1 e^{-i\Omega(t-t_1)} |E\rangle e^{-\pi\gamma t_1} \otimes |\mathbf{v}\rangle, \\
U_{11} &= e^{-i\Omega t} \otimes A - \iint_{0<t_1<t_2<t} dt_1 dt_2 e^{-i\Omega(t-t_2)} |E\rangle e^{-\pi\gamma(t_2-t_1)} \langle E|e^{-i\Omega t_1} \otimes |\mathbf{v}\rangle\langle \mathbf{v}| \\
&\quad + 1 \otimes (1-A).
\end{aligned}$$

We have

$$H_0 = \begin{pmatrix} 0 & 0 \\ 0 & A\Omega A \end{pmatrix}$$

and

$$e^{-iH_0 t} = \begin{pmatrix} 1 & 0 \\ 0 & Ae^{-i\Omega t} A + 1 - A \end{pmatrix}.$$

Then

$$V(t) = e^{iH_0 t} U(t)$$

is given by

$$\begin{aligned}
V_{00} &= e^{-\pi\gamma t}, \\
V_{01} &= -i\int_0^t dt_1 e^{-\pi\gamma(t-t_1)} \langle E|e^{-i\Omega t_1} \otimes \langle \mathbf{v}|, \\
V_{10} &= -i\sqrt{2\pi}\int_0^t dt_1 e^{i\Omega t_1} |E\rangle e^{-\pi\gamma t_1} \otimes |\mathbf{v}\rangle, \\
V_{11} &= 1 - \iint_{0<t_1<t_2<t} dt_1 dt_2 e^{-i\Omega(t-t_2)} |E\rangle e^{-\pi\gamma(t_2-t_1)} \langle E|e^{-i\Omega t_1} \otimes |\mathbf{v}\rangle\langle \mathbf{v}|.
\end{aligned}$$

That reads in the formal time representation, as explained in Sect. 4.2.4,

$$V_{00} = \mathrm{e}^{-\pi\gamma t},$$

$$V_{01}|\tau\rangle = -\mathrm{i}\sqrt{2\pi}\int_0^t \mathrm{d}t_1 \mathrm{e}^{-\pi\gamma(t-t_1)}\delta(t-t_1)\langle\mathbf{v}|,$$

$$\langle\tau|V_{10} = -\mathrm{i}\int_0^t \mathrm{d}t_1\delta(\tau-t_1)|E\rangle\mathrm{e}^{-\pi\gamma t_1}\otimes|\mathbf{v}\rangle,$$

$$\langle\tau_2|V_{11}\tau_1\rangle = \delta(\tau_1-\tau_2) - 2\pi\iint_{0<t_1<t_2<t} \mathrm{d}t_1\mathrm{d}t_2\delta(\tau_2-t_1)\mathrm{e}^{-\pi\gamma(t_2-t_1)}\delta(t_1-\tau_1)$$
$$\otimes|\mathbf{v}\rangle\langle\mathbf{v}|.$$

The matrix element U_{00} describes the decay of the upper state. The transition probability is

$$2\pi\gamma = e^2\frac{4\omega_0^3}{3}\big|\langle\psi_0|\mathbf{x}|\psi_1\rangle\big|^2,$$

in agreement with Landau-Lifschitz [28]. The element U_{10} represents the spontaneous emission. The integrated emitted tensor intensity, in direction $\mathbf{n}$ and with frequency ω, for all times between 0 and ∞, is

$$\begin{aligned}\mathfrak{I}(\omega,\mathbf{n}) &= \frac{1}{\omega^2+\pi^2\gamma^2}\big|\mathbf{v}(\mathbf{n})\big\rangle\big\langle\mathbf{v}(\mathbf{n})\big| \\ &= \frac{1}{\omega^2+\pi^2\gamma^2}\frac{e^2\omega_0^3}{4\pi^2}\big|\Pi(\mathbf{n})\langle\psi_0|\mathbf{x}|\psi_1\rangle\big\rangle\big\langle\langle\psi_1|\mathbf{x}|\psi_0\rangle\Pi(\mathbf{n})\big|.\end{aligned}$$

The fraction of the emitted total intensity in direction $\mathbf{n}$ is

$$\int \mathrm{d}\omega\,\mathrm{trace}\big(\mathfrak{I}(\omega,\mathbf{n})\big) = \frac{3}{8\pi}\left(1-\frac{|\langle\psi_1|\Pi(\mathbf{n})\mathbf{x}|\psi_0\rangle|^2}{|\langle\psi_1|\mathbf{x}|\psi_0\rangle|^2}\right) = \frac{3}{8\pi}\sin^2\vartheta$$

where ϑ is the angle between $\mathbf{n}$ and $\langle\psi_1|\mathbf{x}|\psi_0\rangle$.

The element U_{10} describes absorption, and U_{11} describes undisturbed transmission and scattering.

4.3.3 The Hamiltonian and the Eigenvalue Problem

The Hamiltonian corresponding to the resolvent $R(z)$ is

$$H = \begin{pmatrix} 0 & \langle\hat{E}\mathbf{v}| \\ |\mathbf{v}\hat{E}\rangle & A\hat{\Omega}A \end{pmatrix},$$

where the definitions of $\hat{E}$ and $\hat{\Omega}$ have to be adapted from Sect. 4.2.2 to the vector case here. The domain of H is

$$\begin{aligned} D &= R(z)\mathfrak{H} \\ &= \left\{ c \begin{pmatrix} 1 \\ R_\Omega(z)E\mathbf{v} \end{pmatrix} + \begin{pmatrix} 0 \\ R_\Omega(z)A\mathbf{f}_1 \end{pmatrix} \right. \\ &\quad \left. + \begin{pmatrix} 0 \\ (1-A)\mathbf{f}_2 \end{pmatrix} : c \in \mathbb{C}, \mathbf{f}_1, \mathbf{f}_2 \in L^2(\mathbb{R} \times \mathbb{S}^2, \mathbb{C}^3) \right\}. \end{aligned}$$

One checks immediately that

$$HR(z) = R(z)H = -1 + zR(z).$$

We discuss the eigenvalue problem in the same way as in the previous section. One calculates in a similar way, using the fact that for

$$\xi_i = \begin{pmatrix} c_i \\ \mathbf{f}_i \end{pmatrix}, \qquad r_i \in \mathbb{C}, \qquad \mathbf{f}_i \in C_c^1, \qquad h \in C_c^1, \qquad x \in \mathbb{R}$$

the expression

$$\int \mathrm{d}x\, h(x)(\xi_1|R(x \pm \mathrm{i}0)|\xi_2)$$

is well defined.

Proposition 4.3.1 *We have, given*

$$\xi_i = \begin{pmatrix} c_i \\ \mathbf{f}_i \end{pmatrix}, \quad c_i \in \mathbb{C}, \ \mathbf{f}_i \in C_c^1,$$

that, for $z = x + \mathrm{i}y$, the spectral Schwartz distribution

$$\frac{1}{\pi}\bar{\partial}_z(\xi_1|R(z)|\xi_2) = (\xi_1|\mu(x)|\xi_2)\delta(y)$$

with

$$\mu(x) = p_x^1 + p_x^2 + p_x^3$$

and

$$p_x^1 = |\alpha_x\rangle\langle\alpha_x|, \qquad |\alpha_x\rangle = \frac{1}{\sqrt{x^2 + \pi^2\gamma^2}} \begin{pmatrix} \sqrt{\gamma} \\ x|\mathbf{v}\delta_x\rangle/\sqrt{\gamma} + \sqrt{\gamma}\dfrac{\mathscr{P}}{x - \Omega}|E\rangle \end{pmatrix},$$

$$p_x^2 = \begin{pmatrix} 0 & 0 \\ 0 & q|\delta_x\rangle\langle\delta_x| \end{pmatrix}, \qquad q = A - \frac{|\mathbf{v}\rangle\langle\mathbf{v}|}{\langle\mathbf{v}|\mathbf{v}\rangle},$$

$$p_x^3 = (1 - A)\delta(x).$$

In the same way as before, we obtain the orthonormality relations

$$\langle \alpha_x | \alpha_y \rangle = \delta(x-y)$$

and

$$p_x^i p_y^j = \delta(x-y)\delta_{ij} p_i(x).$$

We also have completeness expressed by

$$\int \mathrm{d}x\, \mu(x) = 1.$$

4.4 The Heisenberg Equation of the Amplified Oscillator

4.4.1 Physical Considerations

Consider a quantum harmonic oscillator, with the usual creation and annihilation operators b^+ and b, in a heat bath of oscillators given by $a_\lambda^+, a_\lambda,\ \lambda \in \Lambda$, with the Hamiltonian

$$H_0 = -\omega_0 b^+ b + \sum_{\lambda\in\Lambda} (\omega_0 + \omega_\lambda) a_\lambda^+ a_\lambda + \sum_{\lambda\in\Lambda} (g_\lambda a_\lambda b + \overline{g}_\lambda a_\lambda^+ b^+).$$

This Hamiltonian, however, is not bounded below, so it cannot describe a real physical system. Nevertheless, it does enable one to discuss the initial behaviour of superradiance, and can be used as the model of a photon multiplier. We now sketch these ideas.

We consider N two-level atoms coupled to a heat bath. The Hilbert space of the atoms is $(\mathbb{C}^2)^{\otimes N}$. The Hamiltonian is

$$H_N = \sigma_3^{(N)} \omega_0 + \sum (\omega_0 + \omega_\lambda) a_\lambda^+ a_\lambda + \sum (N^{-1/2} g_\lambda \sigma_+^{(N)} + N^{-1/2} \overline{g}_\lambda \sigma_-^{(N)} a_\lambda^+)$$

with

$$\sigma_i^{(N)} = \sigma_i \otimes 1 \otimes \cdots \otimes 1 + \cdots + 1 \otimes \cdots \otimes 1 \otimes \sigma_i,$$

the sum of terms with σ_i in all possible positions in the N-fold tensor product, and the spin matrices are as usual given by

$$\sigma_1 = \frac{1}{2}\begin{pmatrix} 0 & 1 \\ 1 & 0 \end{pmatrix}, \qquad \sigma_1 = \frac{1}{2}\begin{pmatrix} 0 & \mathrm{i} \\ -\mathrm{i} & 0 \end{pmatrix}, \qquad \sigma_3 = \frac{1}{2}\begin{pmatrix} -1 & 0 \\ 0 & 1 \end{pmatrix},$$

$$\sigma_+ = \sigma_1 + \mathrm{i}\sigma_2 = \begin{pmatrix} 0 & 0 \\ 1 & 0 \end{pmatrix}, \qquad \sigma_- = \sigma_1 - \mathrm{i}\sigma_2 = \begin{pmatrix} 0 & 1 \\ 0 & 0 \end{pmatrix}.$$

The operators $\sigma_i^{(N)}$ obey the spin commutation relations, and $(\mathbb{C}^2)^{\otimes N}$ can be considered as a "spin representation space", or, in other words, as a representation space of the group U(2). Any irreducible representation space is invariant under the operator H.

In the case of superradiance, at $t = 0$ all atoms are initially in the upper state $\binom{0}{1}$. Then, due to spontaneous emission, one atom emits a photon, the radiation increases the probability that another atom emits a second single photon, etc. Thus an avalanche is created, which dies out when the atoms of a majority of the N atoms are in the lower state $\binom{1}{0}$.

For $t = 0$ the state of the atomic system is $\binom{0}{1}^{\otimes N} = \psi_{N/2}$, the highest weight vector of the representation, and successive applications of $\sigma_-^{(N)}$ create an irreducible invariant subspace spanned by ψ_m, $m = -N/2, -N/2+1, \ldots, N/2$. One has

$$\sigma_3^{(N)}\psi_m = m\psi_m, \qquad \sigma_\pm^{(N)}\psi_m = \left(\frac{N}{2}\left(\frac{N}{2}+1\right) - m(m\pm 1)\right)^{1/2}\psi_{m\pm 1}.$$

Put $\varphi_k = \psi_{N/2-k}$; then

$$N^{-1/2}\sigma_+^{(N)}\varphi_k = N^{-1/2}\left(Nk - k^2 + k\right)^{1/2}\varphi_{k-1} \to \sqrt{k}\,\varphi_{k-1},$$
$$N^{-1/2}\sigma_-^{(N)}\varphi_k = N^{-1/2}\left(N(k+1) - k^2 + k\right)^{1/2}\varphi_{k+1} \to \sqrt{k+1}\,\varphi_{k+1}.$$

For $N \to \infty$ the operator $N^{-1/2}\sigma_-^{(N)}$ becomes the creation operator b^+ and the operator $N^{-1/2}\sigma_+^{(N)}$ becomes the annihilation operator b. Shifting the operator H_N by adding $\omega_0 N/2$ we obtain H_0. By choosing, for $t = 0$, the vector $\psi_{-N/2} = \binom{1}{0}^{\otimes N}$ we would have arrived at the same irreducible representation, and an analogous procedure would have ended with the Hamiltonian for the damped oscillator.

We split H_0 into two commuting operators $H_0 = H_0' + H_0''$ with

$$H_0' = \sum_{\lambda\in\Lambda}\omega_\lambda a_\lambda^+ a_\lambda + \sum_{\lambda\in\Lambda}\left(g_\lambda a_\lambda b + \overline{g}_\lambda a_\lambda^+ b^+\right),$$

$$H_0'' = \omega_0\left(-b^+b + \sum_{\lambda\in\Lambda} a_\lambda^+ a_\lambda\right).$$

The time dependence due to H_0'' is trivial: it describes a fast oscillation modulated by the time development due to H_0'. We disregard it.

Put

$$\eta_t(b^+) = \exp\left(\mathrm{i}H_0't\right)b^+\exp\left(-\mathrm{i}H_0't\right),$$
$$\eta_t(a_\lambda) = \exp\left(\mathrm{i}H_0't\right)a_\lambda\exp\left(-\mathrm{i}H_0't\right).$$

Then

$$\frac{1}{\mathrm{i}}\frac{\mathrm{d}}{\mathrm{d}t}\begin{pmatrix}\eta_t(b^+)\\ \eta_t(a_\lambda)\end{pmatrix} = \sum_{\lambda'} H_{\lambda,\lambda'}\begin{pmatrix}\eta_t(b^+)\\ \eta_t(a_{\lambda'})\end{pmatrix}$$

with

$$H = \begin{pmatrix} 0 & \langle g| \\ -|g\rangle & \Omega \end{pmatrix},$$

where $|g\rangle$ is the column vector in $\mathbb{C}^\Lambda$ with the elements g_λ, $\langle g|$ is the row vector with the entries $\overline{g}_\lambda$, and Ω is the $\Lambda \times \Lambda$-matrix with entries $\omega_\lambda \delta_{\lambda,\lambda'}$. As in the first example of Sect. 4.2.1, we introduce a continuous set of frequencies. Then $|g\rangle$ becomes an L^2-function and Ω the multiplication operator.

4.4.2 The Singular Coupling Limit, Its Hamiltonian and Eigenvalue Problem

We recall the discussions of Sect. 4.2.2. We again have the Hilbert space

$$\mathfrak{H} = \mathbb{C} \oplus L^2(\mathbb{R})$$

with the scalar product

$$\big\langle (c, f) | (c', g) \big\rangle = \overline{c} c' + \int \mathrm{d}x\, \overline{f}(x) g(x).$$

In the last subsection we ended up with the block matrix

$$H_g = \begin{pmatrix} 0 & \langle g| \\ -|g\rangle & \Omega \end{pmatrix},$$

where $|g\rangle$ is an L^2-function and Ω is the multiplication operator. The matrix H is not symmetric but does satisfy the equation

$$J H_g J = H_g^+$$

with

$$J = \begin{pmatrix} -1 & 0 \\ 0 & 1 \end{pmatrix}.$$

Using Krein's formula we obtain the resolvent $R_g(z)$ of H_g as

$$\frac{1}{z - H_g} = R_g(z) = \begin{pmatrix} 0 & 0 \\ 0 & R_\Omega(z) \end{pmatrix} + \begin{pmatrix} 1 \\ -R_\Omega(z)|g\rangle \end{pmatrix} \frac{1}{C_g(z)} \big(1, \langle g| R_\Omega(z)\big)$$

with

$$C_g(z) = z - \langle g | R_\Omega(z) | g \rangle = z - \int \frac{|g(\omega)|^2}{z - \omega} \mathrm{d}\omega.$$

We perform the so-called singular coupling limit. We consider a sequence g_n of square-integrable functions, converging pointwise to E, and uniformly bounded by some constant function with the property

$$g_n(\omega) = \overline{g_n(-\omega)}.$$

Then, for fixed z with $\operatorname{Im} z \neq 0$, the resolvents $R_{g_n}(z)$ converge in operator norm to

$$R(z) = \begin{pmatrix} 0 & 0 \\ 0 & R_\Omega(z) \end{pmatrix} + \begin{pmatrix} 1 \\ -R_\Omega(z)|E\rangle \end{pmatrix} \frac{1}{z - \mathrm{i}\pi\sigma(z)} \big(1, \langle E|R_\Omega(z)\big).$$

Recall the spaces $\mathfrak{L}$ and $\mathfrak{L}^\dagger$, the functionals $\langle \hat{E}|$ and $|\hat{E}\rangle$, and the operator $\hat{\Omega}$ from Sect. 4.4.2. Define the operator

$$\hat{H} : \mathbb{C} \oplus \mathfrak{L} \to \mathbb{C} \oplus \mathfrak{L}^\dagger$$

$$\hat{H} = \begin{pmatrix} 0 & \langle \hat{E}| \\ -|\hat{E}\rangle & \hat{\Omega} \end{pmatrix}.$$

We have to distinguish between right and left domains D_l, resp. D_r, of the operator H corresponding to $R(z)$:

$$D_l = \mathfrak{H}R(z) = \big\{\xi \in \mathbb{C} \oplus \mathfrak{L} : \xi = c\big(1, \langle E|R(z)\big) + \big(0, \langle f|R(z)\big)\big\},$$

$$D_r = R(z)\mathfrak{H} = \left\{\xi \in \mathbb{C} \oplus \mathfrak{L} : \xi = c\begin{pmatrix} 1 \\ -R(z)|E\rangle \end{pmatrix} + \begin{pmatrix} 0 \\ R(z)f \end{pmatrix}\right\},$$

with $c \in \mathbb{C}$, $f \in L^2$. The Hamiltonian H is the restriction of $\hat{H}$ to D_l, resp. D_r. So

$$\langle \xi|H = \langle \xi|\hat{H},$$

$$H|\xi\rangle = \hat{H}|\xi\rangle,$$

for $\xi \in D_l$, resp. for $\xi \in D_r$.

The time development operator corresponding to $R(z)$ is for $t > 0$

$$U(t) = \begin{pmatrix} U_{00} & U_{01} \\ U_{10} & U_{11} \end{pmatrix}$$

with

$$U_{00} = \mathrm{e}^{\pi t},$$

$$U_{01} = \mathrm{i} \int_0^t \mathrm{d}t_1 \mathrm{e}^{\pi(t-t_1)} \langle E|\mathrm{e}^{-\mathrm{i}\Omega t_1},$$

$$U_{10} = -\mathrm{i} \int_0^t \mathrm{d}t_1 \mathrm{e}^{-\mathrm{i}\Omega(t-t_1)} |E\rangle \mathrm{e}^{\pi t_1},$$

$$U_{11} = e^{-i\Omega t} + \iint_{0<t_1<t_2<t} dt_1 dt_2 e^{-i\Omega(t-t_2)} |E\rangle e^{\pi(t_2-t_1)} \langle E| e^{-i\Omega t_1}.$$

Put

$$H_0 = \begin{pmatrix} 0 & 0 \\ 0 & \Omega \end{pmatrix}$$

and

$$V(t) = e^{iH_0 t} U(t).$$

Then

$$\begin{aligned}
V_{00} &= e^{\pi t}, \\
V_{01} &= i \int_0^t dt_1 e^{\pi(t-t_1)} \langle E| e^{-i\Omega t_1}, \\
V_{10} &= -i \int_0^t dt_1 e^{i\Omega t_1} |E\rangle e^{\pi t_1}, \\
V_{11} &= 1 + \iint_{0<t_1<t_2<t} dt_1 dt_2 e^{i\Omega t_2} |E\rangle e^{\pi(t_2-t_1)} \langle E| e^{-i\Omega t_1}
\end{aligned}$$

and, in the formal time representation of Sect. 4.2.2,

$$\begin{aligned}
V_{00}(t) &= e^{\pi t}, \\
\big(V_{01}(t)|\tau\big) &= i(2\pi)^{1/2} \int_0^t dt_1 e^{\pi(t-t_1)} \delta(\tau - t_1), \\
\big(\tau|V_{10}(t)\big) &= -i(2\pi)^{1/2} \int_0^t dt_1 \delta(t_1 - \tau) e^{\pi t_1}, \\
(\tau_2|V_{11}(t)|\tau_1) &= \delta(\tau_1 - \tau_2) - 2\pi \iint_{0<t_1<t_2<t} dt_1 dt_2 \delta(\tau_2 - t_2) e^{\pi(t_2-t_1)} \delta(t_1 - \tau_1).
\end{aligned}$$

Proposition 4.4.1 *The resolvent $R(z)$ is holomorphic outside the real line and away from the two simple poles $\pm i\pi$. The spectral Schwartz distribution $M(z) = (1/\pi)\bar{\partial} R(z)$ has the form*

$$M(x + iy) = \mu(x)\delta(y) + p_{i\pi}\delta(z - i\pi) + p_{-i\pi}\delta(z + i\pi)$$

with

$$\begin{aligned}
\mu(x) &= \frac{1}{2\pi i}\big(R(x - i0) - R(x + i0)\big) = |\alpha_x\rangle\langle\beta_x|, \\
|\alpha_x\rangle &= \big(x^2 + \pi^2\big)^{-1/2} \left(\begin{pmatrix} 1 \\ -\frac{\mathcal{P}}{x-\Omega}|E\rangle \end{pmatrix} + x \begin{pmatrix} 0 \\ |\delta_x\rangle \end{pmatrix} \right),
\end{aligned}$$

$$\langle\beta_x| = (x^2+\pi^2)^{-1/2}\left(-\left(1, \langle E|\frac{\mathscr{P}}{x-\Omega}\right) + x(0, \langle\delta_x|)\right),$$

and

$$p_{\pm i\pi} = |\alpha_{\pm i\pi}\rangle\langle\beta_{\pm i\pi}|,$$

$$|\alpha_{\pm i\pi}\rangle = \begin{pmatrix} 1 \\ -\frac{1}{\pm i\pi - \Omega}|E\rangle \end{pmatrix}, \qquad \langle\beta_{\pm i\pi}| = \left(1, \langle E|\frac{1}{\pm i\pi - \Omega}\right).$$

Proof Write

$$\begin{pmatrix} 1 \\ -R_\Omega(x \pm i0)|E\rangle \end{pmatrix} = a \pm i\pi b$$

with

$$a = \begin{pmatrix} 1 \\ -\frac{\mathscr{P}}{x-\Omega}|E\rangle \end{pmatrix}, \qquad b = \begin{pmatrix} 0 \\ |\delta_x\rangle \end{pmatrix},$$

and

$$\big(1, \langle E|R(x \pm i0)\big) = a' \mp i\pi b'$$

with

$$a' = \left(1, \langle E|\frac{\mathscr{P}}{x-\Omega}\right), \qquad b' = \big(0, \langle\delta_x|\big).$$

Then

$$\begin{aligned}
&\frac{1}{2\pi i}\big(R(x-i0) - R(x+i0)\big) \\
&\quad = bb' + \frac{1}{2\pi i}\left((a - i\pi b)\frac{1}{x+i\pi}(a' + i\pi b') - (a + i\pi b)\frac{1}{x-i\pi}(a' - i\pi b')\right) \\
&\quad = \frac{1}{x^2+\pi^2}(a + xb)(-a' + xb') = |\alpha_x\rangle\langle\beta_x|.
\end{aligned}$$

The terms $p_{\pm i\pi}$ are the residues of $R(z)$ at the points $\pm i\pi$, so, e.g.,

$$p_{i\pi} = \lim_{z\to i\pi}(z - i\pi)R(z) = \begin{pmatrix} 1 \\ -\frac{1}{\pi-\Omega}|E\rangle \end{pmatrix}\left(1, \langle E|\frac{1}{i\pi - \Omega}\right). \qquad \square$$

It is easy to check the bi-orthonormality relations

$$\begin{aligned}
&\langle\alpha_x|\beta_y\rangle = \delta(x-y), \\
&\langle\alpha_x|\beta_{\pm i\pi}\rangle = 0, \qquad \langle\alpha_{\pm i\pi}|\beta_x\rangle = 0, \\
&\langle\alpha_{\pm\, i\pi}|\beta_{\pm i\pi}\rangle = 1, \qquad \langle\alpha_{\pm i\pi}|\beta_{\mp i\pi}\rangle = 0.
\end{aligned}$$

Similarly to the discussions in Sect. 4.2.5, one proves the completeness condition

$$\int \mathrm{d}z\, M(z) = 1.$$

4.5 The Pure Number Process

The pure number quantum stochastic process restricted to the one-particle case is mathematically the easiest of the four examples, but it does not seem to have a direct physical meaning. We consider the Hamiltonian

$$H = \int \mathrm{d}\omega\, \omega\, a^\dagger(\omega) a(\omega) + \left(\int \mathrm{d}\omega\, \overline{g}(\omega) a^\dagger(\omega)\right)\left(\int \mathrm{d}\omega\, g(\omega) a(\omega)\right),$$

with $g \in L^2(\mathbb{R})$.

The underlying Hilbert space is the Fock space. The number operator

$$N = \int \mathrm{d}\omega\, a^\dagger(\omega) a(\omega)$$

commutes with H. The restriction of the Hamiltonian to the one-particle space yields the operator defined in L^2,

$$H_g = \Omega + |g\rangle\langle g|.$$

A slight modification of Krein's formula is needed, and yields

$$R_g(z) = \frac{1}{z - H_g} = R_\Omega(z) + \frac{1}{1 - \langle g|R_\Omega(z)|g\rangle} R_\Omega(z)|g\rangle\langle g|R_\Omega(z).$$

We perform the so-called singular coupling limit. We consider a sequence g_n of square-integrable functions, converging to E pointwise, uniformly bounded by a constant function, with the property $g_n(\omega) = \overline{g_n(-\omega)}$. Then, for fixed z with $\mathrm{Im}\, z \neq 0$, the resolvents $R_{g_n}(z)$ converge in operator norm to

$$R(z) = R_\Omega(z) + \frac{1}{1 + \mathrm{i}\pi\sigma(z)} R_\Omega(z)|E\rangle\langle E|R_\Omega(z)$$

with $\sigma(z) = \mathrm{sign}\, \mathrm{Im}\, z$. The corresponding unitary evolution has, for $t > 0$, the form

$$U(t) = \mathrm{e}^{-\mathrm{i}\Omega t} - \mathrm{i}\frac{1}{1 + \mathrm{i}\pi} \int_0^t \mathrm{d}t_1 \mathrm{e}^{-\mathrm{i}\Omega(t - t_1)}|E\rangle\langle E|\mathrm{e}^{-\mathrm{i}\Omega t_1}.$$

Put, for $t > 0$,

$$U(t) = \mathrm{e}^{-\mathrm{i}\Omega t} V(t),$$

so that

$$V(t) = 1 - \mathrm{i}\frac{1}{1+\mathrm{i}\pi}\int_0^t \mathrm{d}t_1 \mathrm{e}^{\mathrm{i}\Omega(t_1)}|E\rangle\langle E|\mathrm{e}^{-\mathrm{i}\Omega t_1}.$$

In the formal time representation we have

$$\langle h|V(t)|f\rangle = \langle h|f\rangle - \mathrm{i}\frac{2\pi}{1+\mathrm{i}\pi}\int_0^t \mathrm{d}t_1 \overline{h}(t_1) f(t_1),$$

or, in other words, $V(t)$ becomes the multiplication operator

$$\big(V(t)f\big)(\tau) = \left(1 - \frac{2\pi\mathrm{i}}{1+\mathrm{i}\pi}\mathbf{1}_{[0,t]}(\tau)\right) f(\tau).$$

This unitary group was found by Chebotarev [14].

The domain of the selfadjoint operator H is a subspace of the space $\mathfrak{L}$ defined in Sect. 4.2.2. It is

$$D = R(z)L^2(\mathbb{R}) = \left\{ R_\Omega(z)\left(|f\rangle + \frac{\langle E|R_\Omega(z)|f\rangle}{1+\mathrm{i}\sigma(z)\pi}|E\rangle \right) : f \in L^2 \right\}.$$

The Hamiltonian H is the restriction of

$$\hat{H} = \hat{\Omega} + |\hat{E}\rangle\langle\hat{E}|$$

to that domain [42]. With the methods used before we calculate the spectral Schwartz distribution $M(x+\mathrm{i}y) = \mu(x)\delta(y)$ with

$$\mu(x) = \frac{1}{2\pi\mathrm{i}}\big(R(x-\mathrm{i}0) - R(x+\mathrm{i}0)\big) = |\alpha_x\rangle\langle\alpha_x|,$$

$$|\alpha_x\rangle = \big(1+\pi^2\big)^{-1/2}\left(\frac{\mathcal{P}}{x-\Omega}|E\rangle + |\delta_x\rangle\right).$$

Chapter 5
White Noise Calculus

Abstract The creation and annihilation operators cannot be multiplied arbitrarily. Only special monomials can be formed, which are colled *admissible*. Normal ordered monomials are admissible and products of several normal ordered monomials depending on different variables are admissible, too. By a variant of Wick's theorem it can be shown, that any admissible monomial is the linear combination of normal ordered monomials: The coefficients are products of point measures. We prove the representation of unity by monomials of creation and annihilation operators and investigate the duality, which changes creators in annihilators and vice versa.

5.1 Multiplication of Diffusions

Before introducing white noise, we have to offer some preliminary explanations. We define for any locally compact space X, $\mathscr{M}_+(X)$ to be its set of positive measures. Let X and Y be two locally compact spaces. A continuous diffusion is (following Bourbaki, Intégration, Chap. 5 [11]) a vaguely continuous mapping

$$\kappa : X \to \mathscr{M}_+(Y) : x \mapsto \kappa_x.$$

Using the old-fashioned way of writing we have

$$\kappa = \kappa_x(\mathrm{d}y) = \kappa(x, \mathrm{d}y).$$

Vaguely continuous means that the mapping $x \in X \mapsto \int \kappa_x(\mathrm{d}y) f(y)$ is continuous for any $f \in \mathscr{K}(Y)$.

We consider three types of multiplication of diffusions:

1. Let X_1, X_2, Y_1, Y_2 be four locally compact spaces, and let

$$\kappa_1 : X_1 \to \mathscr{M}_+(Y_1),$$
$$\kappa_2 : X_2 \to \mathscr{M}_+(Y_2)$$

be continuous diffusions, then we can have as the product

$$\kappa : X_1 \times X_2 \to \mathscr{M}_+(Y_1 \times Y_2),$$

W. von Waldenfels, *A Measure Theoretical Approach to Quantum Stochastic Processes*, Lecture Notes in Physics 878, DOI 10.1007/978-3-642-45082-2_5,

$$\kappa(x_1, x_2; \mathrm{d}y_1, \mathrm{d}y_2) = \kappa_1(x_1, \mathrm{d}y_1)\kappa_2(x_2, \mathrm{d}y_2)$$

or

$$\kappa_{(x_1,x_2)} = \kappa_{1,x_1} \otimes \kappa_{2,x_2},$$

2. Let X, Y, Z be three locally compact spaces and

$$\kappa_1 : X \to \mathcal{M}_+(Y),$$
$$\kappa_2 : Y \to \mathcal{M}_+(Z)$$

be continuous diffusions, then we can take as a second alternative product

$$\kappa : X \to \mathcal{M}_+(Y \times Z),$$
$$\kappa(x; \mathrm{d}y, \mathrm{d}z) = \kappa_1(x, \mathrm{d}y)\kappa_2(y, \mathrm{d}z).$$

So

$$\iint \kappa(x; \mathrm{d}y, \mathrm{d}z) f(x, y) = \int \kappa_1(x, \mathrm{d}y) \int \kappa_2(y, \mathrm{d}z) f(y, z).$$

This product is familiar from probability theory. If $\kappa_1(x, \mathrm{d}y)$ is the probability of transition from x to y and $\kappa_2(y, \mathrm{d}z)$ is the probability of transition from y to z, then $\kappa(x; \mathrm{d}x, \mathrm{d}y)$ is the transition probability from x to y *and* z.

3. Let X, Y, Z be three locally compact spaces and

$$\kappa_1 : X \to \mathcal{M}_+(Y),$$
$$\kappa_2 : X \to \mathcal{M}_+(Z)$$

be continuous diffusions, then we can take as the third product

$$\kappa : X \to \mathcal{M}_+(Y \times Z),$$
$$\kappa(x; \mathrm{d}y, \mathrm{d}z) = \kappa_1(x, \mathrm{d}y)\kappa_2(x, \mathrm{d}z).$$

So

$$\kappa_x = \kappa_{1,x} \otimes \kappa_{2,x}.$$

Using the positivity of the diffusions it is easy to see, that all three types of multiplications again yield positive continuous diffusions. We shall not introduce different symbols for the multiplications, but rely on the different notations using differentials to make clear which is in play.

5.2 Multiplication of Point Measures

Using Bourbaki's terminology we denote by ε_x the point measure at the point $x \in X$. So for $f \in \mathscr{K}(X)$ we have

$$\int \varepsilon_x(\mathrm{d}y) f(y) = f(x).$$

We consider the diffusion

$$\varepsilon : x \in X \mapsto \varepsilon_x \in \mathscr{M}_+(X).$$

We have the three ways of defining the product of two point measures. If the four variables x_1, x_2, x_3, x_4 are different, then we may first define the tensor product

$$\varepsilon_{x_1}(\mathrm{d}x_2)\varepsilon_{x_3}(\mathrm{d}x_4) = \varepsilon_{x_1} \otimes \varepsilon_{x_3}(\mathrm{d}x_2, \mathrm{d}x_4),$$

$$\iint \varepsilon_{x_1} \otimes \varepsilon_{x_3}(\mathrm{d}x_2, \mathrm{d}x_4) f(x_2, x_4) = f(x_1, x_3).$$

Then a second way is

$$\varepsilon_{x_1}(\mathrm{d}x_2)\varepsilon_{x_2}(\mathrm{d}x_3) =: E_{x_1}(\mathrm{d}x_2, \mathrm{d}x_3) = \varepsilon_{x_1} \otimes \varepsilon_{x_1}(\mathrm{d}x_2, \mathrm{d}x_3),$$

$$\iint E_{x_1}(\mathrm{d}x_2, \mathrm{d}x_3) f(x_2, x_3) = f(x_1, x_1).$$

The third possibility is

$$\varepsilon_{x_1}(\mathrm{d}x_2)\varepsilon_{x_1}(\mathrm{d}x_3) = \varepsilon_{x_1} \otimes \varepsilon_{x_1}(\mathrm{d}x_2, \mathrm{d}x_3).$$

That the last two products amount to the same here is a property of ε_x. We omit the variable x and write only the indices, and use the notation

$$\varepsilon_{x_b}(\mathrm{d}x_c) = \varepsilon(b, c) \quad \text{and} \quad E_{x_1}(\mathrm{d}x_1, \mathrm{d}x_2) = E(1; 2, 3).$$

We want to define the product of

$$\{\varepsilon(b_i, c_i) : i = 1, \dots, n\}.$$

Consider the set

$$S = \{(b_1, c_1), \dots, (b_n, c_n)\},$$

where all the b_i and all the c_i are different and $b_i \neq c_i$. We introduce in S the structure of an oriented graph by defining the relation of being a right neighbor

$$(b, c) \rhd (b', c') \Longleftrightarrow c = b'.$$

An element (b, c) has at most one right neighbor, as $(b_i, c_i) \rhd (b_j, c_j)$ and $(b_i, c_i) \rhd (b_k, c_k)$ implies $b_j = b_k$ and $j = k$. So the components of the graph S are either chains or circuits.

We have to avoid the expression $\varepsilon_x(\mathrm{d}x)$. This notion makes no sense and if one wants to give it a sense, one runs into problems. If X is discrete, then $\varepsilon_x(\mathrm{d}y) = \delta_{x,y}$ and $\varepsilon_x(\mathrm{d}x) = \delta_{x,x} = 1$. If $X = \mathbb{R}$, then $\varepsilon_x(\mathrm{d}y) = \delta(x-y)\mathrm{d}y$, and if one wants to approximate Dirac's delta function one obtains $\varepsilon_x(\mathrm{d}x) = \infty$.

Consider a circuit

$$(1,2), (2,3), \ldots, (k-2, k-1), (k-1, 1).$$

It corresponds to a product

$$\varepsilon(1,2)\varepsilon(2,3)\cdots\varepsilon(k-2,k-1)\varepsilon(k-1,1).$$

Integrating over $x_2, \ldots, x_{k-1}$ one obtains $\varepsilon(1,1)$, which cannot be defined. So in order that $\prod_{i=1}^n \varepsilon(b_i, c_i)$ can be defined, it is necessary that the graph S contain no circuits.

On the other hand, if $(1,2), (2,3), \ldots, (k-2, k-1), (k-1, k)$ is a chain, then using the second form of multiplication we have

$$\begin{aligned}
&\varepsilon(1,2)\varepsilon(2,3)\cdots\varepsilon(k-2,k-1)\varepsilon(k-1,k) \\
&\quad = E(1; 2, 3, \ldots, k) = \varepsilon_{x_1}^{\otimes(k-1)}(\mathrm{d}x_2, \ldots, \mathrm{d}x_k), \\
&\int\cdots\int_{2,3,\ldots,k} E(1; 2, 3, \ldots, k) f(2, 3, \ldots, k) = f(x_1, \ldots, x_1).
\end{aligned}$$

Use the notation $S_- = \{b_1, \ldots, b_n\}$ and $S_+ = \{c_1, \ldots, c_n\}$. If S contains no circuits, then any $p \in S_- \setminus S_+$ is the starting point of a (maximal) chain

$$(p, c_{p,1}), (c_{p,1}, c_{p,2}), \ldots, (c_{p,k-1}, c_{p,k}).$$

Use the notation $\pi_p = \{c_{p,1}, \ldots, c_{p,k}\}$. We have

$$\varepsilon(p, c_{p,1})\varepsilon(c_{p,1}, c_{p,2})\cdots\varepsilon(c_{p,k-1}, c_{p,k}) = E(p; \pi_p)$$

where explicitly

$$E(p; \pi_p) = \varepsilon_{x_p}^{\otimes \# \pi_p}(\mathrm{d}x_{\pi_p}).$$

Finally we adopt

Definition 5.2.1 If S contains no circuits, then

$$E_S = \prod_{i=0}^{n} \varepsilon(b_i, c_i) = \prod_{p \in S_- \setminus S_+} E(p; \pi_p).$$

5.3 White Noise Operators

Recall the generalization of the creation operator a^+ to the diffusion $\varepsilon : x \mapsto \varepsilon_x$ and the definition

$$\big(a^+(\varepsilon(\mathrm{d}y))f\big)(x_\alpha) = \sum_{c\in\alpha} \varepsilon_{x_c}(\mathrm{d}y) f(x_{\alpha\setminus c}).$$

We write for short, if b is an index,

$$a^+\big(\varepsilon(\mathrm{d}x_b)\big) = a^+(\mathrm{d}x_b) = a_b^+.$$

The annihilation operator $a(x) = a(\varepsilon_x)$ is the special case for the annihilation operator $a(\nu)$ (defined in Sect. 2.3)

$$\big(a(\varepsilon_{x_b})f\big)(x_\alpha) = \big(a(x_b)f\big)(x_\alpha) = f(x_{\alpha+b}).$$

We write for short

$$a(\varepsilon_{x_b}) = a_b.$$

If $\alpha = \{b_1, \dots, b_n\}$ is a set, then

$$a_\alpha^+ = a_{b_1}^+ \cdots a_{b_n}^+, \qquad a_\emptyset^+ = 1,$$

$$a_\alpha = a_{b_1} \cdots a_{b_n}, \qquad a_\emptyset = 1.$$

We shall be dealing with functions on the space $\mathfrak{X}$, which we recall is the space of all tuples of elements of X:

$$\mathfrak{X} = \{\emptyset\} + X + X^2 + \cdots.$$

Write for short

$$\mathscr{K} = \mathscr{K}_s(\mathfrak{X}).$$

Recall the function

$$\Phi \in \mathscr{K}; \quad \Phi(x) = \begin{cases} \Phi(x) = 1 & \text{for } x = \emptyset, \\ \Phi(x) = 0 & \text{for } x \neq \emptyset \end{cases}$$

and the measure

$$\Psi \in \mathscr{M}_s(\mathfrak{X}); \quad \Psi(f) = f(\emptyset).$$

We define

$$\Phi_\alpha = a_\alpha^+ \Phi,$$

and then

$$\Phi_\emptyset = \Phi$$

and, for $\alpha \neq \emptyset$,

$$\Phi_\alpha(x_\upsilon) = \varepsilon(\upsilon, \alpha) \quad \text{for } \upsilon \cap \alpha = \emptyset,$$

where

$$\varepsilon(\upsilon, \alpha) = \begin{cases} \sum_{h:\alpha \twoheadrightarrow \upsilon} \prod_{i=1}^{n} \varepsilon(h(b_i), b_i) & \text{if } \#\alpha = \#\upsilon, \\ 0 & \text{otherwise.} \end{cases}$$

More explicitly the last expression could have been written

$$\sum_{h:\alpha \twoheadrightarrow \upsilon} \prod_{i=1}^{n} \varepsilon(x_{h(b_i)}, \mathrm{d}x_{b_i})$$

showing the dependence on the variables of $\mathfrak{X}$. Here the sign $\twoheadrightarrow$ signifies a bijective mapping. So the sum runs over all bijections from α to υ. We call Φ_α a *measure-valued finite-particle vector*. So Φ_α is a continuous diffusion

$$\Phi_\alpha : \mathfrak{X} \to X^\alpha.$$

Extending Ψ we have

$$\Psi a_\upsilon a_\alpha^+ \Phi = \varepsilon(\upsilon, \alpha).$$

Assume we are given a set $\sigma = \{s_1, \ldots, s_m\}$ and a set $S = \{(b_i, c_i) : i = 1, \ldots, n\}$, where all the elements b_i and c_i are different. Use, as above, the notation $S_- = \{b_1, \ldots, b_n\}$ and $S_+ = \{c_1, \ldots, c_n\}$, and assume that $\sigma \cap S_+ = \emptyset$. We extend the relation $\triangleright$ of right neighbor from S to the pair (σ, S) by defining

$$s \triangleright (b, c) \Longleftrightarrow s = b.$$

If the graph (σ, S) is without circuits and $(\sigma \cup S_+ \cup S_-) \cap \upsilon = \emptyset$, then for any $f : \upsilon \twoheadrightarrow \sigma$, the graph

$$S \cup \{(c, f(c)), c \in \upsilon\}$$

is cycle-free, so there are no problems in defining $E_S \Phi_\sigma = \Phi_\sigma E_S$.

The graph naturally is made up of a collection of chains, some of which begin with an element in σ and some of which do not. We break up the nodes in the graph into groups according to the chains in which they are. We carry along the first element in the case of chains that begin in σ. All the rest of the elements in a chain must be target elements in some edge for the relation $\triangleright$, i.e., in S_+. Formally, we set this out in a lemma.

Lemma 5.3.1 *The set of components of the graph* (σ, S) *is*

$$\Gamma = \Gamma(\sigma, S) = \Gamma_1 + \Gamma_2,$$

$$\Gamma_1 = \{\{s, (s, c_{s,1}), (c_{s,1}, c_{s,2}), \ldots, (c_{s,k_s-1}, c_{s,k_s})\}; s \in \sigma\},$$

$$\Gamma_2 = \{\{(t, c_{t,1}), (c_{t,1}, c_{t,2}), \ldots, (c_{t,k_t-1}, c_{t,k_t})\}; t \in S_- \setminus (S_+ + \sigma)\}.$$

Put

$$\begin{aligned}
\xi_s &= \{s, c_{s,1}, \ldots, c_{s,k_s}\} \quad \textit{for } s \in \sigma, \\
\pi_t &= \{c_{t,1}, \ldots, c_{t,k_t}\} \quad \textit{for } t \in S_- \setminus (S_+ + \sigma), \\
\pi &= S_+ + \sigma, \\
\varrho &= S_- \setminus (S_+ + \sigma).
\end{aligned}$$

Then

$$\pi = \sum_{s \in \sigma} \xi_s + \sum_{t \in \varrho} \pi_t.$$

Note that π is made up of the nodes which are in σ, or are second components of pairs; ρ are those nodes which are not connected to σ by a chain. This partitions the chains into two types. The physical reason for these considerations is that there are the chains of interactions connected to the vacuum and those which are not.

So $\Phi_\sigma E_S$ is a continuous diffusion

$$\begin{aligned}
&\Phi_\sigma E_S : \mathfrak{X} \times X^\varrho \to \mathscr{M}_+(X^\pi), \\
&\Phi_\sigma E_S(x_\upsilon + x_\varrho) = \sum_{f:\upsilon \twoheadrightarrow \sigma} \prod_{c \in \upsilon} E(c, \xi_{f(c)}) \prod_{t \in \varrho} E(t, \pi_t).
\end{aligned}$$

Definition 5.3.1 We denote by $\mathscr{G}_{n,\pi,\varrho}$ the additive monoid generated by the elements of the form $\Phi_\sigma E_S$, such that $\sigma \cap S_+ = \emptyset$ and the graph (σ, S) is circuit-free, and that

$$\varrho = S_- \setminus (S_+ + \sigma), \qquad \pi = S_+ + \sigma, \qquad n = \#\sigma.$$

We use the corresponding notation

$$\mathscr{G}_{\pi,\varrho} = \bigoplus_n \mathscr{G}_{n,\pi,\varrho}.$$

We define for $c \notin \sigma$, using $a_c \Phi = 0$,

$$\begin{aligned}
a_c \Phi_\sigma &= \sum_{b \in \sigma} \varepsilon(c, b) \Phi_{\sigma \setminus b}, \\
a_c^+ \Phi_\sigma &= \Phi_{\sigma + c},
\end{aligned}$$

and obtain for $b \neq c$, $b, c \notin \sigma$,

$$a_b^+ a_c^+ \Phi_\sigma = a_c^+ a_b^+ \Phi_\sigma,$$

$$a_b a_c \Phi_\sigma = a_c a_b \Phi_\sigma,$$
$$a_b a_c^+ \Phi_\sigma = \varepsilon(b,c)\Phi_\sigma + a_c^+ a_b \Phi_\sigma.$$

Proposition 5.3.1 *Assume*

$$f = \Phi_\sigma E_S \in \mathcal{G}_{n,\pi,\varrho}.$$

Then, for $c \notin \pi$, we have

$$\big(a_c^+ \Phi_\sigma\big) E_S \in \mathcal{G}_{n+1,\pi+c,\varrho\setminus c}$$

and we can define

$$a_c^+ f = \big(a_c^+ \Phi_\sigma\big) E_S.$$

If $c \notin \pi + \varrho$, then

$$(a_c \Phi_\sigma) E_S \in \mathcal{G}_{n-1,\pi,\varrho+c}$$

and we can define

$$a_c f = (a_c \Phi_\sigma) E_S.$$

Proof We only have to prove that there are no circuits created for the definitions to be good ones.

The graph of $\Phi_\sigma E_S$ is (σ, S). Its set of components is $\Gamma = \Gamma_1 + \Gamma_2$ as above. Assume $c \notin \pi$ and consider $(a_c^+ \Phi_\sigma) E_S$. The corresponding graph is $(S', \sigma') = (\sigma + c, S)$. Denote by $\Gamma' = \Gamma_1' + \Gamma_2'$ the corresponding set of components of (S', σ'). There are two cases:

(a) $c \notin S_-$, in which case $\Gamma_1' = \Gamma_1 + \{c\}$, $\Gamma_2' = \Gamma_2$, and $\pi' = \pi + c$ and $\varrho' = \varrho$.
(b) $c = t \in S_-$, so that

$$\Gamma_1' = \Gamma_1 + \big\{t, (t, c_{t,1}), (c_{t,1}, c_{t,2}), \ldots, (c_{t,k_t-1}, c_{t,k_t})\big\},$$
$$\Gamma_2' = \Gamma_2 \setminus \big\{(t, c_{t,1}), (c_{t,1}, c_{t,2}), \ldots, (c_{t,k_t-1}, c_{t,k_t})\big\}$$

and $\pi' = \pi + c$ and $\varrho' = S_-' \setminus (\sigma' + S_+') = \varrho \setminus \{c\}$.

In both cases the graph $(\sigma + c, S)$ contains no circuits and $(a_c^+ \Phi_\sigma) E_S$ is defined; we set

$$a_c^+ (\Phi_\sigma E_S) = \big(a_c^+ \Phi_\sigma\big) E_S.$$

Assume $c \notin \pi + \varrho$ and consider $(a_c \Phi_\sigma) E_S$. It consists of a sum of terms with a graph of the form $(S'', \sigma'') = (\sigma \setminus b, S + (c, b))$. Denote the corresponding sets of components by Γ_1'', Γ_2''. Then we have

$$\Gamma_1'' = \Gamma_1 \setminus \big\{b, (b_{c_1}, b_{c_2}), \ldots, (b_{c_{k-1}}, b_{c_k})\big\},$$
$$\Gamma_2'' = \Gamma_2 + \big\{(c, b), (b_{c_1}, b_{c_2}), \ldots, (b_{c_{k-1}}, b_{c_k})\big\}$$

and $\pi'' = \pi$, $\varrho'' = \varrho + \{c\}$. The graph (σ'', S'') has no circuits. □

Definition 5.3.2 A finite sequence

$$W = (a_{c_n}^{\vartheta_n}, \ldots, a_{c_1}^{\vartheta_1})$$

with indices $c_1, \ldots, c_n$ and $\vartheta_i = \pm 1$ and the usual

$$a_c^{\vartheta} = \begin{cases} a_c^+ & \text{for } \vartheta = +1, \\ a_c & \text{for } \vartheta = -1 \end{cases}$$

is called *admissible* if

$$i > j \implies \{c_i \neq c_j \text{ or } \{c_i = c_j \text{ and } \vartheta_i = 1, \vartheta_j = -1\}\}.$$

So W is admissible if it contains only pairs (not necessarily neighbors) of the form $(a_c^{\vartheta}, a_{c'}^{\vartheta'})$ with $c \neq c'$, or (a_c^+, a_c) and no pairs of the form (a_c, a_c), (a_c^+, a_c^+) or (a_c, a_c^+).

If W is an admissible sequence, define

$$\omega(W) = \{c_1, \ldots, c_n\},$$

$$\omega_+(W) = \{c_i, 1 \leq i \leq n : \vartheta_i = +1\},$$

$$\omega_-(W) = \{c_i, 1 \leq i \leq n : \vartheta_i = -1\}.$$

If

$$W = (a_{c_n}^{\vartheta_n}, \ldots, a_{c_1}^{\vartheta_1})$$

is an admissible sequence we call

$$M = a_{c_n}^{\vartheta_n} \cdots a_{c_1}^{\vartheta_1}$$

an *admissible monomial.*

The following proposition shows that iterated creators and annihilators can be defined in a suitable way.

Proposition 5.3.2 *Assume*

$$W = (a_{c_n}^{\vartheta_n}, \ldots, a_{c_1}^{\vartheta_1})$$

to be an admissible sequence. Assume disjoint index sets π and ϱ are given and that

$$\omega_+(W) \cap \pi = \emptyset,$$

$$\omega_-(W) \cap (\pi + \varrho) = \emptyset.$$

Define, for $k = 1, \ldots, n$,

$$W_k = (a_{c_k}^{\vartheta_k}, \ldots, a_{c_1}^{\vartheta_1}).$$

Set $\pi_0 = \pi$, $\varrho_0 = \varrho$ *and*

$$\pi_k = \pi + \omega_+(W_k),$$
$$\varrho_k = \big(\varrho + \omega_-(W_k)\big) \setminus \omega_+(W_k)$$

where for the sets α *and* β

$$\alpha \setminus \beta = \alpha \setminus (\alpha \cap \beta).$$

Then we have for the maps of Proposition 5.3.1

$$a_{c_k}^{\vartheta_k} : \mathcal{G}_{\pi_{k-1},\varrho_{k-1}} \to \mathcal{G}_{\pi_k,\varrho_k}$$

and for the corresponding iterated maps

$$M = a_{c_n}^{\vartheta_n} \cdots a_{c_1}^{\vartheta_1} : \mathcal{G}_{\pi,\varrho} \to \mathcal{G}_{\pi',\varrho'}$$

with

$$\pi' = \pi + \omega_+(W),$$
$$\varrho' = \big(\varrho + \omega_-(W)\big) \setminus \omega_+(W).$$

Proof For $l = 1, \ldots, n$ use the shorter notation $\omega_l = \omega(W_l)$ and $\omega_{\pm,k} = \omega_\pm(W_k)$.

We carry out the proof by induction. The case of one operator is trivial. Assume that we have proven the theorem for up to $k-1$ operators. Assume $\vartheta_k = +1$. In order that $a_{c_k}^+$ be defined, $c_k \notin \pi_{k-1}$, in the notation given in the theorem's statement. But $c_k \notin \pi$ by assumption and $c_k \notin \omega_{+,k-1}$, since W_k is admissible. So we have a mapping

$$a_{c_k}^+ : \mathcal{G}_{\pi_{k-1},\varrho_{k-1}} \to \mathcal{G}_{\pi_{k-1}+c_k,\varrho_{k-1}\setminus c_k}.$$

Now $\pi_{k-1} + c_k = \pi + \omega_{+,k} = \pi_k$ and

$$\varrho_{k-1} \setminus c_k = (\varrho + \omega_{-,k-1}) \cap \complement\omega_{+,k-1} \cap \complement\{c_k\} = (\varrho + \omega_{-,k}) \cap \complement\omega_{+,k} = \varrho_k$$

as $\omega_{-,k-1} = \omega_{-,k}$ and $\omega_{+,k-1} + c_k = \omega_{+,k}$.

Assume now, that $\vartheta_k = -1$. In order that a_{c_k} be defined,

$$c_k \notin \pi_{k-1} + \varrho_{k-1} \subset \pi + \varrho + \omega_{k-1}.$$

But $c_k \notin \pi + \varrho$ by assumption and $c_k \notin \omega_{k-1}$, as W_k is admissible. So we have the mapping

$$a_{c_k} : \mathcal{G}_{\pi_{k-1},\varrho_{k-1}} \to \mathcal{G}_{\pi_{k-1},\varrho_{k-1}+c_k}.$$

But $\omega_{+,k} = \omega_{+,k-1}$ and

$$\pi_k = \pi + \omega_{k-1} = \pi_{k-1}$$

and

$$\varrho_{k-1} + c_k = \big((\varrho + \omega_{-,k-1}) \cap \complement\omega_{+,k-1}\big) \cup \{c_k\} = (\varrho + \omega_{-,k}) \cap \complement\omega_{+,k} = \varrho_k$$

as $c_k \notin \omega_{+,k-1}$. □

5.4 Wick's Theorem

We prove a theorem analogous to that of Sect. 1.3 and to Proposition 1.7.2. The general theorem of Sect. 1.3 cannot be applied, as the multiplication is not always defined. But the ideas of our proof are borrowed from there.

Assume two finite index sets σ, τ and a finite set of pairs $S = \{(b_i, c_i) : i \in I\}$, such that all b_i and all c_i are different and $b_i \neq c_i$. We extend the relation of right neighbor to the triple (σ, S, τ) by putting for $(b, c) \in S, t \in \tau$

$$(b, c) \rhd t \Longleftrightarrow c = t.$$

Consider a triple (σ, S, τ), $\sigma \cap \tau = \emptyset$, and two finite sets υ, β such that the three sets $\sigma \cup S_+ \cup S_- \cup \tau$ and υ and β are pairwise disjoint. As

$$\big(a_\sigma^+ a_\tau \Phi_\upsilon\big)(\beta) = \sum_{\upsilon_1 + \upsilon_2 = \upsilon} \varepsilon(\tau, \upsilon_1)\varepsilon(\beta, \sigma + \upsilon_2)$$

we find that the product $(a_\sigma^+ a_\tau \Phi_\upsilon)(\beta) E_S$ is defined if the graph (σ, S, τ) is free of circuits and we define the operator

$$a_\sigma^+ a_\tau E_S = a_\sigma^+ E_S a_\tau = E_S a_\sigma^+ a_\tau$$

that way.

Consider an admissible sequence $W = (a_{c_n}^{\vartheta_n}, \ldots, a_{c_1}^{\vartheta_1})$ and the associated sets ω_+, ω_-. We define the set $\mathfrak{P}(W)$ of all decompositions of $[1, n]$, i.e., all sets of subsets, of the form

$$\mathfrak{p} = \big\{\mathfrak{p}_+, \mathfrak{p}_-, \{q_i, r_i\}_{i \in I}\big\},$$

$$[1, n] = \mathfrak{p}_+ + \mathfrak{p}_- + \sum_{i \in I} \{q_i, r_i\},$$

$$\mathfrak{p}_+ \subset \omega_+,\ \mathfrak{p}_- \subset \omega_-,\ q_i \in \omega_-,\ r_i \in \omega_+,\ q_i > r_i.$$

Lemma 5.4.1 *Assume W to be admissible and $\mathfrak{p} \in \mathfrak{P}(W)$. Then the graph (σ, S, τ) with*

$$\sigma = \{c_s : s \in \mathfrak{p}_+\}, \qquad S = \big\{(c_{q_i}, c_{r_i}) : i \in I\big\}, \qquad \tau = \{c_t : t \in \mathfrak{p}_-\}$$

has no circuits.

Proof Let $(c_q, c_r) \rhd (c_{q'}, c_{r'})$, then $c_r = c_{q'}$ and $r > q'$ as W is admissible. By the definition of $\mathfrak{p}$ we have $q > r$ and $q' > r'$, so $q > r > q' > r'$, and if we have a sequence $(c_{q_1}, c_{r_1}) \rhd \cdots \rhd (c_{q_k}, c_{r_k})$, then $q_1 > r_1 > \cdots q_k > r_k$ and as $\vartheta_1 = -1$, $\vartheta_k = +1$ we have $c_{q_1} \neq c_{r_k}$ as W is admissible. This proves that S is without circuits. For the other components of the graph one uses similar arguments. □

Definition 5.4.1 For $\mathfrak{p} \in \mathfrak{P}(W)$ we define

$$\lfloor W \rfloor_{\mathfrak{p}} = \prod_{s \in \mathfrak{p}_+} a^+_{c_s} \prod_{i \in I} \varepsilon(c_{q_i}, c_{r_i}) \prod_{t \in \mathfrak{p}_-} a_{c_t}.$$

Theorem 5.4.1 (Wick's theorem) *If W is admissible and if M is the corresponding monomial, then*

$$M = \sum_{\mathfrak{p} \in \mathfrak{P}(W)} \lfloor W \rfloor_{\mathfrak{p}}.$$

Proof We proceed by induction. The case $n = 1$ is clear. We write for short $\mathfrak{p}_i = (q_i, r_i)$, $\varepsilon(c(\mathfrak{p}_i)) = \varepsilon(c_{q_i}, c_{r_i})$. Assume

$$V = \left(a^{\vartheta_n}_{c_n}, \ldots, a^{\vartheta_1}_{c_1}\right)$$

to be admissible and set

$$N = a^{\vartheta_n}_{c_n} \cdots a^{\vartheta_1}_{c_1}.$$

Consider $W = (a_{c_{n+1}}, V)$ and define a mapping $\varphi_- : \mathfrak{P}(W) \to \mathfrak{P}(V)$ consisting in erasing $n+1$. Then $n+1$ may occur in one of the $\mathfrak{p}_i$, say in $\mathfrak{p}_{i_0}$, or in $\mathfrak{p}_-$. In the first case

$$\varphi_- \mathfrak{p} = \left\{\mathfrak{p}_+ + \{r_{i_0}\},\ (\mathfrak{p}_i)_{i \in I \setminus i_0},\ \mathfrak{p}_-\right\}$$

in the second case

$$\varphi_- \mathfrak{p} = \left\{\mathfrak{p}_+,\ (\mathfrak{p}_i)_{i \in I},\ \mathfrak{p}_- \setminus \{n+1\}\right\}.$$

Assume

$$\mathfrak{q} = \left\{\mathfrak{q}_+,\ (\mathfrak{q}_j)_{j \in J},\ \mathfrak{q}_-\right\} \in \mathfrak{P}(V)$$

then

$$\begin{aligned}
\varphi_-^{-1} \mathfrak{q} &= \{\mathfrak{p} : \varphi_- \mathfrak{p} = \mathfrak{q}\} = \left\{\mathfrak{p}^{(0)}, \mathfrak{p}^{(l)}, l \in \mathfrak{q}_+\right\}, \\
\mathfrak{p}^{(0)} &= \left\{\mathfrak{q}_+, (\mathfrak{q}_j)_{j \in J}, \mathfrak{q}_- + \{n+1\}\right\}, \\
\mathfrak{p}^{(l)} &= \left\{\mathfrak{q}_+ \setminus l, (\mathfrak{q}_j)_{j \in J}, (n+1, l), \mathfrak{q}_-\right\}.
\end{aligned}$$

Consider

$$a_{c_{n+1}} \lfloor V \rfloor_{\mathfrak{q}} = a_{c_{n+1}} \prod_{s \in \mathfrak{q}_+} a^+_{c_s} \prod_{j \in J} \varepsilon(c(\mathfrak{q}_j)) \prod_{t \in \mathfrak{q}_-} a_{c_t}$$

$$= \prod_{s\in\mathfrak{q}_+} a^+_{c_s} \prod_{j\in J} \varepsilon(c(\mathfrak{q}_j)) \prod_{t\in\mathfrak{q}_-+\{n+1\}} a_{c_t}$$

$$+ \sum_{l\in\mathfrak{q}_+} \prod_{s\in\mathfrak{q}_+\setminus l} a^+_{c_s} \prod_{j\in J} \varepsilon(c(\mathfrak{q}_j)\varepsilon(c_{n+1}, c_l) \prod_{t\in\mathfrak{q}_-} a_{c_t}$$

$$= \sum_{\mathfrak{p}\in\varphi_-^{-1}(\mathfrak{q})} \lfloor W \rfloor_{\mathfrak{p}}.$$

Finally

$$a_{c_{n+1}} N = \sum_{\mathfrak{q}\in\mathfrak{P}(V)} a_{c_{n+1}} \lfloor V \rfloor = \sum_{\mathfrak{q}\in\mathfrak{P}(V)} \sum_{\mathfrak{p}\in\varphi_-^{-1}(\mathfrak{q})} \lfloor W \rfloor_{\mathfrak{p}} = \sum_{\mathfrak{p}\in\mathfrak{P}(W)} \lfloor W \rfloor_{\mathfrak{p}}.$$

Consider now $W = (a^+_{c_{n+1}}, V)$ and define a map $\varphi_+ : \mathfrak{P}(W) \to \mathfrak{P}(V)$ consisting in erasing $n+1$ then

$$\varphi_+\mathfrak{p} = \{\mathfrak{p}_+ \setminus \{n+1\}, (\mathfrak{p}_i)_{i\in I}, \mathfrak{p}_-\},$$

$$\varphi_+^{-1}\mathfrak{q} = \{\mathfrak{q}_+ + \{n+1\}, (\mathfrak{q}_j)_{j\in J}, \mathfrak{q}_- +\},$$

$$a^+_{c_{n+1}} \lfloor V \rfloor_{\mathfrak{q}} = \lfloor W \rfloor_{\varphi_+^{-1}\mathfrak{q}}.$$

By the same reasoning as above one finishes the proof. □

5.5 Representation of Unity

We extend the functional Ψ to $\mathcal{G}_{n,\pi,\varrho}$ by putting

$$\Psi\Phi_\sigma = \begin{cases} 1 & \text{for } \sigma = \emptyset, \\ 0 & \text{otherwise} \end{cases}$$

and

$$\Psi\Phi_\sigma E_S = (\Psi\Phi_\sigma) E_S.$$

Definition 5.5.1 Assume

$$W = (a^{\vartheta_n}_{c_n}, \dots, a^{\vartheta_1}_{c_1})$$

to be admissible and

$$M = a^{\vartheta_n}_{c_n} \cdots a^{\vartheta_1}_{c_1}.$$

Then we define

$$\langle M \rangle = \sum_{\mathfrak{p}\in\mathfrak{P}_0(W)} \lfloor W \rfloor_{\mathfrak{p}}.$$

Here $\mathfrak{P}_0(W)$ is the set of partitions of $[1,n]$ into pairs $\{q_i, r_i\}_{i=1,\dots,n/2}$ such that $q_1 > r_i, \vartheta_{q_i} = -1, \vartheta_{r_i} = +1$, and

$$\lfloor W \rfloor_{\mathfrak{p}} = \prod_i \varepsilon(b_i, c_i).$$

If n is odd or $\mathfrak{P}_0(W)$ is empty, then $\langle M \rangle = 0$.

As a consequence of Wick's Theorem 5.4.1 we obtain

Proposition 5.5.1 *We obtain*

$$(M\Phi)(\emptyset) = \Psi M \Phi = \begin{cases} 0 & \text{if } \vartheta_1 + \cdots + \vartheta_n \neq 0, \\ \langle M \rangle & \text{if } \vartheta_1 + \cdots + \vartheta_n = 0. \end{cases}$$

If M is admissible, then

$$(M\Phi_\beta)(\alpha) = \Psi a_\alpha M a_\beta^+ \Phi = \langle a_\alpha M a_\beta^+ \rangle.$$

We shall use this notation very often.

Theorem 5.5.1 *If $M = M_2 M_1$ is admissible, then*

$$\langle M \rangle = \int_\alpha \Delta\alpha \langle M_2 a_\alpha^+ \rangle \langle a_\alpha M_1 \rangle.$$

Proof Assume

$$M = a_{c_n}^{\vartheta_n} \cdots a_{c_1}^{\vartheta_1},$$
$$M_2 = a_{c_n}^{\vartheta_n} \cdots a_{c_k}^{\vartheta_k},$$
$$M_1 = a_{c_{k-1}}^{\vartheta_{k-1}} \cdots a_{c_1}^{\vartheta_1}.$$

We prove the theorem by induction with respect to k. For $k = n$ we have

$$\Psi a_\alpha^+ a_\alpha M \Phi = \begin{cases} \langle M \rangle & \text{for } \alpha = \emptyset, \\ 0 & \text{otherwise.} \end{cases}$$

Integration yields the result. Put $M_2' = a_{c_n}^{\vartheta_n} \cdots a_{c_{k+1}}^{\vartheta_{k+1}}$. Assume $\vartheta_k = -1$. Then

$$\begin{aligned} \int_\alpha \Delta\alpha \langle M_2 a_\alpha^+ \rangle \langle a_\alpha M_1 \rangle &= \int_\alpha \Delta\alpha \langle M_2' a_{c_k} a_\alpha^+ \rangle \langle a_\alpha M_1 \rangle \\ &= \int_\alpha \Delta\alpha \sum_{b \in \alpha} \langle M_2' a_{\alpha \setminus b}^+ \rangle \langle a_\alpha M_1 \rangle \varepsilon(c_k, b) \end{aligned}$$

$$= \int_\alpha \Delta\alpha \int_b \langle M_2' a_\alpha^+ \rangle \langle a_{\alpha+b} M_1 \rangle \varepsilon(c_k, b)$$

$$= \int_\alpha \Delta\alpha \langle M_2' a_\alpha^+ \rangle \langle a_\alpha a_{c_k} M_1 \rangle.$$

In a similar way one proves

$$\int_\alpha \Delta\alpha \langle M_2' a_\alpha^+ \rangle \langle a_\alpha a_{c_k}^+ M_1 \rangle = \int_\alpha \Delta\alpha \langle M_2' a_{c_k}^+ a_\alpha^+ \rangle \langle a_\alpha M_1 \rangle.$$

□

5.6 Duality

We fix a positive measure λ on X, and instead of writing $\mathrm{e}(\lambda)$ we shall just continue to write λ when there are indexed variables like x_α, and so by abuse of notation

$$\mathrm{e}(\lambda)(\mathrm{d}x_\alpha) = \lambda^{\otimes\alpha}(\mathrm{d}x_\alpha) = \lambda(\mathrm{d}x_\alpha) = \lambda(\alpha) = \lambda_\alpha.$$

We define the measure Λ on X^k given by

$$\int \Lambda(1,\dots,k) f(1,\dots,k)$$
$$= \int \Lambda(dx_1,\dots,dx_k) f(x_1,\dots,x_k) = \int \lambda(dx) f(x,\dots,x).$$

So we have

$$\lambda(1)\varepsilon(1,2)\cdots\varepsilon(k-1,k) = \Lambda(1,2,\dots,k).$$

Assume

$$W = \left(a_{c_n}^{\vartheta_n}, \dots, a_{c_1}^{\vartheta_1}\right),$$

to be an admissible sequence with $\vartheta_1 + \cdots \vartheta_n = 0$. Define as usual $\omega_\pm(W) = \{c_i : \vartheta_i = \pm 1\}$. Recall from Theorem 5.4.1 that

$$\langle M \rangle = \sum_{\mathfrak{p} \in \mathfrak{P}_0(W)} \lfloor W \rfloor_{\mathfrak{p}}.$$

Here $\mathfrak{P}_0(W)$ is the set of partitions of $[1,n]$ into pairs $\{q_i, r_i\}_{i=1,\dots,n/2}$ such that $q_1 > r_i, \vartheta_{q_i} = -1, \vartheta_{r_i} = +1$, and

$$\lfloor W \rfloor_{\mathfrak{p}} = \prod_i \varepsilon(b_i, c_i).$$

Call $S(\mathfrak{p})$ the graph related to $\mathfrak{p}$ and $\Gamma(S(\mathfrak{p}))$ the set of components of the graph. To any $s \in S_-(\mathfrak{p}) \setminus S_+(\mathfrak{p})$ there is associated a component. As

$$S_-(\mathfrak{p}) \setminus S_+(\mathfrak{p}) = \omega_-(W) \setminus \omega_+(W) = \varrho$$

for any $\mathfrak{p}$, we obtain

$$\langle M\rangle\lambda(\varrho) = \sum_{\mathfrak{p}\in\mathfrak{P}_0(W)} \lfloor W\rfloor_{\mathfrak{p}}\lambda(\varrho) = \sum_{\mathfrak{p}\in\mathfrak{P}_0(W)} \prod_{\gamma\in\Gamma(S(\mathfrak{p}))} \Lambda(\gamma).$$

Definition 5.6.1 Assume

$$W = \left(a_{c_n}^{\vartheta_n}, \dots, a_{c_1}^{\vartheta_1}\right),$$

to be an admissible sequence, then define the formally adjoint sequence by

$$W^+ = \left(a_{c_1}^{-\vartheta_1}, \dots, a_{c_n}^{-\vartheta_n}\right).$$

If M is the monomial corresponding to W, we denote by M^+ the monomial corresponding to W^+.

W^+ is admissible as well. Using the symmetry of Λ one sees that

Theorem 5.6.1

$$\langle M\rangle\lambda\big(\omega_-(W)\setminus\omega_+(W)\big) = \big\langle M^+\big\rangle\lambda\big(\omega_+(W)\setminus\omega_-(W)\big).$$

Chapter 6
Circled Integrals

Abstract The circled integral will be needed to treat quantum stochastic differential equations. We solve a circled integral equation, introduce the class $\mathscr{C}^1$, which has remarkable analytical properties, and show, that the solution is a $\mathscr{C}^1$ function.

6.1 Definition

We use the notation $\mathfrak{R}$ for

$$\mathfrak{R} = \{\emptyset\} + \mathbb{R} + \mathbb{R}^2 + \cdots.$$

We provide $\mathfrak{R}$ with the measure $\mathrm{e}(\lambda)$ induced by the Lebesgue measure λ, and write for short $\mathrm{e}(\lambda)(\mathrm{d}t_\alpha) = \mathrm{d}t_\alpha = \lambda_\alpha$. So for a symmetric function

$$\begin{aligned}\int \Delta\alpha f(t_\alpha)\mathrm{d}t_\alpha &= f(\emptyset) + \sum_{n=1}^{\infty} \frac{1}{n!} \int \cdots \int_{\mathbb{R}^n} \mathrm{d}t_1 \cdots \mathrm{d}t_n f(t_1, \ldots, t_n) \\ &= f(\emptyset) + \sum_{n=1}^{\infty} \int \cdots \int_{t_1 < \cdots < t_n} \mathrm{d}t_1 \cdots \mathrm{d}t_n f(t_1, \ldots, t_n).\end{aligned}$$

Definition 6.1.1 Assume given a Banach algebra $\mathfrak{B}$ and a function x

$$x : \mathbb{R} \times \mathfrak{R}^k \to \mathfrak{B},$$

$$(t, w_1, \ldots, w_k) \mapsto x_t(w_1, \ldots, w_k)$$

symmetric in any of the variables $w_1, \ldots, w_k$, and locally integrable in norm with respect to the Lebesgue measure on $\mathbb{R} \times \mathfrak{R}^k$. Let there be given a Lebesgue integrable function $f : \mathbb{R} \to \mathbb{C}$. The *circled integral* $\oint^j (f)x$ is defined by

$$\left(\oint^j (f)x\right)(t_{\alpha_1}, \ldots, t_{\alpha_k}) = \sum_{c \in \alpha_j} f(t_c)\, x_{t_c}(t_{\alpha_1}, \ldots, t_{\alpha_{j-1}}, t_{\alpha_j \setminus c}, t_{\alpha_{j+1}}, \ldots, t_{\alpha_k}).$$

The circled integral has been called *Skorohod integral* by P.A. Meyer [34].

W. von Waldenfels, *A Measure Theoretical Approach to Quantum Stochastic Processes*, Lecture Notes in Physics 878, DOI 10.1007/978-3-642-45082-2_6,

Remark 6.1.1 The function

$$(w_1, \ldots, w_k) \in \mathfrak{R}^k \mapsto \left(\oint^j (f)x \right)(t_{\alpha_1}, \ldots, t_{\alpha_k}) \in \mathfrak{B}$$

is symmetric in each of the variables t_{α_i} and locally integrable.

Proof The symmetry is trivial; for local integrability it is sufficient that all functions have values ≥ 0. Let $g(t_{\alpha_1}, \ldots, t_{\alpha_k})$ be a continuous function with $g \geq 0$ and compact support, symmetric in any of the t_{α_i}, then by the sum-integral lemma, Theorem 2.2.1,

$$\int \cdots \int \left(\oint^j (f)x \right)(t_{\alpha_1}, \ldots, t_{\alpha_k}) g(t_{\alpha_1}, \ldots, t_{\alpha_k}) \mathrm{d}t_{\alpha_1} \cdots \mathrm{d}t_{\alpha_k} \Delta\alpha_1 \cdots \Delta\alpha_k$$

$$= \int_{\mathbb{R}} \int \cdots \int f(t_c) x_{t_c}(t_{\alpha_1}, \ldots, t_{\alpha_k})$$

$$\times g(t_{\alpha_1}, \ldots, \ldots, t_{\alpha_j + c}, \ldots, t_{\alpha_k}) \mathrm{d}t_c \mathrm{d}t_{\alpha_1} \cdots \mathrm{d}t_{\alpha_k} \Delta\alpha_1 \cdots \Delta\alpha_k < \infty. \qquad \square$$

6.2 A Circled Integral Equation

Definition 6.2.1 Consider the subset

$$\left\{ (t_{\alpha_1}, \ldots, t_{\alpha_k}) \in \mathfrak{R}^k : \text{ all } t_i \text{ for } i \in \alpha_1 + \cdots + \alpha_k \text{ are different} \right\};$$

this differs from the set $\mathfrak{R}^k$ by a null set. We define on this set a mapping Ξ onto $\mathfrak{S}(\mathbb{R} \times \{1, \ldots, k\})$, where $\mathfrak{S}$ denotes the set of finite subsets, by mapping

$$(t_{\alpha_1}, \ldots, t_{\alpha_k}) \mapsto \xi = \left\{ (s_1, i_1), \ldots, (s_n, i_n) \right\}$$

where

$$t_{\alpha_1} + \cdots + t_{\alpha_k} = \{s_1, \ldots, s_n\},$$
$$i_l = j \Leftrightarrow s_l \in t_{\alpha_j}.$$

That is we list all the variables occurring in the t_{α_j} as s_l's and add a second index, showing in which block j a variable occurs, to make an entry (s_l, j).

Definition 6.2.2 We are given the Banach algebra $\mathfrak{B}$; assume $A_1, \ldots, A_k, B \in \mathfrak{B}$ and that all points in the following subset of $\mathbb{R}$

$$\{s, t\} \cup \{t_i : i \in \alpha_1 + \cdots + \alpha_k\}$$

are different, which holds a.e., and define

$$u(A_1,\ldots,A_k,B):\{s,t\in\mathbb{R}^2, s<t\}\times\mathfrak{R}^k$$
$$\mapsto u_s^t(A_1,\ldots,A_k,B)(t_{\alpha_1},\ldots,t_{\alpha_k})\in\mathfrak{B}$$

by

$$\begin{aligned} &u_s^t(A_1,\ldots,A_k,B)(t_{\alpha_1},\ldots,t_{\alpha_k}) \\ &\quad=\mathbf{1}\{s<s_1<\cdots<s_n<t\}\exp\bigl((t-s_n)B\bigr)A_{i_n}\exp\bigl((s_n-s_{n-1})B\bigr)A_{i_{n-1}} \\ &\qquad\times\cdots\times A_{i_2}\exp\bigl((s_2-s_1)B\bigr)A_{i_1}\exp\bigl((s_1-s)B\bigr) \end{aligned}$$

where the renumbering of variables defined above is

$$\Xi(t_{\alpha_1},\ldots,t_{\alpha_k})=\{(s_1,i_1),\ldots,(s_n,i_n)\}$$

with

$$s_1<\cdots<s_n.$$

Define the unit function

$$\mathbf{e}:\mathfrak{R}^k\to\mathfrak{B},$$

$$\mathbf{e}(t_{\alpha_1},\ldots,t_{\alpha_k})=\begin{cases}1 & \text{if } t_{\alpha_1}=\cdots=t_{\alpha_k}=\emptyset,\\ 0 & \text{otherwise.}\end{cases}$$

Write, for short,

$$\oint_{s,t}^{j}=\oint^{j}(\mathbf{1}_{]s,t[}).$$

Theorem 6.2.1 *Assume $A_1,\ldots,A_k,B\in\mathfrak{B}$ and that*

$$x:(t,t_{\alpha_1},\ldots,t_{\alpha_k})\in\mathbb{R}\times\mathfrak{R}^k\mapsto x_t(t_{\alpha_1},\ldots,t_{\alpha_k})\in\mathfrak{B}$$

is a symmetric function in each of the variables t_{α_i} and locally integrable. Consider for $t>s$ the equation

$$x_t=\mathbf{e}+\sum_{j=1}^{k}A_j\oint_{s,t}^{j}x+\int_s^t Bx_u\,du.$$

Then

$$x_t=u_s^t(A_1,\ldots,A_k,B)$$

is the unique solution of that equation.

Proof The proof is very similar to that of [41, Lemma 6.1]. We include it for completeness. Using the renumbering Ξ we rewrite the equation in terms of

$$\xi = \{(s_1, i_1), \dots, (s_n, i_n)\}, \quad s_1 < \cdots < s_n,$$

and to spare ourselves further heavy notation we view a pair (s_j, i_j) as also denoting the variable in the t_{α_i} to which it corresponds, namely $\Xi^{-1}(\{(s_j, i_j)\})$; we extend this then to all of ξ. With this convention we obtain a rewritten form of the equation to be solved, with a sum running now to n over the list of all the variables in the k different t_{α_i},

$$x_t(\xi) = \mathbf{e}(\xi) + \sum_{l=1}^{n} A_{i_l} x_{s_l}\big(\xi \setminus (s_l, i_l)\big)\mathbf{1}\{s < s_l < t\} + B \int_s^t x_u(\xi)du.$$

Then we can make use of the equation

$$x_t(\emptyset) = 1 + B \int_s^t x_u(\emptyset)du$$

whose solution is

$$x_t(\emptyset) = \exp\big((t - s)B\big).$$

We want to prove by induction that $x_t(\xi) = 0$ if $\{s_1, \dots, s_n\} \not\subset]s, t[$.

Assume $n = 1$ and $s_1 \notin]s, t[$; then, looking at the equation above for $\xi = x_t(\{(s_1, i_1)\})$ we see the $\mathbf{e}(\xi)$ term vanishes since $\xi \neq \emptyset$, the second term vanishes because the set in $\{s < s_i < t\}$ is empty, and we are left with

$$x_t\big(\{(s_1, i_1)\}\big) = B \int_s^t x_u\big(\{(s_1, i_1)\}\big)du$$

which has only the solution, namely $x_t(\{(s_1, i_1)\}) = 0$.

With $n > 1$, if $\{s_1, \dots, s_n\} \not\subset]s, t[$, since the s_j were chosen ordered, then at least one of the s_i, either s_1 or s_n, is not in $]s, t[$. Assume $s_1 \notin]s, t[$, then

$$x_t(\xi) = \sum_{l=2}^{n} A_{i_l} \mathbf{1}\{s < s_l < t\} x_{s_l}\big(\xi \setminus (s_l, i_l)\big) + B \int_s^t x_u(\xi)du.$$

The first sum vanishes, since, for each contribution, $s_1 < s$ is still contained in the shorter set of indices $\xi \setminus (s_l, i_l)$ so the induction hypothesis applies; the integral contribution vanishes as argued above; therefore $x_t(\xi) = 0$.

Now if $\{s_1, \dots, s_n\} \subset]s, t[$, then $x_{s_l}(\xi \setminus (s_l, i_l)) = 0$ for $l < n$, since

$$\{s_1, \dots, s_n\} \setminus s_l \not\subset]s, s_l[;$$

similarly $x_u(\xi) = 0$ for $u < s_n$. So we are left with the final contribution

$$x_t(\xi) = A_{i_n} x_{s_n}\big(\{(s_1, i_1), \dots, (s_{n-1}, i_{n-1})\}\big) + B \int_{s_n}^t x_u(\xi)du.$$

But it is known how to solve this integral equation, and we get

$$x_t(\xi) = e^{B(t-s_n)} A_{i_n} x_{s_n}\big((s_1, i_1), \dots, (s_{n-1}, i_{n-1})\big).$$

Repeating this procedure to pull out all the exponential terms we finally obtain, as asserted,

$$\begin{aligned} x_t(\xi) &= \mathbf{1}\{s < s_1 < \cdots < s_n < t\} \exp\big((t - s_n)B\big) A_{i_n} \exp\big((s_n - s_{n-1})B\big) A_{i_{n-1}} \\ &\quad \times \cdots \times A_{i_2} \exp\big((s_2 - s_1)B\big) A_{i_1} \exp\big((s_1 - s)B\big) \\ &= u_s^t(A_1, \dots, A_k, B)(t_{\alpha_1}, \dots, t_{\alpha_k}). \end{aligned}$$

□

In a similar way one proves, for the lower variable s of the evolution,

Proposition 6.2.1 *For $s < t$, the function*

$$s \mapsto y_s = u_s^t(A_1, \dots, A_k, B)$$

is the unique solution of the equation

$$y_s = \mathbf{e} + \sum_{j=1}^{k} \left(\oint_{s,t}^{j} y \right) A_{i_j} - \int_s^t y_u \mathrm{d}u\, B.$$

Proof Similar to the previous theorem's proof. □

Remark 6.2.1 Again use the representation Ξ, and write

$$u_t^s(A_1, \dots, A_k, B)(t_{\alpha_1}, \dots, t_{\alpha_k}) = u_s^t(\xi)$$

with

$$\xi = \big\{(s_1, i_1), \dots, (s_n, i_n)\big\} \quad \text{and} \quad s_1 < \cdots < s_n,$$

and assuming $s < r < t$ and $s_{j-1} < r < s_j$; then

$$u_s^t(\xi) = u_r^t(\xi_2) u_s^r(\xi_1)$$

with

$$\begin{aligned} \xi_1 &= \big\{(s_1, i_1), \dots, (s_{j-1}, i_{j-1})\big\}, \\ \xi_2 &= \big\{(s_j, i_j), \dots, (s_n, i_n)\big\}. \end{aligned}$$

6.3 Functions of Class $\mathscr{C}^1$

It will be important for later calculations that we are working with what are called $\mathscr{C}^1$-functions.

Definition 6.3.1 Assume a function

$$x : (t, t_{\alpha_1}, \dots, t_{\alpha_k}) \in \mathbb{R} \times \mathfrak{R}^k \mapsto x_t(t_{\alpha_1}, \dots, t_{\alpha_k}) \in \mathfrak{B}$$

is symmetric in each of $t_{\alpha_1}, \dots, t_{\alpha_k}$. Then x is called of class $\mathscr{C}^0$ if the function is locally integrable and continuous in the subspace where all points $t, t_i,\ i \in \alpha_1 + \cdots \alpha_k$, are different. We call x of class $\mathscr{C}^1$ if it is of class $\mathscr{C}^0$ and if, on the same subspace, the functions

$$\begin{aligned}
\left(\partial^c x\right)_t(t_{\alpha_1}, \dots, t_{\alpha_k}) &= \frac{\mathrm{d}}{\mathrm{d}t} x_t(t_{\alpha_1}, \dots, t_{\alpha_k}),\\
\left(R^j_{\pm} x\right)_t(t_{\alpha_1}, \dots, t_{\alpha_k}) &= x_{t\pm 0}\left(t_{\alpha_1}, \dots, t_{\alpha_{j-1}}, t_{\alpha_j} + \{t\}, t_{\alpha_{j+1}}, \dots, t_{\alpha_k}\right)
\end{aligned}$$

exist for $j = 1, \dots, k$, and are of class $\mathscr{C}^0$. Here $\mathrm{d}/\mathrm{d}t = \partial^c$ is the usual derivative at the points of ordinary differentiability, and $R^j_{\pm}$ denote respectively the limits at t from above and below, which are assumed to exist where the function is not continuous. Put

$$D^j x = R^j_+ x - R^j_- x.$$

Proposition 6.3.1 *If x_t is of class $\mathscr{C}^1$, then on the subspace*

$$S \subset \mathfrak{R}^k = \left\{(t_{\alpha_1}, \dots, t_{\alpha_k})\right\},$$

where all points $t_i, i \in \alpha_1, \dots, \alpha_k$ are different, the function $x_t(t_{\alpha_1}, \dots, t_{\alpha_k})$ has left and right limits at every point t, so $x_{t\pm 0}(t_{\alpha_1}, \dots, t_{\alpha_k})$ are well defined and we have for $s < t$

$$x_{t-0} = x_{s+0} + \int_s^t \mathrm{d}t' \partial^c x_{t'} + \sum_{j=1}^k \oint_{s,t}^j D^j x.$$

Conversely, if $k + 1$ functions $f_0, \dots, f_k$ of type $\mathscr{C}^0$ are given, and g is locally integrable and continuous on S, then

$$x_t = g + \int_s^t \mathrm{d}t' f_0(t') + \sum_{j=1}^k \oint_{s,t}^j f_j$$

is of type $\mathscr{C}^1$, and

$$\begin{aligned}
\left(\partial^c x\right)_t(t_{\alpha_1}, \dots, t_{\alpha_k}) &= f_0(t_{\alpha_1}, \dots, t_{\alpha_k}),\\
\left(D^j x\right)_t(t_{\alpha_1}, \dots, t_{\alpha_k}) &= f_j(t_{\alpha_1}, \dots, t_{\alpha_k}).
\end{aligned}$$

Hence

$$\begin{aligned}
\left(R^j_- x\right)_t(t_{\alpha_1}, \dots, t_{\alpha_k}) &= x(t)(t_{\alpha_1}, \dots, t_{\alpha_k}),\\
\left(R^j_+ x\right)_t(t_{\alpha_1}, \dots, t_{\alpha_k}) &= x(t)(t_{\alpha_1}, \dots, t_{\alpha_k}) + \left(D^j x\right)_t(t_{\alpha_1}, \dots, t_{\alpha_k}).
\end{aligned}$$

Proof We have that, on S, the function x_t is continuous if t is not one of the variables in $t_{\alpha_1+\cdots+\alpha_k}$. If t is a variable in $t_{\alpha_1+\cdots+\alpha_k}$, e.g., t is a variable in t_{α_j}, we obtain

$$x_{t\pm 0}(t_{\alpha_1},\ldots,t_{\alpha_k}) = \left(R^j_{\pm}x\right)_t\left(t_{\alpha_1},\ldots,t_{\alpha_{j-1}},t_{\alpha_j}\setminus\{t\},t_{\alpha_{j+1}},\ldots,t_{\alpha_k}\right);$$

but the right-hand side is well defined, because t is not amongst the variables of $t_{\alpha_1+\cdots+\alpha_{j-1}+\alpha_j+\alpha_{j+1}+\cdots+\alpha_k}\setminus\{t\}$. So $x_{t\pm 0}$ is well defined on S.

To finish the proof we discuss only the case $k=1$, since for general k we can use analogous reasoning using the representation Ξ. Assume then we have $\alpha=\alpha_1$, so

$$t_\alpha \cap]s,t[= \{s_1 < \cdots < s_n\}$$

and put $s_0 = s$, $s_{n+1} = t$; then

$$\begin{aligned}
x_{t-0}(t_\alpha) - x_{s+0}(t_\alpha) &= \sum_{i=0}^{n}\int_{s_i}^{s_{i+1}} \mathrm{d}t'\,\partial^{\mathrm{c}}x\left(t'\right)(t_\alpha) + \sum_{i=1}^{n}\left(x_{s_i+0}(t_\alpha) - x_{s_i-0}(t_\alpha)\right)\\
&= \int_s^t \mathrm{d}t'\,\partial^{\mathrm{c}}x\left(t'\right)(t_\alpha) + \sum_{i=1}^{n}(Dx)_{s_i}(t_\alpha\setminus s_i)\\
&= \int_s^t \mathrm{d}t'\,\partial^{\mathrm{c}}x\left(t'\right)(t_\alpha) + \oint_{s,t}(Dx)(t_\alpha)
\end{aligned}$$

since we naturally write $D^1 = D$ and $\oint^1 = \oint$. □

Proposition 6.3.2 *For fixed s, the function $u^{\cdot}_s : t \mapsto u^t_s(A_i,B)$, and for fixed t, the functions $u^t_{\cdot} : s \mapsto u^t_s(A_i,B)$, are each of class $\mathscr{C}^1$, and one has*

$$\begin{aligned}
\partial^{\mathrm{c}}_t u^t_s &= Bu^t_s,\\
\left(R^j_+ u^{\cdot}_s\right)_t &= A_j u^t_s,\\
\left(R^j_- u^{\cdot}_s\right)_t &= 0,\\
\partial^{\mathrm{c}}_s u^t_s &= -u^t_s B,\\
\left(R^j_+ u^t_{\cdot}\right)_s &= 0,\\
\left(R^j_- u^t_{\cdot}\right)_s &= u^t_s A_j
\end{aligned}$$

for $j = 1,\ldots,k$.

Proof By straight-forward calculation. □

We recall the definition of the Schwartz test functions on the real line. They make up the space $C^\infty_{\mathrm{c}}(\mathbb{R})$ of infinitely differentiable functions of compact support.

Definition 6.3.2 For f locally integrable on $\mathbb{R}$, the Schwartz derivative is the functional given by

$$(\partial f)(\varphi) = -\int f(t)\varphi'(t)\mathrm{d}t$$

for Schwartz test functions φ. If the functional is given by

$$(\partial f)(\varphi) = \int g(t)\varphi(t)\mathrm{d}t,$$

where g is locally integrable, we write

$$g = \partial f.$$

If f is continuously differentiable except at a finite set of points $\{t_1, \dots, t_n\}$, then its Schwartz differential is the measure

$$\partial f = \partial^{\mathrm{c}} f + \sum_{i=1}^{n} \big(f(t_i + 0) - f(t_i - 0)\big)\varepsilon_{t_i}(\mathrm{d}t),$$

where $\partial^{\mathrm{c}} f$ is the usual derivative outside the jump points, and ε_t is the point measure in the point t.

We extend the notion of the circled integral to the vaguely continuous measure-valued function $\varepsilon : x \mapsto \varepsilon_x$ by defining

$$\left(\oint^j \varepsilon(\mathrm{d}t)x\right)(t_{\alpha_1}, \dots, t_{\alpha_k}) = \sum_{c\in\alpha_j} \varepsilon_{t_c}(\mathrm{d}t)\, x_{t_c}(t_{\alpha_1}, \dots, t_{\alpha_{j-1}}, t_{\alpha_j\setminus c}, t_{\alpha_{j+1}}, \dots, t_{\alpha_k}).$$

This expression is scalarly defined, i.e., for any function f with compact support in $\mathbb{R}$ we have

$$\int \left(\oint^j \varepsilon(\mathrm{d}t)x\right) f(t) = \oint^j (f)x.$$

Proposition 6.3.3 *If x is of class $\mathscr{C}^1$, then its Schwartz derivative is*

$$(\partial x_t)(\mathrm{d}t) = \big(\partial^{\mathrm{c}} x\big)_t \mathrm{d}t + \sum_{j=1}^{k} \oint^j \varepsilon(\mathrm{d}t)\big(D^j x\big).$$

Proof We calculate

$$-\int \cdots \int \Delta\alpha_1 \cdots \Delta\alpha_k \mathrm{d}t_{\alpha_1} \cdots \mathrm{d}t_{\alpha_k} g(t_{\alpha_1}, \dots, t_{\alpha_k}) \int \mathrm{d}t\, \varphi'(t) x_t(t_{\alpha_1}, \dots, t_{\alpha_k})$$

where g is a continuous function of compact support. It is sufficient to calculate the integral outside the null set where all the t_i, for $i \in \alpha_1 + \cdots + \alpha_k$, are different.

Using the representation (Definition 6.2.1)

$$\Xi(t_{\alpha_1},\dots,t_{\alpha_k}) = \xi = \{(s_1,i_1),\dots,(s_n,i_n)\}$$

with $s_1 < \dots < s_n$, we may write

$$-\int \mathrm{d}t\,\varphi'(t)x_t(t_{\alpha_1},\dots,t_{\alpha_k}) = -\int \mathrm{d}t\,\varphi'(t)x_t(\xi)$$

$$= \int \mathrm{d}t\,\varphi(t)\partial^{\mathrm{c}}x_t(\xi) + \sum_{j=1}^{n}\varphi(t_j)\big(x_{s_j+0}(\xi) - x_{s_j-0}(\xi)\big).$$

The second term equals

$$\sum_{j=1}^{k}\sum_{c\in\alpha_j}\varphi(t_c)\big(x_{t_c+0}(t_{\alpha_1},\dots,t_{\alpha_k}) - x_{t_c-0}(t_{\alpha_1},\dots,t_{\alpha_k})\big)$$

$$= \sum_{j=1}^{k}\sum_{c\in\alpha_j}\varphi(t_c)\big(D^j x\big)_{t_c}(t_{\alpha_1},\dots,t_{\alpha_j\setminus c},\dots,t_{\alpha_k})$$

$$= \sum_{j=1}^{k}\left(\oint^{j}(\varphi)D^j x\right)(t_{\alpha_1},\dots,t_{\alpha_k}).$$

From there one obtains the proposition immediately. □

Chapter 7
White Noise Integration

Abstract We define integrals of normal ordered monomials. These integrals are scalarly defined as sesquilinear forms over $\mathscr{K}_s(\mathfrak{X}, \mathfrak{k})$, the space of all symmetric, continuous functions of compact support with values in a Hilbert space $\mathfrak{k}$. We can define products of those objects as scalarly defined integrals. We define $\mathscr{C}^1$-processes and calculate their Schwartz derivatives. We prove Ito's theorem for $\mathscr{C}^1$-processes.

7.1 Integration of Normal Ordered Monomials

In the following we shall, if not otherwise stated, skip $\Delta\alpha$ etc. in the integrals. So we write, e.g.,

$$\int \mu(\mathrm{d}\alpha) \quad \text{for} \int \mu(\mathrm{d}\alpha)\Delta\alpha.$$

Recall that this expression stands for

$$\int \mu(\mathrm{d}\alpha) = \mu(\emptyset) + \sum_{n=1}^{\infty} \frac{1}{n!} \int \cdots \int \mu(\mathrm{d}x_1, \ldots, \mathrm{d}x_n).$$

With this simplified notation the sum-integral lemma, Theorem 2.2.1, reads

$$\int_{\alpha_1} \cdots \int_{\alpha_k} \mu(\mathrm{d}x_{\alpha_1}, \ldots, \mathrm{d}x_{\alpha_k}) = \int_{\alpha} \sum_{\alpha_1 + \cdots + \alpha_n = \alpha} \mu(\mathrm{d}x_{\alpha_1}, \ldots, \mathrm{d}x_{\alpha_k})$$

or, by neglecting the $\mathrm{d}x$,

$$\int_{\alpha_1} \cdots \int_{\alpha_k} \mu(\alpha_1, \ldots, \alpha_k) = \int_{\alpha} \sum_{\alpha_1 + \cdots + \alpha_k = \alpha} \mu(\alpha_1, \ldots, \alpha_k).$$

Recall an admissible monomial is of the form (Definition 5.3.2)

$$M = a_{c_n}^{\vartheta_n} \cdots a_{c_1}^{\vartheta_1}.$$

W. von Waldenfels, *A Measure Theoretical Approach to Quantum Stochastic Processes*, Lecture Notes in Physics 878, DOI 10.1007/978-3-642-45082-2_7,

Let S_+ be the set of all i, such that $\vartheta_i = +1$, and S_- the set of all i, such that $\vartheta_i = -1$. If λ is the base measure, we will use the fact, that

$$\langle M\rangle \lambda_{S_- \setminus S_+}$$

is a positive measure in the usual sense on an appropriate space. We shall denote by $\langle \Phi|$ the measure, concentrated on $\emptyset$, which we denoted by Ψ in Sect. 2.1 So

$$\langle \Phi | f\rangle = f(\emptyset).$$

A monomial

$$M = a_{c_n}^{\vartheta_n} \cdots a_{c_1}^{\vartheta_1}$$

is called *normal ordered* if all the creators a_c^+ are to the left of the annihilators a_c, i.e.,

$$\vartheta_i = +1, \vartheta_j = -1 \Longrightarrow i > j.$$

Using the commutation relations it is clear that any normal ordered monomial can be brought into the form

$$a^+(\mathrm{d}x_{s_1}) \cdots a^+(\mathrm{d}x_{s_l}) a^+(\mathrm{d}x_{t_1}) \cdots a^+(\mathrm{d}x_{t_m})$$
$$a(x_{t_1}) \cdots a(x_{t_m}) a(x_{u_1}) \cdots a(x_{u_n}) = a^+_{\sigma+\tau} a_{\tau+\upsilon},$$

with

$$\sigma = \{s_1, \ldots, s_l\}, \qquad \tau = \{t_1, \ldots, t_m\}, \qquad \upsilon = \{u_1, \ldots, u_n\}.$$

Assume five finite, pairwise disjoint, index sets $\pi, \sigma, \tau, \upsilon, \rho$ and consider the admissible monomial $a_\pi a^+_{\sigma+\tau} a_{\tau+\upsilon} a^+_\rho$. The indices of creators make up the set $S_+ = \sigma + \tau + \rho$, and the indices of annihilators $S_- = \pi + \tau + \upsilon$. So $S_- \setminus S_+ = \pi + \upsilon$. Following Sect. 5.6, $\langle a_\pi a^+_{\sigma+\tau} a_{\tau+\upsilon} a^+_\rho\rangle \lambda_{\pi+\upsilon}$ is for fixed $\#\pi, \#\sigma, \#\tau, \#\upsilon, \#\rho$, a measure on $X^{\#(\pi+\sigma+\tau+\upsilon+\rho)}$. Letting the numbers $\#\pi, \#\sigma, \#\tau, \#\upsilon, \#\rho$ run from 0 to ∞ we arrive at a measure $\mathfrak{m}$ on $\mathfrak{X}^5$

$$\mathfrak{m} = \mathfrak{m}(\pi, \sigma, \tau, \upsilon, \rho) = \big\langle a_\pi a^+_{\sigma+\tau} a_{\tau+\upsilon} a^+_\rho \big\rangle \lambda_{\pi+\upsilon}.$$

Using Theorem 5.5.1 and Theorem 5.6.1, we obtain (forgetting about the $\Delta\omega$),

$$\begin{aligned}\mathfrak{m} &= \int_\omega \big\langle a_\omega a_{\sigma+\tau} a^+_\pi \big\rangle \big\langle a_\omega a_{\tau+\upsilon} a^+_\rho \big\rangle \lambda_{\omega+\sigma+\tau+\upsilon} \\ &= \int_\omega \varepsilon(\sigma+\tau+\omega, \pi)\varepsilon(\tau+\upsilon+\omega, \rho)\lambda_{\omega+\sigma+\tau+\upsilon}.\end{aligned}$$

If $\varphi \in \mathscr{K}_s(\mathfrak{X}^5)$ then

$$\int \mathfrak{m}(\pi, \sigma, \tau, \upsilon, \rho)\varphi(\pi, \sigma, \tau, \upsilon, \rho) = \int \varphi(\sigma+\tau+\omega, \sigma, \tau, \upsilon, \tau+\upsilon+\omega)\lambda_{\omega+\sigma+\tau+\upsilon},$$

(forgetting about the $\Delta\sigma, \Delta\tau, \ldots$).

Assume we have a Hilbert space $\mathfrak{k}$ with a countable basis. We often write the scalar product $x, y \mapsto \langle x, y\rangle$ in the form x^+y by introducing the dual vector x^+. We denote by $B(\mathfrak{k})$ the space of bounded linear operators on $\mathfrak{k}$. We provide $B(\mathfrak{k})$ with the operator norm topology. If $A \in B(\mathfrak{k})$, then A^+ denotes the adjoint operator.

Assume the function $F : \mathfrak{X}^3 \to B(\mathfrak{k})$ is locally λ-integrable, i.e., locally integrable with respect to $\mathrm{e}(\lambda)^{\otimes 3}$, and $f, g \in \mathscr{K}_s(\mathfrak{X}, \mathfrak{k})$ (continuous in the norm topology of $\mathfrak{k}$). The integral

$$\int \mathfrak{m}(\pi,\sigma,\tau,\upsilon,\rho) f^+(\pi) F(\sigma,\tau,\upsilon) g(\rho)$$

$$= \int f^+(\sigma+\tau+\omega) F(\sigma,\tau,\upsilon) g(\tau+\upsilon+\omega)\lambda_{\omega+\sigma+\tau+\upsilon} = \langle f|\mathscr{B}(F)|g\rangle$$

exists and defines a sesquilinear form on $\mathscr{K}_s(\mathfrak{X}, \mathfrak{k})$. We may say that

$$\mathscr{B}(F) = \int F(\sigma,\tau,\upsilon) a^+_{\sigma+\tau} a_{\tau+\upsilon} \lambda_\upsilon$$

is scalarly defined as a sesquilinear form in f, g by using

$$\langle f| = \int f^+(\pi)\langle\Phi|a_\pi\lambda_\pi, \qquad |g\rangle = \int g(\rho) a^+_\rho|\Phi\rangle;$$

note that a^+_ρ is a measure but a_π has to be multiplied with the base measure λ_π. We shall use the following formulas, which can be established easily.

Lemma 7.1.1

$$a_\omega a^+_\rho|\Phi\rangle = \sum_{\alpha\subset\omega} \varepsilon(\omega,\alpha) a^+_{\rho\setminus\alpha}|\Phi\rangle,$$

$$a^+_\omega a_\omega a^+_\rho|\Phi\rangle = \sum_{\alpha\subset\omega} \varepsilon(\omega,\alpha) a^+_\rho|\Phi\rangle,$$

$$\langle\Phi|a_\pi a^+_\omega = \sum_{\alpha\subset\pi} \varepsilon(\alpha,\omega)\langle\Phi|a_{\pi\setminus\alpha},$$

$$\langle\Phi|a_\pi a^+_\tau a_\tau = \sum_{\alpha\subset\pi} \varepsilon(\alpha,\tau)\langle\Phi|a_{\omega\setminus\alpha}.$$

Proposition 7.1.1 *The sesquilinear form $\langle f|\mathscr{B}(F)|g\rangle$ induces a mapping $\mathscr{O}(F)$ from $\mathscr{K}_s(\mathfrak{X})$ into the locally λ-integrable functions on $\mathfrak{X}$, and we have*

$$\langle f|\mathscr{B}(F)|g\rangle = \int f^+(\omega)\big(\mathscr{O}(F)g\big)(\omega)\lambda_\omega = \big\langle f|\mathscr{O}(F)g\big\rangle_\lambda,$$

$$\big(\mathscr{O}(F)g\big)(\omega) = \sum_{\alpha\subset\omega}\sum_{\beta\subset\omega\setminus\alpha}\int_\upsilon \lambda_\upsilon F(\alpha,\beta,\upsilon) g(\omega\setminus\alpha+\upsilon).$$

If we define

$$F^+(\sigma,\tau,\upsilon)=F(\upsilon,\sigma,\tau)^+,$$

we obtain

$$\langle f|\mathscr{O}(F)g\rangle_\lambda=\langle\mathscr{O}(F^+)f|g\rangle_\lambda.$$

Proof We have

$$\mathfrak{m}=\langle a_\pi a^+_{\sigma+\tau}a_{\tau+\upsilon}a^+_\rho\rangle\lambda_{\pi+\upsilon}=\langle\Phi|a_\pi a^+_{\sigma+\tau}a_{\tau+\upsilon}a^+_\rho|\Phi\rangle\lambda_{\pi+\upsilon}.$$

Now

$$\langle\Phi|a_\pi a^+_\sigma=\sum_{\alpha\subset\pi}\varepsilon(\alpha,\sigma)\langle\Phi|a_{\pi\setminus\alpha}$$

and

$$\langle\Phi|a_\omega a^+_\tau a_\tau=\sum_{\beta\subset\omega}\varepsilon(\beta,\tau)\langle\Phi|a_{\omega\setminus\beta}.$$

From there one obtains the first formula. Using the results of Sect. 5.6, we have

$$\mathfrak{m}(\pi,\sigma,\tau,\upsilon,\rho)=\mathfrak{m}(\rho,\upsilon,\tau,\sigma,\pi)$$

and obtain

$$\langle f|\mathscr{O}(F)g\rangle_\lambda=\overline{\langle g|\mathscr{O}(F^+)f\rangle_\lambda}=\langle\mathscr{O}(F^+)f|g\rangle_\lambda.\qquad\square$$

Consider a new longer similar expression, a measure on $\mathfrak{X}^8$,

$$\mathfrak{m}(\pi,\sigma_1,\tau_1,\upsilon_1,\sigma_2,\tau_2,\upsilon_2,\rho)=\langle a_\pi a^+_{\sigma_1+\tau_1}a_{\tau_1+\upsilon_1}a^+_{\sigma_2+\tau_2}a_{t_2+\upsilon_2}a^+_\rho\rangle\lambda_{\pi+\upsilon_1+\upsilon_2}.$$

Assume $F,G:\mathfrak{X}^3\to B(\mathfrak{k})$ to be λ-measurable and define

$$\begin{aligned}&\langle f|\mathscr{B}(F,G)|g\rangle\\&\quad=\int\mathfrak{m}(\pi,\sigma_1,\tau_1,\upsilon_1,\sigma_2,\tau_2,\upsilon_2,\rho)f^+(\pi)F(\sigma_1,\tau_1,\upsilon_1)G(\sigma_2,\tau_2,\upsilon_2)g(\rho)\end{aligned}$$

provided the integral exists in norm. So the bilinear form $\mathscr{B}(F,G)$ in F and G, whose values are sesquilinear forms in f and g, can be written as the scalarly defined integral

$$\mathscr{B}(F,G)=\int F(\sigma_1,\tau_1,\upsilon_1)G(\sigma_2,\tau_2,\upsilon_2)a^+_{\sigma_1+\tau_1}a_{\tau_1+\upsilon_1}a^+_{\sigma_2+\tau_2}a_{\tau_2+\upsilon_2}\lambda_{\upsilon_1+\upsilon_2}.$$

One obtains

Proposition 7.1.2

$$\langle f|\mathscr{B}(F,G)|g\rangle=\langle\mathscr{O}(F^+)f|\mathscr{O}(G)g\rangle_\lambda.$$

Proof Use the representation of unity from Sect. 5.6, and obtain

$$\begin{aligned}\mathfrak{m} &= \int_\omega \langle a_\pi a^+_{\sigma_1+\tau_1} a_{\tau_1+\upsilon_1} a^+_\omega \rangle \langle a_\omega a^+_{\sigma_2+\tau_2} a_{t_2+\upsilon_2} a^+_\rho \rangle \lambda_{\pi+\upsilon_1+\upsilon_2} \\ &= \int_\omega \langle a_\omega a^+_{\tau_1+\upsilon_1} a_{\tau_1+\sigma_1} a^+_\pi \rangle \langle a_\omega a^+_{\sigma_2+\tau_2} a_{t_2+\upsilon_2} a^+_\rho \rangle \lambda_{\omega+\sigma_1+\upsilon_2}.\end{aligned}$$

From there one obtains the result. □

Therefore a sufficient condition for the existence of $\langle f|\mathscr{B}(F,G)|g\rangle$ is that $\mathscr{O}(F^+)$ and $\mathscr{O}(G)$ are bounded operators from $\mathscr{K}_s(\mathfrak{X},\mathfrak{k})$, provided with the $L^2(\mathfrak{X},\mathfrak{k},\lambda)$-norm, into $L^2(\mathfrak{X},\mathfrak{k},\lambda)$.

7.2 Meyer's Formula

As might be guessed from Wick's theorem, there exists an H such that $\mathscr{B}(F,G) = \mathscr{B}(H)$. In fact we have the following theorem, basically due to P.A. Meyer [34].

Theorem 7.2.1 (Meyer's formula) *If F, G are locally λ-integrable on $\mathfrak{X}^3$, symmetric in each variable, such that*

$$\mathscr{B}(F,G) = \int F(\sigma_2,\tau_2,\upsilon_2)G(\sigma_1,\tau_1,\upsilon_1) a^+_{\sigma_2+\tau_2} a_{\tau_2+\upsilon_2} a^+_{\sigma_1+\tau_1} a_{\tau_1+\upsilon_1} \lambda_{\upsilon_1+\upsilon_2}$$

exists, then there exists a locally λ-integrable function H on $\mathfrak{X}^3$, symmetric in each variable, such that

$$\mathscr{B}(F,G) = \mathscr{B}(H)$$

and H is given by the formula

$$\begin{aligned}H(\sigma,\tau,\upsilon) = \sum \int_\kappa \lambda_\kappa F(&\alpha_1, \alpha_2+\beta_1+\beta_2, \gamma_1+\gamma_2+\kappa) \\ &\times G(\kappa+\alpha_2+\alpha_3, \beta_2+\beta_3+\gamma_2, \gamma_3)\end{aligned}$$

where the sum runs through all indices $\alpha_1, \ldots, \gamma_3$ with

$$\begin{aligned}\alpha_1+\alpha_2+\alpha_3 &= \sigma, \\ \beta_1+\beta_2+\beta_3 &= \tau, \\ \gamma_1+\gamma_2+\gamma_3 &= \upsilon.\end{aligned}$$

That is essentially Meyer's formula [34, p. 92]. The difference is mainly, that his formula is formulated for sets of coordinates, whereas our formula deals with sets of indices of coordinates; in addition, our formula holds for any locally compact set and for any base measure λ; Meyer considers only $X = \mathbb{R}$ and the Lebesgue measure. This formula was proven in [43] for C_c-functions. In order to generalize it to more complicated functions one must use the extension theorems of measure theory.

Proof We prove the theorem only for positive C_c-functions and leave the generalization to the reader. Recall from Sect. 5.3 that

$$\varepsilon(\alpha, \beta) = \langle a_\alpha, \overset{+}{a}_\beta \rangle = \sum_{\varphi \in B(\alpha,\beta)} \prod_{c \in \alpha} \varepsilon\big(c, \varphi(c)\big) \qquad (*)$$

where $B(\alpha, \beta)$ is the set of all bijections $\varphi : \alpha \to \beta$. If $\#\alpha \neq \#\beta$, then $B(\alpha, \beta) = \emptyset$ and $\varepsilon(\alpha, \beta) = 0$.

One shows easily that

$$\varepsilon(\alpha_1 + \alpha_2, \beta) = \sum_{\beta_1+\beta_2=\beta} \varepsilon(\alpha_1, \beta_1)\varepsilon(\alpha_2, \beta_2),$$

$$\varepsilon(\alpha, \beta_1 + \beta_2) = \sum_{\alpha_1+\alpha_2=\alpha} \varepsilon(\alpha_1, \beta_1)\varepsilon(\alpha_2, \beta_2).$$

From there one concludes that

$$(*) \quad \varepsilon(\alpha_1 + \alpha_2, \beta_1 + \beta_2) = \sum \varepsilon(\alpha_{11}, \beta_{11})\varepsilon(\alpha_{12}, \beta_{21})\varepsilon(\alpha_{21}, \beta_{12})\varepsilon(\alpha_{22}, \beta_{22})$$

where the sum runs through all indices $\alpha_{11}, \dots, \beta_{22}$ with

$$\alpha_{11} + \alpha_{12} = \alpha_1, \qquad \alpha_{21} + \alpha_{22} = \alpha_2,$$

$$\beta_{11} + \beta_{12} = \beta_1, \qquad \beta_{21} + \beta_{22} = \beta_2.$$

We have

$$\mathscr{B}(F, G) = \int \lambda_{\upsilon_1+\upsilon_2} F(\sigma_2, \tau_2, \tau_1) G(\sigma_1, \tau_1, \upsilon_1) \overset{+}{a}_{\sigma_2+\tau_2} a_{\tau_2+\upsilon_2} \overset{+}{a}_{\sigma_1+\tau_1} a_{\tau_1+\upsilon_1},$$

where the integral runs over all (mutually disjoint) index sets $\sigma_1, \dots, \upsilon_2$. Calculate

$$\overset{+}{a}_{\sigma_2+\tau_2} a_{\tau_2+\upsilon_2} \overset{+}{a}_{\sigma_1+\tau_1} a_{\tau_1+\upsilon_1}$$
$$= \sum \overset{+}{a}_{\sigma_2+\tau_2+\sigma_{11}+\tau_{11}} a_{\tau_{21}+\upsilon_{21}+\tau_1+\upsilon_1} \varepsilon(\tau_{22} + \upsilon_{22}, \sigma_{12} + \tau_{12})$$

where the indices obey the conditions

$$\tau_{21} + \tau_{22} = \tau_2, \qquad \upsilon_{21} + \upsilon_{22} = \upsilon_2,$$

$$\sigma_{11} + \sigma_{12} = \sigma_1, \qquad \tau_{11} + \tau_{12} = \tau_1.$$

Following $(*)$

$$\varepsilon(\tau_{22} + \upsilon_{22}, \sigma_{12} + \tau_{12}) = \sum \varepsilon(\tau_{221}, \sigma_{121})\varepsilon(\tau_{222}, \tau_{121})\varepsilon(\upsilon_{221}, \sigma_{122})\varepsilon(\upsilon_{222}, \tau_{122})$$

with

$$\tau_{221} + \tau_{222} = \tau_{22}, \qquad \upsilon_{221} + \upsilon_{222} = \upsilon_{22},$$
$$\sigma_{121} + \sigma_{122} = \sigma_{12}, \qquad \tau_{121} + \tau_{122} = \tau_{12}.$$

Using the sum-integral lemma

$$\begin{aligned}\mathscr{B}(F,G) = \int &\lambda_{\upsilon_{21}+\upsilon_{221}+\upsilon_{222}+\upsilon_1} F(\sigma_2, \tau_{21}+\tau_{221}+\tau_{222}, \upsilon_{21}+\upsilon_{221}+\upsilon_{222}) \\ &\times G(\sigma_{11}+\sigma_{121}+\sigma_{122}, \tau_{11}+\tau_{121}+\tau_{122}, \upsilon_1)\varepsilon(\tau_{221},\sigma_{121})\varepsilon(\tau_{222},\tau_{121}) \\ &\times \varepsilon(\upsilon_{221},\sigma_{122})\varepsilon(\upsilon_{222},\tau_{122}) a^+_{\sigma_2+\tau_{21}+\tau_{221}+\tau_{222}+\sigma_{11}+\tau_{11}} \\ &\times a_{\tau_{21}+\upsilon_{21}+\tau_{11}+\tau_{121}+\tau_{122}+\upsilon_1}\end{aligned}$$

where the integral runs over all indices. Put

$$\begin{aligned} &\sigma_2 = \alpha_1, && \sigma_{121} = \tau_{221} = \alpha_2, && \sigma_{11} = \alpha_3, \\ &\tau_{21} = \beta_1, && \tau_{222} = \tau_{121} = \beta_2, && \tau_{11} = \beta_3, \\ &\upsilon_{21} = \gamma_1, && \upsilon_{222} = \tau_{122} = \gamma_2, && \upsilon_1 = \gamma_3, \\ & && \sigma_{122} = \upsilon_{221} = \kappa, && \end{aligned}$$

where the equalities in the second column hold after integration. Define

$$\begin{aligned}\alpha_1 + \alpha_2 + \alpha_3 &= \sigma, \\ \beta_1 + \beta_2 + \beta_3 &= \tau, \\ \gamma_1 + \gamma_2 + \gamma_3 &= \upsilon,\end{aligned}$$

and obtain the theorem using the sum-integral lemma again. □

7.3 Quantum Stochastic Processes of Class $\mathscr{C}^1$: Definition and Fundamental Properties

Recall the definition of functions of class $\mathscr{C}^1$ from Definition 6.3.1, and use, instead of the index sets $\alpha_1, \ldots, \alpha_k$, the index sets σ, τ, υ; set $\mathfrak{B} = B(\mathfrak{k})$, where $\mathfrak{k}$ is a Hilbert space. Assume $x_t(\sigma, \tau, \upsilon)$ of class $\mathscr{C}^1$, and use the notation

$$\begin{aligned}\left(R^1_\pm x\right)_t(t_\sigma, t_\tau, t_\upsilon) &= x_{t\pm 0}\big(t_\sigma + \{t\}, t_\tau, t_\upsilon\big), \\ \left(R^0_\pm x\right)_t(t_\sigma, t_\tau, t_\upsilon) &= x_{t\pm 0}\big(t_\sigma, t_\tau + \{t\}, t_\upsilon\big), \\ \left(R^{-1}_\pm x\right)_t(t_\sigma, t_\tau, t_\upsilon) &= x_{t\pm 0}\big(t_\sigma, t_\tau, t_\upsilon + \{t\}\big), \\ \left(D^i x\right)_t &= \left(R^i_+ x\right)_t - \left(R^i_- x\right)_t.\end{aligned}$$

Note we are using $1, 0, -1$ as indices where a while before we used $1, 2, 3$ analogously. Recall, from Sect. 7.1, the definition of the measure

$$\mathfrak{m}(\pi, \sigma, \tau, \upsilon, \rho) = \langle a_\pi a^+_{\sigma+\tau} a_{\tau+\upsilon} a^+_\rho \rangle \lambda_{\pi+\upsilon}$$

and the sesquilinear form

$$\langle f | \mathscr{B}(F) | g \rangle = \int \mathfrak{m}(\pi, \sigma, \tau, \upsilon, \rho) f^+(\pi) F(\sigma, \tau, \upsilon) g(\rho).$$

Definition 7.3.1 If x_t is of class $\mathscr{C}^1$, we call $\mathscr{B}(x_t)$ a quantum stochastic process of class $\mathscr{C}^1$.

Theorem 7.3.1 *If x_t is of class $\mathscr{C}^1$, then the Schwartz derivative of $\langle f | \mathscr{B}(x_t) | g \rangle$ for $f, g \in \mathscr{K}_s(\mathfrak{R}, \mathfrak{k})$ is a locally integrable function*

$$\begin{aligned} \partial \langle f | \mathscr{B}(x_t) | g \rangle &= \langle f | \mathscr{B}(\partial^{\mathrm{c}} x_t) | g \rangle \\ &\quad + \langle a(t) f | \mathscr{B}(\mathrm{D}^1 x_t) | g \rangle + \langle a(t) f | \mathscr{B}(\mathrm{D}^0 x_t) | a(t) g \rangle \\ &\quad + \langle f | \mathscr{B}(\mathrm{D}^{-1} x_t) | a(t) g \rangle \end{aligned}$$

and we have, for $s < t$,

$$\langle f | \mathscr{B}(x_t) | g \rangle - \langle f | \mathscr{B}(x_s) | g \rangle = \int_s^t \mathrm{d}t' \partial \langle f | \mathscr{B}(x_{t'}) | g \rangle.$$

Using the notation of Sect. 2.4, we may write

$$\partial \mathscr{B}(x_t) = \mathscr{B}(\partial^{\mathrm{c}} x_t) + a^\dagger(t) \mathscr{B}(\mathrm{D}^1 x_t) + a^\dagger(t) \mathscr{B}(\mathrm{D}^0 x_t) a(t) + \mathscr{B}(\mathrm{D}^{-1} x_t) a(t).$$

Proof From Proposition 6.3.1 we have, with $\varphi = \mathbf{1}_{]s,t[}$,

$$\begin{aligned} &\langle f | \mathscr{B}(x_t) | g \rangle - \langle f | \mathscr{B}(x_s) | g \rangle \\ &\quad = \langle f | \mathscr{B}(x_{t-0}) | g \rangle - \langle f | \mathscr{B}(x_{s+0}) | g \rangle \\ &\quad = \int \mathfrak{m} f^+(\pi) \partial^{\mathrm{c}}_t x_t(\sigma, \tau, \upsilon) g(\rho) \varphi(t) \mathrm{d}t \\ &\qquad + \int \mathfrak{m} f^+(\pi) \sum_{c \in \sigma} (\mathrm{D}^0 x)_c x_{t_c}^-(\sigma \setminus c, \tau, \upsilon) g(\rho) \varphi(t_c) \\ &\qquad + \int \mathfrak{m} f^+(\pi) \sum_{c \in \tau} (\mathrm{D}^1 x)_c x_{t_c}(\sigma, \tau \setminus c, \upsilon) g(\rho) \varphi(t_c) \\ &\qquad + \int \mathfrak{m} f^+(\pi) \sum_{c \in \upsilon} (\mathrm{D}^{-1} x)_c x_{t_c}(\sigma, \tau, \upsilon \setminus c) g(\rho) \varphi(t_c). \end{aligned}$$

By using the sum-integral lemma, we obtain for the last three terms

$$\int \langle a_\pi a^+_{\sigma+c+\tau} a_{\tau+\upsilon} a^+_\rho \rangle \lambda_{\pi+\upsilon} f^+(\pi) \left(D^1 x\right)_c g(\rho)\varphi(c)$$

$$+ \int \langle a_\pi a^+_{\sigma+\tau+c} a_{\tau+c+\upsilon} a^+_\rho \rangle \lambda_{\pi+\upsilon} f^+(\pi) \left(D^0 x\right)_c g(\rho)\varphi(c)$$

$$+ \int \langle a_\pi a^+_{\sigma+c+\tau} a_{\tau+\upsilon+c} a^+_\rho \rangle \lambda_{\pi+\upsilon+c} f^+(\pi) \left(D^{-1} x\right)_c g(\rho)\varphi(c)$$

where the integration is over all indices $\pi, \sigma, \tau, \upsilon, \rho, c$. From there we deduce the result. □

7.4 Ito's Theorem

Recall Sect. 7.1, and consider the measure

$$\mathfrak{m}(\pi, \sigma_1, \tau_1, \upsilon_1, \sigma_2, \tau_2, \upsilon_2, \rho) = \langle a_\pi a^+_{\sigma_1+\tau_1} a_{\tau_1+\upsilon_1} a^+_{\sigma_2+\tau_2} a_{\tau_2+\upsilon_2} a^+_\rho \rangle \lambda_{\pi+\upsilon_1+\upsilon_2}.$$

Assume $F, G : \mathfrak{R}^3 \to B(\mathfrak{k})$ to be λ-measurable, and define

$$\begin{aligned} &\langle f | \mathscr{B}(F, G) | g \rangle \\ &\quad = \int \mathfrak{m}(\pi, \sigma_1, \tau_1, \upsilon_1, \sigma_2, \tau_2, \upsilon_2, \rho) f^+(\pi) F(\sigma_1, \tau_1, \upsilon_1) G(\sigma_2, \tau_2, \upsilon_3) g(\rho) \end{aligned}$$

provided the integral exists in norm.

Theorem 7.4.1 *Assume x_t, y_t to be of class $\mathscr{C}^1$, and that for $f, g \in \mathscr{K}_s(\mathfrak{R}, \mathfrak{k})$ the sesquilinear forms $\langle f | \mathscr{B}(F_t, G_t) | g \rangle$ exist in norm, and $t \in \mathbb{R} \mapsto \langle f | \mathscr{B}(F_t, G_t) | g \rangle$ is locally integrable, where F_t can be any function in $\{x_t, \partial^c x_t, R^1_\pm x_t, R^0_\pm x_t, R^{-1}_\pm x_t\}$ and G_t can be any function in $\{y_t, \partial^c y_t, R^1_\pm y_t, R^0_\pm y_t, R^{-1}_\pm y_t\}$.*

Then $\langle f | \mathscr{B}(x_t, y_t) | g \rangle$ is a continuous function, its Schwartz derivative is a locally integrable function, and a formula for it is

$$\begin{aligned} \partial \langle f | \mathscr{B}(x_t, y_t) | g \rangle = &\ \langle f | \mathscr{B}(\partial^c x_t, y_t) + \mathscr{B}(f, \partial^c y_t) + I_{-1,+1,t} | g \rangle \\ &+ \langle a(t) f | \mathscr{B}(D^1 x_t, y_t) + \mathscr{B}(f, D^1 y_t) + I_{0,+1,t} | g \rangle \\ &+ \langle a(t) f | \mathscr{B}(D^0 x_t, y_t) + \mathscr{B}(f, D^0 y_t) + I_{0,0,t} | a(t) g \rangle \\ &+ \langle f | \mathscr{B}(D^{-1} x_t, y_t) + \mathscr{B}(f, D^{-1} y_t) + I_{-1,0,t} | a(t) g \rangle \end{aligned}$$

with

$$I_{i,j,t} = \mathscr{B}(R^i_+ x_t, R^j_+ y_t) - \mathscr{B}(R^i_- x_t, R^j_- y_t).$$

So, for $s < t$,

$$\langle f|\mathscr{B}(x_t, y_t)|g\rangle - \langle f|\mathscr{B}(x_s, y_s)|g\rangle = \int_s^t \mathrm{d}s'\partial\langle f|\mathscr{B}(x_{s'}, y_{s'})|g\rangle.$$

Again using the notation $a^\dagger$, we may write

$$\begin{aligned}\partial\mathscr{B}(x_t, y_t) = &\big(\mathscr{B}(\partial^{c}x_t, y_t) + \mathscr{B}(f, \partial^{c}y_t) + I_{-1,+1,t}\big)\\ &+ a^\dagger(t)\big(\mathscr{B}(D^1x_t, y_t) + \mathscr{B}(f, D^1y_t) + I_{0,+1,t}\big)\\ &+ a^\dagger(t)\big(\mathscr{B}(D^0x_t, y_t) + \mathscr{B}(f, D^0y_t) + I_{0,0,t}\big)a(t)\\ &+ \big(\mathscr{B}(D^{-1}x_t, y_t) + \mathscr{B}(f, D^{-1}y_t) + I_{-1,0,t}\big)a(t).\end{aligned}$$

We start with a lemma.

Lemma 7.4.1 *Assume x_t be of class $\mathscr{C}^1$, and define the function N on $\mathfrak{X}^3$ by*

$$N(\sigma, \tau, \upsilon) = \begin{cases} 1 & \text{if } \{t_{\sigma+\tau+\upsilon}\}^\bullet \text{ has a repeated point,}\\ 0 & \text{otherwise.}\end{cases}$$

Then the functions

$$x_{t\pm 0}(\sigma, \tau, \upsilon)\big(1 - N(\sigma, \tau, \upsilon)\big)$$

are everywhere defined Borel functions, and we consider

$$\int \big\langle a_\pi a_c a^+_{\sigma+\tau} a_{\tau+\upsilon} a^+_\rho\big\rangle f^+(\pi)h(c)x_{t_c+0}(\sigma, \tau, \upsilon)\big(1 - N(\sigma, \tau, \upsilon)\big)g(\rho).$$

Understand this expression as a scalarly defined integral and obtain

$$\begin{aligned}&\int a_{t_c} a^+_{\sigma+\tau} a_{\tau+\upsilon}\lambda_\upsilon x_{t_c+0}(\sigma, \tau, \upsilon)\big(1 - N(\sigma, \tau, \upsilon)\big)\\ &= \mathscr{O}(x_{t_c})a_c + \mathscr{O}\big((R^1_+x)_{t_c}\big) + \mathscr{O}\big((R^0_+)_{t_c}\big)a_c.\end{aligned}$$

Proof The function

$$\begin{aligned}&x_{t+0}(\sigma, \tau, \upsilon)\big(1 - N(\sigma, \tau, \upsilon)\big)\\ &= \big(1 - N(\sigma, \tau, \upsilon)\big)\begin{cases} x_t(\sigma, \tau, \upsilon) & \text{if } t_c \notin t_{\sigma+\tau+\upsilon},\\ (R^1_+x)_t(\sigma\setminus b, \tau, \upsilon) & \text{if } t = t_b, b\in\sigma,\\ (R^0_+x)_t(\sigma, \tau\setminus b, \upsilon) & \text{if } t = t_b, b\in\tau,\\ (R^{-1}_+x)_t(\sigma, \tau, \upsilon\setminus b) & \text{if } t = t_b, b\in\upsilon\end{cases}\end{aligned}$$

is defined everywhere. We calculate

$$\int a_{t_c} a^+_{\sigma+\tau} a_{\tau+\upsilon}\lambda_\upsilon x_{t_c+0}(\sigma, \tau, \upsilon)\big(1 - N(\sigma, \tau, \upsilon)\big)$$

$$= \int x_{t_c+0}(\sigma,\tau,\upsilon) a^+_{\sigma+\tau} a_{\tau+\upsilon+c} \lambda_\upsilon \big(1 - N(\sigma,\tau,\upsilon)\big)$$

$$+ \int x_{t_c+0}(\sigma,\tau,\upsilon)\big[a_{t_c}, a^+_{\sigma+\tau} a_{\tau+\upsilon}\big] \lambda_\upsilon \big(1 - N(\sigma,\tau,\upsilon)\big).$$

Because, in the first term on the right-hand side upon insertion of f, g, h all measures are λ-based, we may neglect N and replace $t_c + 0$ by t_c. The second term equals, with the help of the sum-integral lemma and integrating over t_b,

$$\int \Big(\sum_{b\in\sigma} \varepsilon(c,b) a^+_{\sigma\backslash b+\tau} a_{\tau+\upsilon}\lambda_\upsilon + \sum_{b\in\sigma} \varepsilon(c,b) a^+_{\sigma+\tau\backslash b} a_{\tau+\upsilon}\lambda_\upsilon \Big) x_{t_c+0}(\sigma,\tau,\upsilon)$$

$$\times \big(1 - N(\sigma,\tau,\upsilon)\big)$$

$$= \int a^+_{\sigma+\tau} a_{\tau+\upsilon} \lambda_\upsilon$$

$$\times \big(1 - N(\sigma+c,\tau,\upsilon)\big)\big(x_{t_c+0}(\sigma+c,\tau,\upsilon)$$

$$+ \big(1 - N(\sigma,\tau+c,\upsilon)\big) x_{t_c+0}(\sigma,\tau+c,\upsilon) a_c\big).$$

If we insert the functions f, g, h into the expressions, we see that we have to deal with integrals over λ-based measures; we may neglect N. We use the expressions R^1_+, R^0_+ introduced in Sect. 7.3, and arrive at

$$\int a_{t_c} a^+_{\sigma+\tau} a_{\tau+\upsilon} \lambda_\upsilon x_{t_c+0}(\sigma,\tau,\upsilon)\big(1 - N(\sigma,\tau,\upsilon)\big)$$

$$= \int a^+_{\sigma+\tau} a_{\tau+\upsilon} \lambda_\upsilon \big(a_c x_{t_c}(\sigma,\tau,\upsilon) + \big(R^1_+ x\big)_{t_c}(\sigma,\tau,\upsilon) + a_c\big(R^0_+\big)_{t_c}(\sigma,\tau,\upsilon)\big)$$

$$= \mathscr{O}(x_{t_c}) a_c + \mathscr{O}\big(\big(R^1_+ x\big)_{t_c}\big) + \mathscr{O}\big(\big(R^0_+\big)_{t_c}\big) a_c. \qquad \square$$

Proof of Ito's Theorem By the formulae in Sect. 7.1, the sesquilinear form $\mathscr{B}(F, G)$ vanishes if one of the functions F or G is a Lebesgue null function. So for fixed t

$$\mathscr{B}(x_t, y_t) = \mathscr{B}(x_{t\pm0}, y_{t\pm0}).$$

Define

$$N(\sigma,\tau,\upsilon) = \begin{cases} 1 & \text{if } \{t_{\sigma+\tau+\upsilon}\}^\bullet \text{ has a repeated point,} \\ 0 & \text{otherwise.} \end{cases}$$

As N is a Lebesgue null function, we have, for $t_0 < t_1$,

$$\langle f|\mathscr{B}(x_{t_1}, y_{t_1})|g\rangle - \langle f|\mathscr{B}(x_{t_0}, y_{t_0})|g\rangle$$

$$= \int \mathfrak{m}(\pi,\sigma_1,\tau_1,\upsilon_1,\sigma_2,\tau_2,\upsilon_2,\pi) f^+(\pi)\big(1 - N(\sigma_1,\tau_1,\upsilon_1)\big)$$

$$\times \left(x_{t_1-0}(\sigma_1,\tau_1,\upsilon_1)y_{t_1-0}(\sigma_2,\tau_2,\upsilon_2) - x_{t_0+0}(\sigma_1,\tau_1,\upsilon_1)y_{t_0+0}(\sigma_2,\tau_2,\upsilon_2)\right)$$
$$\times \left(1 - N(\sigma_2,\tau_2,\upsilon_2)\right)g(\rho).$$

We consider the set

$$(t_{\sigma_1+\tau_1+\upsilon_1} \cup t_{\sigma_2+\tau_2+\upsilon_2}) \cap]t_0,t_1[= \left\{t^1 < \cdots < t^{n-1}\right\},$$

and put $t_0 = t^0$ and $t_1 = t^n$ to obtain

$$\begin{aligned}
&\left(x_{t_1-0}(\sigma_1,\tau_1,\upsilon_1)y_{t_1-0}(\sigma_2,\tau_2,\upsilon_2) - x_{t_0+0}(\sigma_1,\tau_1,\upsilon_1)y_{t_0+0}(\sigma_2,\tau_2,\upsilon_2)\right)\\
&\quad \times \left(1 - N(\sigma_1,\tau_1,\upsilon_1)\right)\left(1 - N(\sigma_2,\tau_2,\upsilon_2)\right)\\
&\quad = \sum_{i=1}^{n} \int_{t^{i-1}}^{t^i} \mathrm{d}t \left(\partial^{\mathrm{c}} x_t(\sigma_1,\tau_1,\upsilon_1)y_t(\sigma_2,t_2,\upsilon_2) + x_t(\sigma_1,\tau_1,\upsilon_1)\partial^{\mathrm{c}} y_t(\sigma_2,t_2,\upsilon_2)\right)\\
&\quad \times \left(1 - N(\sigma_1,\tau_1,\upsilon_1)\right)\left(1 - N(\sigma_2,\tau_2,\upsilon_2)\right)\\
&\quad + \sum_{i=1}^{n-1} \left(x_{t^i+0}(\sigma_1,\tau_1,\upsilon_1)y_{t^i+0}(\sigma_2,\tau_2,\upsilon_2) - x_{t^i-0}(\sigma_1,\tau_1,\upsilon_1)y_{t^i-0}(\sigma_2,\tau_2,\upsilon_2)\right)\\
&\quad \times \left(1 - N(\sigma_1,\tau_1,\upsilon_1)\right)\left(1 - N(\sigma_2,\tau_2,\upsilon_2)\right).
\end{aligned}$$

The first sum equals

$$\int_{t_0}^{t_1} \mathrm{d}t \left(\partial^{\mathrm{c}} x_t(\sigma_1,\tau_1,\upsilon_1)y_t(\sigma_2,t_2,\upsilon_2) + x_t(\sigma_1,\tau_1,\upsilon_1)\partial^{\mathrm{c}} y_t(\sigma_2,t_2,\upsilon_2)\right).$$

Remark that the points of each of $t_{\sigma_1+\tau_1+\upsilon_1}$ and $t_{\sigma_2+\tau_2+\upsilon_2}$ are all different, but there may be points common to both. The second sum equals

$$\begin{aligned}
&\sum_{c\in\sigma_1+\tau_1+\upsilon_1,\ t_c\in]t_0,t_1[} \left(x_{t_c+0}(\sigma_1,\tau_1,\upsilon_1) - x_{t_c-0}(\sigma_1,\tau_1,\upsilon_1)\right)y_{t_c-0}(\sigma_2,\tau_2,\upsilon_2)\\
&\quad \times \left(1 - N(\sigma_1,\tau_1,\upsilon_1)\right)\left(1 - N(\sigma_2,\tau_2,\upsilon_2)\right)\\
&\quad + \sum_{c\in\sigma_2+\tau_2+\upsilon_2,\ t_c\in]t_0,t_1[} (x_{t_c-0}(\sigma_1,\tau_1,\upsilon_1)\left(y_{t_c+0}(\sigma_2,\tau_2,\upsilon_1) - y_{t_c-0}(\sigma_2,\tau_2,\upsilon_2)\right)\\
&\quad \times \left(1 - N(\sigma_1,\tau_1,\upsilon_1)\right)\left(1 - N(\sigma_2,\tau_2,\upsilon_2)\right)
\end{aligned}$$

as, for example,

$$x_{t_c+0}(\sigma_1,\tau_1,\upsilon_1) - x_{t_c-0}(\sigma_1,\tau_1,\upsilon_1) = 0$$

for $t_c \notin t_{\sigma_1+\tau_1+\upsilon_1}$. We discuss the integrals of the terms of the form

$$\sum_{c\in\sigma_i+\tau_i+\upsilon_i,\ t_c\in]t_0,t_1[} x_{t_c\pm 0}(\sigma_1,\tau_1,\upsilon_1)\left(1 - N(\sigma_1,\tau_1,\upsilon_1)\right)$$

$$\times\, y_{t_c \pm 0}(\sigma_2, \tau_2, \upsilon_2)\big(1 - N(\sigma_2, \tau_2, \upsilon_2)\big)$$

and assume at first, that f, g, x_t, y_t are ≥ 0 , then define

$$\varphi(t) = \mathbf{1}\{t \in]t_0, t_1[\}$$

and consider

$$\begin{aligned}
&\int f(\pi) \sum_{c \in \sigma_1 + \tau_1 + \upsilon_1} x_{t_c+0}(\sigma_1, \tau_1, \upsilon_1)\big(1 - N(\sigma_1, \tau_1, \upsilon_1)\big) \\
&\qquad \times y_{t_c+0}(\sigma_2, \tau_2, \upsilon_2)\big(1 - N(\sigma_2, \tau_2, \upsilon_2)\big) g(\rho)\varphi(t_c) \\
&\qquad \times \mathfrak{m}(\pi, \sigma_1, \tau_1, \upsilon_1, \sigma_2, \tau_2, \upsilon_2, \rho) \\
&\quad = I + II + III.
\end{aligned}$$

We split up the sum into three parts

$$\sum_{c \in \sigma_1 + \tau_1 + \upsilon_1} = \sum_{c \in \sigma_1} + \sum_{c \in \tau_1} + \sum_{c \in \upsilon_1}.$$

We have, using the sum-integral lemma,

$$\begin{aligned}
I &= \int f(\pi) \sum_{c \in \sigma_1} \big(R^1_+ x\big)_{t_c}(\sigma_1 \setminus c, \tau_1, \upsilon_1)\big(1 - N(\sigma_1, \tau_1, \upsilon_1)\big) \\
&\quad \times y_{t_c+0}(\sigma_2, \tau_2, \upsilon_2)\big(1 - N(\sigma_2, \tau_2, \upsilon_2)\big) g(\rho)\varphi(t_c) \\
&\quad \times \mathfrak{m}(\pi, \sigma_1, \tau_1, \upsilon_1, \sigma_2, \tau_2, \upsilon_2, \rho) \\
&= \int f(\pi)\big(R^1_+ x\big)_{t_c}(\sigma_1, \tau_1, \upsilon_1)\big(1 - N(\sigma_1 + c, \tau_1, \upsilon_1)\big) \\
&\quad \times y_{t_c+0}(\sigma_2, \tau_2, \upsilon_2)\big(1 - N(\sigma_2, \tau_2, \upsilon_2)\big) g(\rho)\varphi(t_c) \\
&\quad \times \big\langle a_\pi a^+_{\sigma_1 + c + \tau_1} a_{\tau_1 + \upsilon_1} a^+_{\sigma_2 + \tau_2} a_{\tau_2 + \upsilon_2} a^+_\rho \big\rangle \lambda_{\pi + \upsilon_1 + \upsilon_2} \\
&= \int \mathrm{d}t \varphi(t) \big\langle a(t) f \big| \mathscr{B}\big(R^1_+ x_t, y_t\big) | g \big\rangle \\
&= \int_{t_0}^{t_1} \mathrm{d}t \big\langle a(t) f \big| \mathscr{B}\big(R^1_+ x_t, y_t\big) | g \big\rangle.
\end{aligned}$$

The integral over $N(\sigma_1 + c, \tau_1, \upsilon_1)$ and $N(\sigma_2, \tau_2, \upsilon_2)$ vanishes, and $y_{t+0} = y_t$ a.e. with respect to the integrating measure.

In the same way

$$\begin{aligned}
II &= \int f(\pi)\big(R^0_+ x\big)_{t_c}(\sigma_1, \tau_1, \upsilon_1)\big(1 - N(\sigma_1, \tau_1 + c, \upsilon_1)\big) \\
&\quad \times y_{t_c+0}(\sigma_2, \tau_2, \upsilon_2)\big(1 - N(\sigma_2, \tau_2, \upsilon_2)\big) g(\rho)\varphi(t_c)
\end{aligned}$$

$$\times \langle a_\pi a^+_{\sigma_1+\tau_1+c} a_{\tau_1+c+\upsilon_1} a^+_{\sigma_2+\tau_2} a_{\tau_2+\upsilon_2} a^+_\rho \rangle \lambda_{\pi+\upsilon_1+\upsilon_2}.$$

Using the representation of unity from Sect. 5.5, we obtain

$$\begin{aligned} II = \int & f(\pi)\big(R^0_+ x\big)_{t_c}(\sigma_1,\tau_1,\upsilon_1)\big(1-N(\sigma_1,\tau_1+c,\upsilon_1)\big) \\ & \times y_{t_c+0}(\sigma_2,\tau_2,\upsilon_2)\big(1-N(\sigma_2,\tau_2,\upsilon_2)\big) g(\rho)\varphi(t_c) \\ & \times \int_\omega \langle a_\pi a^+_{\sigma_1+\tau_1+c} a_{\tau_1+\upsilon_1} a^+_\omega \rangle \langle a_\omega a_c a^+_{\sigma_2+\tau_2} a_{\tau_2+\upsilon_2} a^+_\rho \rangle \lambda_{\pi+\upsilon_1+\upsilon_2}. \end{aligned}$$

Now, by the proof of Theorem 7.3.1,

$$\begin{aligned} &\int f(\pi)\big(R^0_+ x\big)_{t_c}(\sigma_1,\tau_1,\upsilon_1)\big(1-N(\sigma_1,\tau_1+c,\upsilon_1)\big)\langle a_\pi a^+_{\sigma_1+\tau_1+c} a_{\tau_1+\upsilon_1} a^+_\omega \rangle \lambda_{\pi+\upsilon_1} \\ &= \big(\mathscr{O}\big(\big(R^0_+ x\big)_{t_c}\big)^+ a_{t_c} f\big)^+(\omega)\lambda_\omega, \end{aligned}$$

and by the last lemma

$$\begin{aligned} &\int y_{t_c+0}(\sigma_2,\tau_2,\upsilon_2)\big(1-N(\sigma_2,\tau_2,\upsilon_2)\big) g(\rho)\langle a_\omega a_c a^+_{\sigma_2+\tau_2} a_{\tau_2+\upsilon_2} a^+_\rho \rangle \lambda_{\upsilon_2} \\ &= \langle a_\omega a_c a^+_{\sigma_2+\tau_2} a_{\tau_2+\upsilon_2} a^+_\rho \rangle \lambda_{\pi+\upsilon_1+\upsilon_2} g(\omega) \\ &= \big(\big(\mathscr{O}(y_{t_c}) a_c + \mathscr{O}\big(\big(R^1_+ y\big)_{t_c}\big) + \mathscr{O}\big(\big(R^0_+ y\big)_{t_c} a_c\big)\big) g\big)(\omega), \end{aligned}$$

where $N(\sigma_1,\tau_1+c,\upsilon_1)$ and $N(\sigma_2,\tau_2,\upsilon_2)$ can be safely neglected. So finally

$$\begin{aligned} II &= \int \mathrm{d}t_c \varphi(t_c) \langle \big(\mathscr{O}\big(\big(R^0_+ x\big)_{t_c}\big)^+ a_{t_c} f\big) \big| \big(\mathscr{O}(x_{t_c}) a_c + \mathscr{O}\big(\big(R^1_+ x\big)_{t_c}\big) + \mathscr{O}\big(\big(R^0_+\big)_{t_c} a_c\big)\big) g \rangle_\lambda \\ &= \int \mathrm{d}t \varphi(t)\big(\langle a_t f | \mathscr{B}\big(R^0_+ x\big)_t, y_t\big) \big| a(t) g\rangle + \langle a_t f | \mathscr{B}\big(R^0_+ x_t, R^1_+ y_t\big) | g\rangle \\ &\quad + \langle a_t f | \mathscr{B}\big(R^0_+ x_t, R^0_+ y_t | a_t g\rangle\big). \end{aligned}$$

We calculate

$$\begin{aligned} III = \int & f(\pi)\big(R^{-1}_+ x\big)_{t_c}(\sigma_1,\tau_1,\upsilon_1)\big(1-N(\sigma_1,\tau_1,\upsilon_1+c)\big) \\ & \times y_{t_c+0}(\sigma_2,\tau_2,\upsilon_2)\big(1-N(\sigma_2,\tau_2,\upsilon_2)\big) g(\rho)\varphi(t_c) \\ & \times \langle a_\pi a^+_{\sigma_1+\tau_1} a_{\tau_1+\upsilon_1+c} a^+_{\sigma_2+\tau_2} a_{\tau_2+\upsilon_2} a^+_\rho \rangle \lambda_{\pi+\upsilon_1+c+\upsilon_2} \\ = \int & f(\pi)\big(R^{-1}_+ x\big)_{t_c}(\sigma_1,\tau_1,\upsilon_1)\big(1-N(\sigma_1,\tau_1,\upsilon_1+c)\big) \\ & \times y_{t_c+0}(\sigma_2,\tau_2,\upsilon_2)\big(1-N(\sigma_2,\tau_2,\upsilon_2)\big) g(\rho)\varphi(t_c) \\ & \times \int_\omega \langle a_\pi a^+_{\sigma_1+\tau_1} a_{\tau_1+\upsilon_1} a^+_\omega \rangle \langle a_\omega a_c a^+_{\sigma_2+\tau_2} a_{\tau_2+\upsilon_2} a^+_\rho \rangle \lambda_{\pi+\upsilon_1+c+\upsilon_2}. \end{aligned}$$

By calculations similar to those for II one obtains

$$III = \int \mathrm{d}\varphi(t)\big(\langle f|\mathscr{B}(R_{\mp}^{-1}x_t, y_t)|a(t)g\rangle + \langle f|\mathscr{B}(R_{\mp}^{-1}x_t, R_{+}^{1}y_t)|g\rangle$$
$$+ \langle f|\mathscr{B}(R_{\mp}^{-1}x_t, R_{+}^{0}y_t)|a(t)g\rangle\big).$$

The assumptions of our theorem guarantee that all the expressions exist and we may extend the formulas to vector- and operator-valued functions. By analogous calculations,

$$\int f(\pi)^{+} \sum_{c\in\sigma_1+\tau_1+\upsilon_1} x_{t_c\pm 0}(\sigma_1,\tau_1,\upsilon_1)\big(1-N(\sigma_1,\tau_1,\upsilon_1)\big)$$
$$\times\, y_{t_c+0}(\sigma_2,\tau_2,\upsilon_2)\big(1-N(\sigma_2,\tau_2,\upsilon_2)\big)g(\rho)\varphi(t_c)$$
$$\times\, \mathfrak{m}(\pi,\sigma_1,\tau_1,\upsilon_1,\sigma_2,\tau_2,\upsilon_2,\rho)$$
$$= \int_{t_0}^{t_1} \mathrm{d}t\, K_{\pm,+}(t)$$

and

$$\int f(\pi)^{+} \sum_{c\in\sigma_2+\tau_2+\upsilon_2} x_{t_c-0}(\sigma_1,\tau_1,\upsilon_1)\big(1-N(\sigma_1,\tau_1,\upsilon_1)\big)$$
$$\times\, y_{t_c\pm 0}(\sigma_2,\tau_2,\upsilon_2)\big(1-N(\sigma_2,\tau_2,\upsilon_2)g(\rho)\varphi(t_c)\big)$$
$$\times\, \mathfrak{m}(\pi,\sigma_1,\tau_1,\upsilon_1,\sigma_2,\tau_2,\upsilon_2,\rho)$$
$$= \int \mathrm{d}t\varphi(t)K_{-,\pm}(t)$$
$$= \int_{t_0}^{t_1} \mathrm{d}t\, K_{-,\pm}(t).$$

We have

$$K_{\pm,+} = K_{\pm,+}^{(1)} + K_{\pm,+}^{(2)},$$
$$K_{-,\pm} = K_{-,\pm}^{(1)} + K_{-,\pm}^{(2)}$$

with

$$K_{\pm,+}^{(1)} = \langle a(t)f|\mathscr{B}(R_{\pm}^{1}x_t, y_t)|g\rangle + \langle a(t)f|\mathscr{B}(R_{\pm}^{0}x_t, y_t)|a(t)g\rangle$$
$$+ \langle f|\mathscr{B}(R_{\pm}^{-1}x_t, y_t)|a(t)g\rangle,$$
$$K_{-,\pm}^{(1)} = \langle a(t)f|\mathscr{B}(x_t, R_{\pm}^{1}y_t)|g\rangle + \langle a(t)f|\mathscr{B}(x_t, R_{+}^{0}y_t)|a(t)g\rangle$$
$$+ \langle f|\mathscr{B}(x_t, R_{\mp}^{-1}y_t)|a(t)g\rangle$$

and

$$K^{(2)}_{\pm,\pm} = \langle a(t)f \,|\, \mathscr{B}(R^0_{\pm}x_t, R^1_{\pm}y_t)|g\rangle + \langle a(t)f \,|\, \mathscr{B}(R^0_{\pm}x_t, R^0_{\pm}y_t)\big|a(t)g\rangle$$
$$+ \langle f|\mathscr{B}(R^{-1}_{\mp}x_t, R^0_{\pm}y_t)\big|a(t)g\rangle + \langle f|\mathscr{B}(R^{-1}_{\mp}x_t, R^1_{\pm}y_t)|g\rangle.$$

From there one achieves the final result without great difficulty. □

Chapter 8
The Hudson-Parthasarathy Differential Equation

Abstract The Hudson-Parthasarathy quantum stochastic differential equation can be solved by a classical integral in a high-dimensional space. With the help of an a priori estimate it is possible to show that the solution is unitary, under the usual assumptions. The unitarity allows stronger estimates: the Γ_k-norm is of polynomial growth. This provides the resolvent of the associated one-parameter group with the properties needed for the discussion of the Hamiltonian. An explicit form of the Hamiltonian can be established.

8.1 Formulation of the Equation

We shall investigate the quantum stochastic differential equation that reads in the Hudson-Parthasarathy calculus [34, 36]

$$\mathrm{d}_t U_s^t = A_1 \mathrm{d}B_t^+ U_s^t + A_0 \mathrm{d}\Lambda_t U_s^t + A_{-1} \mathrm{d}B_t U_s^t + B U_s^t \mathrm{d}t, \quad \text{with } U_s^s = 1,$$

where A_1, A_0, A_{-1}, B are operators in $B(\mathfrak{k})$. In his white noise calculus Accardi [3] formulates it as a *normal ordered* equation

$$\frac{\mathrm{d}U_s^t}{\mathrm{d}t} = A_1 a_t^+ U_s^t + A_0 a_t^+ U_s^t a_t + A_{-1} U_s^t a_t + B U_s^t.$$

Our formulation is very similar to Accardi's. We interpret U_s^t as a sesquilinear form over $\mathscr{K}_s(\mathfrak{R})$ given by the classical integrals

$$\langle f | U_s^t | g \rangle = \int f^+(\pi) u_s^t(\sigma, \tau, \upsilon) g(\varrho) \langle a_\pi a_{\sigma+\tau}^+ a_{\tau+\upsilon} a_\varrho^+ \rangle \lambda_{\pi+\upsilon}$$

where u_s^t is locally integrable in all five variables $s, t, \sigma, \tau, \upsilon$. We formulate the differential equation in the weak sense as

$$\frac{\mathrm{d}}{\mathrm{d}t} \langle f | U_s^t | g \rangle = \langle a_t f | A_1 U_s^t | g \rangle + \langle a_t f | A_0 U_s^t | a_t g \rangle + \langle f | A_{-1} U_s^t | a_t g \rangle + \langle f | B U_s^t | g \rangle,$$

$$U_s^s = 1$$

W. von Waldenfels, *A Measure Theoretical Approach to Quantum Stochastic Processes*, Lecture Notes in Physics 878, DOI 10.1007/978-3-642-45082-2_8,

or, using the operator $a^\dagger$ and interpreting the bracket as a weak integral,

$$\frac{dU_s^t}{dt} = a^\dagger(t)A_1U_s^t + a^\dagger(t)A_0U_s^t a(t) + A_{-1}U_s^t a(t) + BU_s^t,$$

which is still more similar to Accardi's formulation. We can write the differential equation better as the integral equation

$$\langle f|U_s^t|g\rangle = \langle f|g\rangle + \int_s^t dr\, \langle a_r f|A_1U_s^r|g\rangle + \int_s^t dr\, \langle a_r f|A_0U_s^r|a_r g\rangle$$
$$+ \int_s^t dr\, \langle f|A_{-1}U_s^r|a_r g\rangle + \int_s^t dr\, \langle f|BU_s^r|g\rangle \qquad (*)$$

for $t \geq s$. We shall show that this equation has a unique solution, which can be given explicitly.

8.2 Existence and Uniqueness of the Solution

Lemma 8.2.1 *The equation* $(*)$ *is equivalent to the circled integral equation* $(**)$

$$u_s^t = \mathbf{e} + A_1 \oint_{s,t}^{1} u_s^\cdot + A_0 \oint_{s,t}^{0} u_s^\cdot + A_{-1} \oint_{s,t}^{-1} u_s^\cdot + B \int_s^t dr\, u_s^r \qquad (**)$$

where

$$\mathbf{e}(\sigma,\tau,\upsilon) = \begin{cases} 1 & \text{if } \sigma+\tau+\upsilon = \emptyset, \\ 0 & \text{otherwise.} \end{cases}$$

Proof Consider, for example, the term

$$\int_s^t dr\, \langle a_r f|A_0U_s^r|a_r g\rangle$$
$$= \int \mathbf{1}_{[s,t]}(t_c) f^+(\omega+\sigma+\tau+c)A_0 u_s^{t_c}(\sigma,\tau,\upsilon) g(\omega+\tau+\upsilon+c)\lambda_{\omega+\sigma+\tau+\upsilon+c}$$
$$= \int \sum_{c\in\tau} \mathbf{1}_{[s,t]}(t_c) f^+(\omega+\sigma+\tau)A_0 u_s^{t_c}(\sigma,\tau\setminus c,\upsilon) g(\omega+\tau+\upsilon)\lambda_{\omega+\sigma+\tau+\upsilon}$$
$$= \int f^+(\omega+\sigma+\tau)A_0\left(\oint_{s,t}^{0} u_s^\cdot\right)(\sigma,\tau,\upsilon) g(\omega+\tau+\upsilon)\lambda_{\omega+\sigma+\tau+\upsilon}.$$

Remark that the function $u_s^t(\sigma,\tau,\upsilon)$ is determined by the sesquilinear form $\langle f|U_s^t|g\rangle$ Lebesgue almost everywhere. □

Applying Theorem 6.2.1, we obtain immediately

Theorem 8.2.1 *Equation* (∗) *has a unique solution, namely*

$$U_s^t = \mathscr{B}\big(u_s^t(A_1, A_0, A_{-1}; B)\big).$$

We recall the definition of u_s^t:

$$\begin{aligned} &u_s^t(\sigma, \tau, \upsilon)\\ &\quad = (-\mathrm{i})^n \mathrm{e}^{B(t-s_n)} A_{i_n} \mathrm{e}^{B(s_n-s_{n-1})} A_{i_{n-1}}\\ &\qquad \cdots A_{i_2} \mathrm{e}^{B(s_2-s_1)} A_{i_1} \mathrm{e}^{B(s_1-s)} \mathbf{1}\{t_{\sigma+\tau+\upsilon} \subset]s,t[\} \end{aligned}$$

if $t_{\sigma+\tau+\upsilon}$ is without a repeated point and

$$t_{\sigma+\tau+\upsilon} = \{s < s_1 < s_2 < \cdots < s_{n-1} < s_n < t\},$$

where the A_{i_j} are numbered accordingly.

If $\mathbb{O}_a$ is the operator inducing the normal ordering of a and a^+, one may write

$$\begin{aligned} U_s^t = 1 + \sum_{n=1}^{\infty} (-\mathrm{i})^n \int \cdots \int_{s<s_1<s_2<\cdots<s_n<t} \\ \mathbb{O}_a\big(\mathrm{e}^{B(t-s_n)}\big(A_1 a^+(\mathrm{d}s_n) + A_0 a^+(\mathrm{d}s_n)a(s_n) + A_{-1}a(s_n)\mathrm{d}s_n\big)\mathrm{e}^{B(s_n-s_{n-1})} \\ \cdots \mathrm{e}^{B(s_2-s_1)}\big(A_1 a^+(\mathrm{d}s_1) + A_0 a^+(\mathrm{d}s_1)a(s_1) + A_{-1}a(s_1)\mathrm{d}s_1\big)\big)\mathrm{e}^{B(s_1-s)}\big). \end{aligned}$$

Using the notation $a^+(\mathrm{d}t) = a^\dagger(t)\mathrm{d}t$, the last equation becomes

$$\begin{aligned} U_s^t = 1 + \sum_{n=1}^{\infty} (-\mathrm{i})^n \int \cdots \int_{s<s_1<s_2<\cdots<s_n<t} \mathrm{d}s_1 \cdots \mathrm{d}s_n \\ \mathbb{O}_a\big(\mathrm{e}^{B(t-s_n)}\big(A_1 a^\dagger(s_n) + A_0 a^\dagger(s_n)a(s_n) + A_{-1}a(s_n)\big)\mathrm{e}^{B(s_n-s_{n-1})} \\ \cdots \mathrm{e}^{B(s_2-s_1)}\big(A_1 a^\dagger(s_1) + A_0 a^\dagger(s_1)a(s_1) + A_{-1}a(s_1)\big)\mathrm{e}^{B(s_1-s)}\big). \end{aligned}$$

8.3 Examples

8.3.1 A Two-Level Atom in a Heatbath of Oscillators

We discuss the four examples introduced in Chap. 4.

We consider the equation

$$(\mathrm{d}/\mathrm{d}t)U_s^t = -\mathrm{i}\sqrt{2\pi}a^\dagger(t)E_{-+}U_s^t - \mathrm{i}\sqrt{2\pi}E_{+-}U_s^t a(t) - \pi E_{++}U_s^t,$$

where the four $E_{\pm\pm}$ are the 2×2-matrix units. Then

$$U_s^t = \mathrm{e}^{-\pi E_{++}(t-s)} + \sum_{n=1}^{\infty}(-\sqrt{2\pi}\mathrm{i})^n \int \cdots \int_{s<s_1<s_2<\cdots<s_n<t} \mathrm{d}s_1 \cdots \mathrm{d}s_n$$

$$\mathbb{O}_a\big(\mathrm{e}^{-\pi E_{++}(t-s_n)}\big(E_{-+}a^\dagger(s_n) + E_{+-}a(s_n)\big)\mathrm{e}^{-\pi E_{++}(s_n-s_{n-1})}$$

$$\cdots \mathrm{e}^{-\pi E_{++}(s_2-s_1)}\big(E_{-+}a^\dagger(s_1) + E_{+-}a(s_1)\big)\mathrm{e}^{-\pi E_{++}(s_1-s)}\big).$$

Use the notation $|+\rangle = \binom{1}{0}$ and $|-\rangle = \binom{0}{1}$. Then we calculate

$$U_s^t|+\rangle \otimes |\emptyset\rangle = \mathrm{e}^{-\pi(t-s)}|+\rangle \otimes |\emptyset\rangle - \mathrm{i}\sqrt{2\pi}\int_s^t \mathrm{d}s_1|-\rangle \otimes a^\dagger(s_1)|\emptyset\rangle \mathrm{e}^{-\pi(s_1-s)}$$

since

$$\big(E_{-+}a^\dagger(s_1) + E_{+-}a(s_1)\big)\big(|+\rangle \otimes |\emptyset\rangle\big) = |-\rangle \otimes a^\dagger(s_1)|\emptyset\rangle,$$

$$\mathbb{O}_a\big(E_{-+}a^\dagger(s_2) + E_{+-}a(s_2)\big)\big(E_{-+}a^\dagger(s_1) + E_{+-}a(s_1)\big)|+\rangle \otimes |\emptyset\rangle = 0.$$

Also

$$U_s^t\big(|-\rangle \otimes a^\dagger(s)|\emptyset\rangle\big) = \big(|-\rangle \otimes a^\dagger(s)|\emptyset\rangle\big) - \mathrm{i}\sqrt{2\pi}\int_s^t \mathrm{d}s_1 \mathrm{e}^{-\pi(t-s_1)}|+\rangle \otimes |\emptyset\rangle$$

$$- 2\pi \iint_{s<s_1<s_2<t} \mathrm{d}s_1\mathrm{d}s_2 \mathrm{e}^{-\pi(s_2-s_1)}\delta(s-s_1)|-\rangle \otimes a^\dagger(s_2)|\emptyset\rangle$$

since

$$\big(E_{-+}a^\dagger(s_1) + E_{+-}a(s_1)\big)\big(|-\rangle \otimes a^\dagger(s)|\emptyset\rangle\big) = \delta(s_1 - s)\big(|+\rangle \otimes |\emptyset\rangle\big),$$

$$\mathbb{O}_a\big(E_{-+}a^\dagger(s_2) + E_{+-}a(s_2)\big)\big(E_{-+}a^\dagger(s_1) + E_{+-}a(s_1)\big)\big(|-\rangle \otimes a^\dagger(s)|\emptyset\rangle\big)$$

$$= \delta(s - s_1)|-\rangle \otimes a^\dagger(s_2)|\emptyset\rangle.$$

The terms of third and higher orders vanish. So the subspace spanned by $|+\rangle \otimes |\emptyset\rangle$ and $|-\rangle \otimes a^\dagger(s)|\emptyset\rangle$, $s \in \mathbb{R}$, stays invariant, and the restriction of U_0^t to this subspace coincides with the matrix $V(t)$ in the formal time representation (see Sect. 4.2.4), as

$$V(t) = \begin{pmatrix} V_{00} & V_{01} \\ V_{10} & V_{11} \end{pmatrix}$$

and

$$V_{00}(t) = \mathrm{e}^{-\pi t},$$

$$\big(V_{01}(t)|\tau\big) = -\mathrm{i}(2\pi)^{1/2}\int_0^t \mathrm{d}t_1 \mathrm{e}^{-\pi(t-t_1)}\delta(\tau - t_1),$$

$$\big(\tau|V_{10}(t)\big) = -\mathrm{i}(2\pi)^{1/2}\int_0^t \mathrm{d}t_1\delta(t_1-\tau)\mathrm{e}^{-\pi t_1},$$

$$(\tau_2|V_{11}(t)|\tau_1) = \delta(\tau_1-\tau_2) - 2\pi\iint_{0<t_1<t_2<t} \mathrm{d}t_1\mathrm{d}t_2\delta(\tau_2-t_2)\mathrm{e}^{-\pi(t_2-t_1)}\delta(t_1-\tau_1).$$

8.3.2 A Two-Level Atom Interacting with Polarized Radiation

We work with the space

$$X = L^2\big(\mathbb{R}\times\mathbb{S}^2\times\{1,2,3\}\big)$$

provided with the measure

$$\langle\lambda|f\rangle = \iint \mathrm{d}t\,\omega_0^2\mathbf{dn}\sum_{i=1,2,3} f(t,\mathbf{n},i),$$

where $\mathbf{dn}$ is the surface element on the unit sphere such that

$$\int_{\mathbb{S}^2}\mathbf{dn} = 4\pi$$

and ω_0 is the transition frequency. Use the notation again

$$\mathfrak{X} = \{\emptyset\} + X + X^2 + \cdots$$

and consider

$$\Gamma = L^2(\mathfrak{X},\mathbb{C}^3).$$

Recall the vector

$$\mathbf{v}(\mathbf{n}) = \Pi(\mathbf{n})\mathbf{q},$$

where $\Pi(\mathbf{n})$ is the projector on the plane perpendicular to $\mathbf{n}$,

$$\Pi(\mathbf{n})_{ij} = \delta_{ij} - \mathbf{n}_i\mathbf{n}_j$$

and $\mathbf{q}$ is a fixed vector given by physics.

One finds

$$\gamma = \int \omega_0^2\mathbf{dn}\big|\mathbf{v}(\mathbf{n})\big|^2 = \frac{8\pi}{3}|\mathbf{q}|^2.$$

We have the annihilation operators $a(t,\mathbf{n},i)$ and the creation operators $a^+(\mathrm{d}(t,\mathbf{n},i))$. Define the vectors

$$\mathbf{a}(t,\mathbf{n}) = \big(a(t,\mathbf{n},i)\big)_{i=1,2,3},\qquad \mathbf{a}^+\big(\mathrm{d}(t,\mathbf{n})\big) = \big(a\big(\mathrm{d}(t,\mathbf{n},i)\big)\big)_{i=1,2,3}.$$

Consider the quantum stochastic differential equation

$$\mathrm{d}_t U_s^t = -\mathrm{i}\sqrt{2\pi}\int_{\mathbb{S}^2} \langle \mathbf{v}(\mathbf{n}), \mathbf{a}^+(\mathrm{d}(\mathbf{n},t))\rangle E_{01} U_s^t$$
$$- \mathrm{i}\sqrt{2\pi} E_{10} U_s^t \mathrm{d}t \int_{\mathbb{S}^2} \omega_0^2 \mathrm{d}\mathbf{n} \langle \mathbf{a}(t,\mathbf{n}), \mathbf{v}(\mathbf{n})\rangle - \pi\gamma E_{11} U_s^t \mathrm{d}t.$$

We use the notation

$$K(\mathrm{d}t) = \int_{\mathbb{S}^2} \langle \mathbf{v}(\mathbf{n}), \mathbf{a}^+(\mathrm{d}(\mathbf{n},t))\rangle E_{01} + \int_{\mathbb{S}^2} E_{10} \mathrm{d}t \omega_0^2 \mathrm{d}\mathbf{n} \langle \mathbf{a}(t,\mathbf{n}), \mathbf{v}(\mathbf{n})\rangle$$

then we assume without proof, that the solution is analogous to the series of Theorem 8.2.1

$$U_s^t = 1 + \sum_{n=1}^{\infty} (-\mathrm{i}\sqrt{2\pi})^n \int \cdots \int_{s<s_1<\cdots<s_n<t}$$
$$\mathbb{O}_a \mathrm{e}^{-\pi\gamma(t-s_n)} K(\mathrm{d}s_n) \cdots \mathrm{e}^{-\mathrm{i}\pi\gamma(s_2-s_1)} K(\mathrm{d}t_1) \mathrm{e}^{-\pi\gamma(s_1-s)}.$$

By a similar calculation to that in Sect. 8.3.1 we obtain that the subspace spanned by $\binom{1}{0} \otimes |\emptyset\rangle$ and $\binom{0}{1} \otimes a^+(\mathrm{d}(t,\mathbf{n},i))|\emptyset\rangle$ stays invariant and that the restriction of U_0^t to that subspace equals $V(t)$ in the formal time representation in Sect. 4.2.3.

8.3.3 The Heisenberg Equation of the Amplified Oscillator

This is formally very similar to the first example in Sect. 8.3.1. We have the stochastic differential equation

$$\frac{\mathrm{d}}{\mathrm{d}t} U_s^t = \mathrm{i}\sqrt{2\pi} a^\dagger(t) E_{-+} U_s^t - \mathrm{i}\sqrt{2\pi} E_{+-} U_s^t a(t) + \pi E_{++} U_s^t.$$

The subspace spanned by $|+\rangle \otimes |\emptyset\rangle$ and by the $|-\rangle \otimes a^+(\mathrm{d}s)|\emptyset\rangle$ stays invariant, and the restriction of U_0^t coincides with the matrix $V(t)$ in Sect. 4.4.2. But the analytical character is very different, as was pointed out there.

8.3.4 A Pure Number Process

The differential equation is of the form

$$\mathrm{d}_t U_s^t = c a^+(\mathrm{d}t) U_s^t a(t).$$

It can be solved by the infinite series of Theorem 4.2.1. The number operator $\int a^+(\mathrm{d}t)a(t)$ is an invariant. If we restrict to the one-particle space, we obtain

$$\begin{aligned}\langle\emptyset|a(s_2)U_s^t a^\dagger(s_1)|\emptyset\rangle &= \delta(s_1-s_2)+c\int_s^t \mathrm{d}t_1\delta(s_1-t_1)\delta(t_1-s_2)\\ &= \big(1+c\mathbf{1}_{[s,t]}(s_1)\big)\delta(s_1-s_2)\end{aligned}$$

in agreement with the formula for $V(t)$ in Sect. 4.5 with

$$c=\frac{-\mathrm{i}2\pi}{1+\mathrm{i}\pi}.$$

8.4 A Priori Estimate and Continuity at the Origin

Definition 8.4.1 We define the Fock space

$$\Gamma=L^2_\mathrm{s}\big(\mathfrak{R},\mathfrak{k},\mathrm{e}(\lambda)\big)$$

of all symmetric square-integrable functions with respect to Lebesgue measure from $\mathfrak{R}$ to $\mathfrak{k}$. If f is a measurable function on $\mathfrak{R}$ define the operator N by $(Nf)(w)=(\#w)f(w)$, and define Γ_k as the space of those measurable symmetric functions from $\mathfrak{R}$ to $\mathfrak{k}$ for which

$$\int \Delta(w)\big\langle f(w)\big|(N+1)^k f(w)\big\rangle \mathrm{d}w<\infty.$$

We denote by $\|.\|_{\Gamma_k}$ the corresponding norm. We write for short

$$\mathscr{K}=\mathscr{K}_\mathrm{s}(\mathfrak{R},\mathfrak{k})$$

for the space of all symmetric continuous functions from $\mathfrak{R}$ to $\mathfrak{k}$ with compact support. Call $\mathscr{K}^{(n)}$, resp. $\Gamma^{(n)}$, the subspaces where $f(w)=0$ for $\#w>n$.

We extend the notions of a and a^+. We define $a(\varphi\lambda)f=a(\varphi)f$ and $a^+(\varphi)f$ for $\varphi\in L^2(\mathbb{R})$ and $f\in\Gamma^{(n)}$. We have the well known relations

$$\begin{aligned} a(\varphi):\Gamma^{(n)}&\to\Gamma^{(n-1)}, &\quad \big\|a(\varphi)f\big\|_\Gamma&\le\sqrt{n}\|\varphi\|_{L^2}\|f\|_\Gamma,\\ a(\varphi)^+:\Gamma^{(n)}&\to\Gamma^{(n+1)}, &\quad \big\|a(\varphi)^+f\big\|_\Gamma&\le\sqrt{n+1}\|\varphi\|_{L^2}\|f\|_\Gamma.\end{aligned}$$

One sees easily

Lemma 8.4.1 *We have for $\varphi\in L^2(\mathbb{R})$ the equations*

$$\int a^+(\sigma)\mathrm{e}(\varphi)(\sigma)=\exp\Big(\int a^+(\mathrm{d}t)\varphi(t)\Big),$$

$$\int \lambda_\upsilon a(\upsilon)\mathrm{e}(\varphi)(\upsilon) = \exp\left(\int \mathrm{d}t\, a(t)\varphi(t)\right),$$

$$\int a^+(\tau)a(\tau)\mathrm{e}(\varphi)(\tau) = \mathbb{O}_a \exp\left(\int a^+(\mathrm{d}t)a(t)\varphi(t)\right),$$

with

$$\mathrm{e}(\varphi)(t_1, \ldots, t_n) = \varphi(t_1)\cdots\varphi(t_n)$$

as usual.

Lemma 8.4.2 *Assume we are given a Lebesgue measurable function* $f : \mathfrak{R} \to \mathfrak{k}$, *then*

$$\int \lambda_{\xi+\omega}\mathbf{1}\{\#\xi = k\}\left\|f(\xi+\omega)\right\|^2 = \langle f|\binom{N}{k}|f\rangle.$$

Proof The left-hand side of the last equation equals

$$\int \lambda_\omega \sum_{\xi\subset\omega}\left\|f(\omega)\right\|^2\mathbf{1}\{\#\xi = k\} = \int \lambda_\omega \binom{\#\omega}{k}\left\|f(\omega)\right\|^2,$$

after a change of variable and using the sum-integral lemma, and the resulting right-hand side is what was needed. □

Lemma 8.4.3 *If*

$$f = \exp\left(\int a^+(\mathrm{d}t)\varphi(t)\right)g,$$

then

$$\langle f|\binom{N}{k}|f\rangle \le \sum_{k_1\le k} \mathrm{e}^{4\|\varphi\|^2}\langle g|2^N\binom{N}{k_1}|g\rangle.$$

Proof We assume $f \ge 0$ and $g \ge 0$. One obtains

$$f(\sigma) = \sum_{\sigma_1+\sigma_2} \mathrm{e}\big(\varphi(\sigma_1)\big)g(\sigma_2)$$

and

$$\int \lambda_{\xi+\omega}\mathbf{1}\{\#\xi = k\}\left\|f(\xi+\omega)\right\|^2$$

$$= \int \lambda_{\xi+\omega}\mathbf{1}\{\#\xi = k\}\left(\sum_{\substack{\omega_1+\omega_2=\omega\\ \xi_1+\xi_2=\xi}} \mathrm{e}(\varphi)(\xi_1+\omega_1)g(\xi_2+\omega_2)\right)^2.$$

Using Cauchy-Schwarz inequality

$$\leq 2^k \int 2^{\#\omega} \lambda_{\xi+\omega} \mathbf{1}\{\#\xi = k\} \sum_{\substack{\omega_1+\omega_2=\omega \\ \xi_1+\xi_2=\xi}} \mathrm{e}(\varphi^2)(\xi_1+\omega_1) g(\xi_2+\omega_2)^2$$

$$= \sum_{k_1+k_2=k} \int \lambda_{\xi_1+\xi_2+\omega_1+\omega_2} \mathbf{1}\{\#\xi_1 = k_1\}\mathbf{1}\{\#\xi_2 = k_2\}$$

$$2^{k_1}\mathrm{e}(\varphi^2)(\xi_1)2^{\#\omega_1}\mathrm{e}(\varphi^2)(\omega_1)2^{k_2+\#\omega_2}g(\xi_2+\omega_2)^2$$

$$\leq \sum_{k_1 \leq k} \mathrm{e}^{4\|\varphi\|^2} \langle g | 2^N \binom{N}{k_1} | g\rangle$$

as

$$\int \lambda_{\xi_1} \mathbf{1}\{\#\xi_1 = k_1\} 2^{k_1} \mathrm{e}(\varphi^2)(\xi_1) = \frac{1}{k_1!} 2^{k_1} \|\varphi\|^{2k_1} < \mathrm{e}^{2\|\varphi\|^2}$$

and

$$\int \lambda_{\omega_1} 2^{\#\omega_1} \mathrm{e}(\varphi^2)(\omega_1) = \mathrm{e}^{2\|\varphi\|^2}$$

and

$$\int \lambda_{\xi_2+\omega_2} \mathbf{1}\{\xi_2 = k_2\} 2^{\#(\xi_2+\omega_2)} g(\xi_2+\omega_2)^2 = \langle g | 2^N \binom{N}{k_1} | g\rangle$$

by the same reasoning as in the proof of the preceding lemma. □

Proposition 8.4.1 *Assume*

$$U_s^t = \mathscr{O}\big(u_s^t(A_1, A_0, A_{-1}; B)\big).$$

Then there exist constants $C_{n,k}(t-s)$ such that, for $f \in \mathscr{K}^{(n)}$,

$$\|U_s^t f\|_{\Gamma_k} \leq C_{n,k}(t-s)\|f\|_{\Gamma}.$$

Furthermore, for $t \downarrow s$ and $f \in \mathscr{K}$,

$$\|U_s^t f - f\|_{\Gamma_k} \to 0.$$

Proof Define

$$C = \max\big(\|A_i\|, i = 1, 0, -1, \|B\|\big);$$

then

$$\big\|u_s^t(\sigma,\tau,\upsilon)\big\| \leq \mathrm{e}^{C(t-s)} C^{\#\sigma+\#\tau+\#\upsilon} \mathbf{1}\big\{t_{\sigma+\tau+\upsilon} \subset [s,t]\big\} = \mathrm{e}^{C(t-s)}\mathrm{e}(\chi)(\sigma+\tau+\upsilon)$$

with $\chi(r) = C\mathbf{1}_{[s,t]}(r)$ and $\mathrm{e}(\chi)(\omega) = \prod_{c\in\omega}\chi(t_c)$.

We have using Proposition 7.1.1

$$\|\mathscr{O}(u_s^t)f(\omega)\| \le \sum_{\sigma\subset\omega}\sum_{\tau\subset\omega\setminus\sigma}\int \lambda_\upsilon e^{C(t-s)}e(\chi)(\sigma+\tau+\upsilon)\|f(\omega\setminus\sigma+\upsilon)\| = e^{C(t-s)}\big(R_s^t\, S_s^t\, T_s^t\,\|f\|\big)(\omega).$$

For $g \in \mathscr{K}^{(n)}(\mathfrak{R},\mathbb{R})$, $g \ge 0$ we have

$$\int \lambda_\upsilon\big(a_\upsilon e(\chi)(\upsilon)g\big)(\omega) = (T_s^t g)(\omega) = \int \lambda_\upsilon e(\chi)(\upsilon)g(\omega+\upsilon) = \big(\exp(a(\chi))g\big)(\omega).$$

As $T_s^t : \mathscr{K}^{(n)} \to \mathscr{K}^{(n)}$, we may estimate the Γ_k-norm by the Γ-norm. We have

$$\|T_s^t g\| \le \sum_{l=0}^{n}(1/l!)\sqrt{n(n-1)\cdots(n-l+1)}C^l(t-s)^{l/2}\|g\|_\Gamma$$

as

$$\|\chi\|_{L^2} = C\sqrt{t-s}.$$

Furthermore we have

$$S_s^t : \mathscr{K}^{(n)} \to \mathscr{K}^{(n)},$$

$$\int \big(a_\tau^+ a_\tau g\big)(\omega) = \big(S_s^t g\big)(\omega) = \sum_{\tau\subset\omega} e(\chi)(\tau)g(\omega) = e(1+\chi)(\omega)g(\omega)$$

and

$$\big\|S_t^s g\big\|_\Gamma \le (1+C)^n\|g\|_\Gamma.$$

Again

$$R_s^t : \mathscr{K}^{(n)} \to \Gamma^k$$

$$\int \big(a_\sigma^+ e(g)\big)(\omega) = \big(R_t^s g\big)(\omega) = \sum_{\sigma\subset\omega} e(\chi)(\sigma)g(\omega\setminus\sigma) = \exp\big(a^+(\chi)g\big)(\omega).$$

Use the inequality of the last lemma and obtain the first assertion.

We investigate the second assertion. As

$$\big(\mathscr{O}(u_s^t)f - f\big)(\omega) = \sum_{\sigma\subset\omega}\sum_{\tau\subset\omega\setminus\sigma}\int \lambda_\upsilon u_s^t(\sigma,\tau,\upsilon)\mathbf{1}\{\sigma+\tau+\upsilon\neq\emptyset\}f(\omega\setminus\sigma+\upsilon)$$

we may estimate the norm by

$$\sum_{\sigma\subset\omega}\sum_{\tau\subset\omega\setminus\sigma}\int \lambda_\upsilon \exp\bigl(C(t-s)\bigr)\mathrm{e}(\chi)(\sigma+\tau+\upsilon)\bigl\|f(\omega\setminus\sigma+\upsilon)\bigr\|\mathbf{1}\{\sigma+\tau+\upsilon\neq\emptyset\}$$

$$=\exp\bigl(C(t-s)\bigr)\bigl((R_s^tS_s^tT_s^t-1)\|f\|\bigr)(\omega).$$

We have

$$\bigl\|T_s^tg-g\bigr\|_\Gamma\leq\sum_{l=1}^n(1/l!)\sqrt{n(n-1)\cdots(n-l+1)}C^l(t-s)^{l/2}\|g\|_\Gamma=O(\sqrt{t-s}),$$

and

$$\begin{aligned}\bigl((S_s^t-1)g\bigr)(\omega)&=\bigl(\mathrm{e}(1+\chi)(\omega)-1\bigr)g(\omega)\\&=\sum_{c\in\omega}\chi(c)\mathrm{e}(1+\chi)(\omega\setminus c)g(\omega)\leq\sum_{c\in\omega}\chi(c)(1+C)^{n-1}g(\omega).\end{aligned}$$

Since $f\in\mathcal{K}^n$, there exists a compact interval $K\subset\mathbb{R}$, $[s,t]\subset K$, such that $g(\omega)\leq\mathrm{e}(\mathbf{1}_K)(\omega)$ for $\#\omega\leq n$, if $g(\omega)\leq 1$ for all ω. We have

$$\sum_{c\in\omega}\chi(c)(1+C)^{n-1}g(\omega)\leq\sum_{c\in\omega}\chi(c)(1+C)^n\mathrm{e}(\mathbf{1}_K)(\omega\setminus c).$$

The norm is bounded above by

$$\sqrt{n+1}\sqrt{t-s}(1+C)^n\exp\bigl(|K|/2\bigr).$$

We have

$$\bigl((R_s^t-1)g\bigr)(\omega)=\sum_{\sigma\subset\omega,\sigma\neq\emptyset}\mathrm{e}(\chi)(\sigma)g(\omega\setminus\sigma).$$

Hence

$$\begin{aligned}&\Bigl\langle R_s^tg-g\Bigl|\binom{N}{k}\Bigr|R_s^tg-g\Bigr\rangle\\&\quad=\int\lambda_{\xi+\omega}\mathbf{1}\{\#\xi=k\}\bigl(R_s^tg-g\bigr)^2(\xi+\omega)\\&\quad=\int\lambda_{\xi+\omega}\mathbf{1}\{\#\xi=k\}\Biggl(\sum_{\substack{\omega_1+\omega_2=\omega\\\xi_1+\xi_2=\xi\\\omega_1+\xi_1\neq\emptyset}}\mathrm{e}(\chi)(\xi_1+\omega_1)g(\xi_2+\omega_2)\Biggr)^2\\&\quad\leq\sum_{k_1+k_2=k}\int_{\xi_1+\omega_1\neq\emptyset}\lambda_{\xi_1+\xi_2+\omega_1+\omega_2}\mathbf{1}\{\#\xi_1=k_1\}\mathbf{1}\{\#\xi_2=k_2\}\end{aligned}$$

$$\times\, 2^{k_1} \mathrm{e}(\chi^2)(\xi_1) 2^{\#\omega_1} \mathrm{e}(\chi^2)(\omega_1) 2^{k_2+\#\omega_2} g(\xi_2+\omega_2)^2$$
$$\le \sum_{k_1 \le k} \left(\mathrm{e}^{4\|\chi\|^2} - 1\right) \langle g | 2^N \binom{N}{k_1} | g \rangle = O(t-s)$$

because

$$\int_{\xi_1+\omega_1 \neq \emptyset} \lambda_{\xi_1+\omega_1} \mathbf{1}\{\#\xi_1 = k_1\} 2^{\#(\xi_1+\omega_1)} \mathrm{e}(\chi^2)(\xi_1+\omega_1) \le \mathrm{e}^{4\|\chi\|^2} - 1.$$

From these results one obtains the second assertion of the proposition easily. □

8.5 Consecutive Intervals in Time

We start with a lemma.

Lemma 8.5.1 *Assume $s < r < t$. Multiply the measure*

$$\mathfrak{m} = \langle a_\pi \overset{+}{a}_{\sigma_2+\tau_2} a_{\tau_2+\upsilon_2} \overset{+}{a}_{\sigma_1+\tau_1} a_{\tau_1+\upsilon_1} \overset{+}{a}_\varrho \rangle \lambda_{\pi+\upsilon_1+\upsilon_2}$$

by the Borel function

$$F = \mathbf{1}\{t_{\sigma_1+\tau_1+\upsilon_1} \subset [s,r]\}\, \mathbf{1}\{t_{\sigma_2+\tau_2+\upsilon_2} \subset [r,t]\}.$$

Then

$$F\mathfrak{m} = F\langle a_\pi \overset{+}{a}_{\sigma_2+\tau_2+\sigma_1+\tau_1} a_{\tau_2+\upsilon_2+\tau_1+\upsilon_1} \overset{+}{a}_\varrho \rangle \lambda_{\pi+\upsilon_1+\upsilon_2}.$$

Proof Integrate against a C_c^∞-function f, considering the integral

$$\int f(\pi, \sigma_1, \ldots, \upsilon_2, \varrho) F\mathfrak{m}.$$

Take $c \in \tau_2 + \upsilon_2$, e.g., $c \in \upsilon_2$, then

$$a_{\tau_2+\upsilon_2} \overset{+}{a}_{\sigma_1+\tau_1} = a_{\tau_2+\upsilon_2 \setminus c} \overset{+}{a}_{\sigma_1+\tau_1} a_c + \sum_{b \in \sigma_2+\tau_2} \varepsilon(c,b) a_{\tau_2+\upsilon_2 \setminus c} \overset{+}{a}_{(\sigma_1+\tau_1)\setminus c}.$$

But

$$\int f^+(\pi, \ldots, \varrho) \varepsilon(c,b) \langle a_\pi \overset{+}{a}_{\sigma_2+\tau_2} a_{\tau_2+\upsilon_2 \setminus c} \overset{+}{a}_{(\sigma_1+\tau_1)\setminus b} a_{\tau_1+\upsilon_1} \overset{+}{a}_\varrho \rangle \lambda_{\pi+\upsilon_1+\upsilon_2} = 0$$

as

$$\int \lambda_c \varepsilon(c,b) \mathbf{1}\{r < t_c < t\} \mathbf{1}\{s < t_b < r\} = \int \lambda_c \mathbf{1}\{r < t_c < t\} \mathbf{1}\{s < t_c < r\} = 0.$$

One proves the lemma by induction. □

Lemma 8.5.2 *Assume*

$$t_0 < t_1 < \cdots < t_n$$

and multiply the measure

$$\mathfrak{m} = \langle a_\pi a^+_{\sigma_n+\tau_N} a_{\tau_n+\upsilon_n} \cdots a^+_{\sigma_1+\tau_1} a_{\tau_1+\upsilon_1} a^+_\varrho \rangle \lambda_{\pi+\upsilon_1+\cdots+\upsilon_n}$$

by the Borel function

$$F = \mathbf{1}\{t_{\sigma_1+\tau_1+\upsilon_1} \subset [t_0, t_1]\} \cdots \mathbf{1}\{t_{\sigma_n+\tau_n+\upsilon_n} \subset [t_{n-1}, t_n]\}.$$

Then

$$F\mathfrak{m} = F\langle a_\pi a^+_{\sigma_n+\tau_n+\cdots+\sigma_1+\tau_1} a_{\tau_n+\upsilon_n+\cdots+\tau_1+\upsilon_1} a^+_\varrho \rangle \lambda_{\pi+\upsilon_1+\cdots+\upsilon_n}.$$

For the proof use the duality theorem in Sect. 5.6.

We consider again $u^t_s(A_1, A_0, A_{-1}; B)$. We have shown, in Sect. 8.4, that the map $\mathscr{O}(u^t_s) : \mathscr{K}^{(n)} \to \Gamma$ is bounded. So $\mathscr{B}(u^t_r, u^r_s)$ exists (see Sect. 7.1).

Proposition 8.5.1 *For $s < r < t$ we have*

$$\mathscr{B}(u^t_r, u^r_s) = \mathscr{B}(u^t_s).$$

Proof We have

$$\langle f | \mathscr{B}(u^t_r, u^r_s) | g \rangle = \int f^+(\pi) u^t_r(\sigma_2, \tau_2, \upsilon_2) u^r_t(\sigma_1, \tau_1, \upsilon_1) g(\varrho)$$
$$\langle a_\pi a^+_{\sigma_2+\tau_2} a_{t_2+\upsilon_2} a^+_{\sigma_1+\tau_1} a_{\tau_1+\upsilon_1} a^+_\varrho \rangle \lambda_{\pi+\upsilon_1+\upsilon_2}.$$

As, e.g., $u^t_r(\sigma_2, \tau_2, \upsilon_2)$ vanishes if $t_{\sigma_2+\tau_2+\upsilon_2} \not\subset [r, t]$, we may apply Lemma 8.5.1 and we obtain

$$\langle f | \mathscr{B}(u^t_r, u^r_s) | g \rangle = \int f^+(\pi) u^t_r(\sigma_2, \tau_2, \upsilon_2) u^r_t(\sigma_1, \tau_1, \upsilon_1) g(\varrho) \mathfrak{m}'$$

with

$$\mathfrak{m}' = \langle a_\pi a^+_{\sigma_2+\tau_2} a^+_{\sigma_1+\tau_1} a_{\tau_2+\upsilon_2} a_{\tau_1+\upsilon_1} a^+_\varrho \rangle \lambda_{\pi+\upsilon_1+\upsilon_2}.$$

If $\{r, s, t, t_{\sigma+\tau+\tau}\}^\bullet$ is without multiple points, then we showed in Remark 6.2.1 that

$$u^t_s(\sigma, \tau, \upsilon) = u^t_r(\sigma_2, \tau_2, \upsilon_2) u^r_s(\sigma_1, \tau_1, \upsilon_1)$$

with

$$\sigma_2 = \{c \in \sigma : r < t_c < t\}, \qquad \sigma_1 = \{c \in \sigma : r < t_c < t\}$$

etc. But $t^\bullet_{\sigma+\tau+\upsilon}$ is without multiple points $\mathfrak{m}'$-a.e. □

8.6 Unitarity

The following theorem is essentially due to Hudson and Parthasarathy [36].

Theorem 8.6.1 *Assume* $u_s^t = u_s^t(A_1, A_0, A_{-1}; B)$. *The mapping*

$$\mathscr{O}\left(u_s^t\right) : \mathscr{K} \to \Gamma$$

can be extended to a unitary mapping

$$U_s^t : \Gamma \to \Gamma,$$

if and only if the operators A_i, $i = 1, 0, -1$ *and* B *fulfill the following conditions: There exists a unitary operator* Υ *such that*

$$\begin{aligned} A_0 &= \Upsilon - 1, \\ A_1 &= -\Upsilon A_{-1}^+, \\ B + B^+ &= -A_1^+ A_1 = -A_{-1} A_{-1}^+. \end{aligned}$$

Proof We recall Proposition 6.3.2. For fixed s, the function $u_s^{\cdot} : t \mapsto u_s^t$, and for fixed t, the function $u_{\cdot}^t : s \mapsto u_s^t$ is of class $\mathscr{C}^1$, and one has

$$\begin{aligned} \partial_t^c u_s^t &= B u_s^t, \\ \left(R_+^j u_s^{\cdot}\right)_t &= A_j u_s^t, \\ \left(R_-^j u_s^{\cdot}\right)_t &= 0, \\ \partial_s^c u_s^t &= -u_s^t B, \\ \left(R_+^j u_{\cdot}^t\right)_s &= 0, \\ \left(R_-^j u_{\cdot}^t\right)_s &= u_s^t A_j \end{aligned}$$

for $j = 1, 0, -1$. We recall Ito's formula from Theorem 7.4.1. Assume x_t, y_t to be of class $\mathscr{C}^1$, and that for $f, g \in \mathscr{K}_s(\mathfrak{R}, \mathfrak{k})$ the sesquilinear forms $\langle f|\mathscr{B}(F_t, G_t)|g\rangle$ exist in norm and $t \in \mathbb{R} \mapsto \langle f|\mathscr{B}(F_t, G_t)|g\rangle$ is locally integrable, where F_t can be any function in $\{x_t, \partial^c x_t, R_\pm^1 x_t, R_\pm^0 x_t, R_\mp^{-1} x_t\}$ and G_t can be any function in $\{y_t, \partial^c y_t, R_\pm^1 y_t, R_\pm^0 y_t, R_\mp^{-1} y_t\}$.

Then $t \mapsto \langle f|\mathscr{B}(x_t, y_t)|g\rangle$ is continuous and its Schwartz derivative is a locally integrable function, and this yields

$$\begin{aligned} \partial\langle f|\mathscr{B}(x_t, y_t)|g\rangle &= \langle f|\mathscr{B}\left(\partial^c x_t, y_t\right) + \mathscr{B}\left(f, \partial^c y_t\right) + I_{-1,+1,t}|g\rangle \\ &\quad + \left\langle a(t) f\middle|\mathscr{B}\left(D^1 x_t, y_t\right) + \mathscr{B}\left(f, D^1 y_t\right) + I_{0,+1,t}\middle|g\right\rangle \end{aligned}$$

$$+\langle a(t)f|\mathscr{B}(D^0x_t,y_t)+\mathscr{B}(f,D^0y_t)+I_{0,0,t}|a(t)g\rangle$$
$$+\langle f|\mathscr{B}(D^{-1}x_t,y_t)+\mathscr{B}(f,D^{-1}y_t)+I_{-1,0,t}|a(t)g\rangle$$

with

$$I_{i,j,t}=\mathscr{B}(R_+^i x_t,R_+^j y_t)-\mathscr{B}(R_-^i x_t,R_-^j y_t).$$

We want to calculate the Schwartz derivatives of the functions

$$t\mapsto\langle f|\mathscr{B}((u_s^t)^+,u_s^t)|g\rangle$$
$$s\mapsto\langle f|\mathscr{B}(u_s^t,(u_s^t)^+)|g\rangle.$$

The derivatives exist, because u and u^+, and the ∂^c, R and D operators, applied to u and u^+, map $\mathscr{K}\to\Gamma$. We obtain

$$\partial_t\langle f|\mathscr{B}((u_s^t)^+,u_s^t)|g\rangle=\langle f|\mathscr{B}(u^+,C_1u)|g\rangle+\langle a(t)f|\mathscr{B}(u^+,C_2u)|g\rangle$$
$$+\langle a(t)f|\mathscr{B}(u^+,C_3u)|a(t)g\rangle+\langle f|\mathscr{B}(u^+,C_4u)|a(t)g\rangle,$$
$$\partial_s\langle f|\mathscr{B}(u_s^t,(u_s^t)^+)|g\rangle=\langle f|\mathscr{B}(u,C_5u^+)|g\rangle+\langle a(t)f|\mathscr{B}(u,C_6u^+)|g\rangle$$
$$+\langle a(t)f|\mathscr{B}(u,C_7u^+)|a(t)g\rangle+\langle f|\mathscr{B}(u,C_8u^+)|a(t)g\rangle$$

with

$$C_1=B+B^++A_1^+A_1,\qquad C_5=B+B^++A_{-1}A_{-1}^+,$$
$$C_2=A_{-1}^++A_1+A_0^+A_1,\qquad C_6=A_1+A_{-1}^++A_0A_{-1}^+,$$
$$C_3=A_0^++A_0+A_0^+A_0,\qquad C_7=A_0+A_0^++A_0A_0^+,$$
$$C_4=A_1^++A_{-1}+A_1^+A_0,\qquad C_8=A_{-1}+A_1^++A_{-1}A_0^+.$$

The operator $\mathscr{O}(u_s^t)$ is unitary if both derivatives vanish, and they vanish if $C_i=0,\ i=1,\dots,8$. The equations $C_3=0$ and $C_7=0$ imply

$$(1+A_0^+)(1+A_0)=(1+A_0)(1+A_0^+)=1.$$

So

$$\Upsilon=1+A_0$$

is unitary. The equations are not independent. We have $C_2^+=C_4$ and $C_6^+=C_8$. Furthermore

$$C_2=A_{-1}^++(1+A_0^+)A_1=A_{-1}^++\Upsilon^+A_1=\Upsilon^+C_6.$$

So $C_2=0$ implies $A_1=-\Upsilon A_{-1}^+$, and we conclude $C_1=C_5$. □

Definition 8.6.1 For $t < s$ we define

$$U_s^t = (U_t^s)^+.$$

Proposition 8.6.1 *For $r, s, t \in \mathbb{R}$ we have*

$$U_r^t U_s^r = U_s^t.$$

Proof For both cases $s < r < t$ and $t < r < s$, the assertion follows from Proposition 8.5.1. For $s < t < r$, we calculate

$$\langle f | U_r^t U_s^r g \rangle = \langle U_t^r f | (U_t^r)^+ (U_s^t g) \rangle = \langle f | U_s^t g \rangle.$$

The other variants can be calculated similarly. □

8.7 Estimation of the Γ_k-Norm

Recall from Sect. 7.1

$$\int \mathfrak{m}(\pi, \sigma, \tau, \upsilon, \varrho) f^+(\omega) F(\sigma, \tau, \upsilon) g(\varrho) = \int f^+(\omega) \big(\mathscr{O}(F) g\big)(\omega) = \langle f, \mathscr{O}(F) g \rangle$$

with

$$\big(\mathscr{O}(F) g\big)(\omega) = \sum_{\alpha \subset \omega} \sum_{\beta \subset \omega \setminus \alpha} \int_\upsilon \lambda_\upsilon F(\alpha, \beta, \upsilon) g(\omega \setminus \alpha + \upsilon)$$

and

$$\mathfrak{m} = \langle a_\omega a_{\sigma+\tau}^+ a_{\tau+\upsilon} a_\varrho^+ \rangle \lambda_{\omega+\upsilon}.$$

So $\mathscr{O}(F)$ is a mapping from $\mathscr{K}_s(\mathfrak{X}) = \mathscr{K}$ into the locally λ-integrable functions on $\mathfrak{X}$. Extend it to those functions g such that the integral exists in norm for almost all ω and yields a locally integrable function in ω.

Recall furthermore

$$F^+(\sigma, \tau, \upsilon) = F(\upsilon, \tau, \sigma)^+$$

and the relation

$$\langle f | \mathscr{O}(F) g \rangle = \langle \mathscr{O}(F^+) f | g \rangle$$

for $f, g \in \mathscr{K}$.

Lemma 8.7.1 *Assume a locally integrable function $F : \mathfrak{R}^3 \to B(\mathfrak{k})$ and a bounded operator $T : \Gamma \to \Gamma$ are given such that*

$$T \upharpoonright \mathscr{K} = \mathscr{O}(F).$$

Then

$$T^+ \restriction \mathscr{K} = \mathscr{O}(F^+).$$

Proof Assume $h, g \in \mathscr{K}$. Then

$$\langle h|Tg\rangle = \big\langle h|\mathscr{O}(F)g\big\rangle = \big\langle \mathscr{O}\big(F^+\big)h|g\big\rangle = \big\langle T^+h|g\big\rangle.$$

As this holds for all $g \in \mathscr{K}$, we have $T^+h = \mathscr{O}(F^+)h$. □

Lemma 8.7.2 *Assume a locally integrable function $F : \mathfrak{R}^3 \to B(\mathfrak{k})$ and a bounded operator $T : \Gamma \to \Gamma$ are given such that*

$$T \restriction \mathscr{K} = \mathscr{O}(F)$$

and there is a function $f \in \Gamma$, such that $\mathscr{O}(F)f$ exists, i.e.,

$$\int \big\|h(\omega)\big\| \big\|F(\sigma,\tau,\upsilon)\big\| \big\|f(\varrho)\big\| \mathfrak{m} < \infty$$

for all $h \in \mathscr{K}$. Then

$$\mathscr{O}(F)f = Tf.$$

Proof We have, for all $h \in \mathscr{K}$,

$$\int h^+(\omega)\big(\mathscr{O}(F)f\big)(\omega)\,\mathrm{d}\omega = \overline{\int f^+(\omega)\big(\mathscr{O}\big(F^+\big)h\big)(\omega)\,\mathrm{d}\omega} = \overline{\big\langle f|T^+h\big\rangle} = \langle h|Tf\rangle.$$

□

Lemma 8.7.3 *Assume we have $u_s^t = u_s^t(A_i, B)$ satisfying the unitarity conditions, and that U_s^t is the corresponding unitary operator. Assume $G_1, \dots, G_k \in B(\mathfrak{k})$ and $s = t_0 < t_1 < \cdots < t_k < t_{k+1} = t$, and also that*

$$\begin{aligned} &F(\sigma,\tau,\upsilon) \\ &\quad = \sum_{\substack{\sigma_0+\sigma_1+\cdots+\sigma_k=\sigma\\ \tau_0+\tau_1+\cdots+\tau_k=\tau\\ \upsilon_0+\upsilon_1+\cdots+\upsilon_k=\upsilon}} u_{t_k}^t(\sigma_k,\tau_k,\upsilon_k)G_k u_{t_{k-1}}^{t_k}(\sigma_{k-1},\tau_{k-1},\upsilon_{k-1})G_{k-1} \\ &\qquad \cdots G_2 u_{t_1}^{t_2}(\sigma_1,\tau_1,\upsilon_1)G_1 u_s^{t_1}(\sigma_0,\tau_0,\upsilon_0). \end{aligned}$$

Then, for $g \in \mathscr{K}$,

$$\mathscr{O}(F)g = U_{t_k}^t G_k U_{t_{k-1}}^{t_k} G_{k-1} \cdots G_2 U_{t_1}^{t_2} G_1 U_s^{t_1} g.$$

Proof The case $k = 0$ is clear. We prove the induction step from $k-1$ to k. Put, for short,

$$u(i) = u_{t_i}^{t_{i+1}}(\sigma_i, \tau_i, \upsilon_i).$$

Then split up F by writing

$$F = \sum u(k)G_k \cdots u(1)G_1 u(0) = \sum_{\substack{\sigma_k+\sigma'=\sigma \\ \tau_k+\tau'=\tau \\ \upsilon_k+\upsilon'=\upsilon}} u(k)G_k F'(\sigma', \tau', \upsilon')$$

with

$$F'(\sigma', \tau', \upsilon') = \sum_{\substack{\sigma_0+\cdots+\sigma_{k-1}=\sigma' \\ \tau_0+\cdots+\tau_{k-1}=\tau' \\ \upsilon_0+\cdots+\upsilon_{k-1}=\upsilon'}} u(k-1)G_{k-1} \cdots u(1)G_1 u(0).$$

Put

$$C = \max\bigl(\|A_i\|,\ i = 1, 0, -1;\ \|B\|;\ \|G_i\|,\ i = 1, \ldots, k\bigr).$$

We have

$$\bigl\|F(\sigma, \tau, \upsilon)\bigr\| \le C^{k+\#(\sigma+\tau+\upsilon)} \mathbf{1}\bigl\{t_{\sigma+\tau+\upsilon} \subset [s, t]\bigr\}.$$

So it is clearly locally integrable. An analogous assertion holds for F'. For $h \in \mathscr{K}$,

$$\int h^+(\omega)\bigl(\mathscr{O}(F)g\bigr)(\omega)\lambda_\omega = \int h^+(\pi)F(\sigma, \tau, \upsilon)g(\varrho)\bigl\langle a_\pi a^+_{\sigma+\tau} a_{\tau+\upsilon} a^+_\varrho\bigr\rangle \lambda_{\pi+\upsilon}.$$

Using Lemma 8.5.1, we see the last term equals

$$\int h^+(\pi)u(k)G_k F'(\sigma', \tau', \upsilon')g(\varrho)\bigl\langle a_\pi a^+_{\sigma_k+\tau_k} a_{\tau_k+\upsilon_k} a^+_{\sigma'+\tau'} a_{\tau'+\upsilon'} a^+_\varrho\bigr\rangle \lambda_{\pi+\upsilon_k+\upsilon'}.$$

Now following Theorem 5.6.1, the representation of unity gives

$$\bigl\langle a_\pi a^+_{\sigma_k+\tau_k} a_{\tau_k+\upsilon_k} a^+_{\sigma'+\tau'} a_{\tau'+\upsilon'} a^+_\varrho\bigr\rangle = \int_\omega \bigl\langle a_\pi a^+_{\sigma_k+\tau_k} a_{\tau_k+\upsilon_k} a^+_\omega\bigr\rangle\bigl\langle a_\omega a^+_{\sigma'+\tau'} a_{\tau'+\upsilon'} a^+_\varrho\bigr\rangle.$$

Put

$$\int G_k F'(\sigma', \tau', \upsilon')g(\varrho)\bigl\langle a_\omega a^+_{\sigma'+\tau'} a_{\tau'+\upsilon'} a^+_\varrho\bigr\rangle \lambda_{\upsilon'} = \bigl(\mathscr{O}(G_k F')g\bigr)(\omega) = f(\omega).$$

Then

$$\begin{aligned}
&\int h^+(\pi)F(\sigma, \tau, \upsilon)g(\varrho)\bigl\langle a_\pi a^+_{\sigma+\tau} a_{\tau+\upsilon} a^+_\varrho\bigr\rangle \lambda_{\pi+\upsilon} \\
&\quad = \int h^+(\pi)u(k)(\sigma_k, \tau_k, \upsilon_k)f(\omega)\bigl\langle a_\pi a^+\sigma_k + \tau_k a_{\tau_k+\upsilon_k}\bigr\rangle \lambda_{\pi+\upsilon_k}.
\end{aligned}$$

By the induction hypothesis

$$f = G_k U_{t_{k-1}}^{t_k} \cdots G_1 U_s^{t_1} g \in \Gamma.$$

We make the estimate

$$\begin{aligned}
&\int \|h(\pi)\| \, \|u(k)\| \, \|f(\omega)\| \langle a_\omega a^+_{\sigma_k+\tau_k} a_{\tau_k+\upsilon_k} \rangle \lambda_{\omega+\upsilon_k} \\
&= \int \|h(\pi)\| \, \|u(k\| \left\| \int G_k F' g(\varrho) \right\| \langle a_\pi a^+_{\sigma+\tau} a_{\tau+\upsilon} a^+_\varrho \rangle \lambda_{\pi+\upsilon} \\
&\leq \int \|h(\omega)\| \, \|F(\sigma,\tau,\upsilon)\| \, \|g(\varrho)\| \mathfrak{m} < \infty
\end{aligned}$$

and note

$$\int h^+(\omega) \mathscr{O}\big(u(k) f\big)(\omega) \lambda_\omega = \langle h | U_{t_k}^t f \rangle.$$

Continue with

$$\mathscr{O}(F) g = U_{t_k}^t f = U_{t_k}^t G_k U_{t_{k-1}}^{t_k} G_{k-1} \cdots G_2 U_{t_1}^{t_2} G_1 U_s^{t_1} g. \qquad \square$$

Lemma 8.7.4 *such that for* $g \in \mathscr{K}$ *we have* $\|\mathscr{O}(F)g\|_\Gamma \leq \text{const}\,\|g\|_\Gamma$ *and* $\|\mathscr{O}(F^+)g\|_\Gamma \leq \text{const}\,\|g\|_\Gamma$. *Let* $T : \Gamma \to \Gamma$ *the operator, such that* $\mathscr{O}(F)$ *is the restriction of* T *to* $\mathscr{K}$. *Then* $\mathscr{O}(F^+)$ *is the restriction of* T^+ *to* $\mathscr{K}$. *Assume* $f \in \Gamma$ *such that* $\mathscr{O}(\|F\|_{B(\mathfrak{k})}) \|f\|_{\mathfrak{k}} \in L^2(\mathbb{R})$, *then*

$$Tf = \mathscr{O}(F) f.$$

Proof Assume $g, h \in \mathscr{K}$, then

$$\begin{aligned}
\langle h | Tg \rangle = \langle h | \mathscr{O}(F) g \rangle &= \int h^+(\omega) F(\sigma,\tau,\upsilon) g(\varrho) \mathfrak{m} \\
&= \overline{\int g^+(\omega) F^+(\sigma,\tau,\upsilon) h(\varrho) \mathfrak{m}} = \overline{\langle \mathscr{O}(F^+) h | g \rangle} = \overline{\langle T^+ h | g \rangle}
\end{aligned}$$

with

$$\mathfrak{m} = \langle a_\omega a^+_{\sigma+\tau} a_{\tau+\upsilon} a^+_\varrho \rangle \lambda_{\omega+\upsilon}.$$

So $\mathscr{O}(F^+)$ is the restriction of T^+ to $\mathscr{K}$.

We have

$$\int \|h(\omega)\| \, \|F(\sigma,\tau,\upsilon)\| \, \|g(\varrho)\| \mathfrak{m} < \infty.$$

Hence

$$\int h^+(\omega) \big(\mathscr{O}(F) f\big)(\omega) \mathrm{d}\omega = \overline{\int f^+(\omega) \big(\mathscr{O}(F^+) h\big)(\omega) \mathrm{d}\omega} = \overline{\langle f | T^+ h \rangle} = \langle h | Tf \rangle.$$

As this holds for any $h \in \mathscr{K}$ the assertion follows. □

Lemma 8.7.5 *Assume $u_s^t = u_s^t(A_i, B)$ satisfying the unitarity conditions and U_s^t the corresponding unitary operator. Assume $G_1, \dots, G_k \in B(\mathfrak{k})$ and $s = t_0 < t_1 < \dots < t_k < t_{k+1} = t$ and*

$$F(\sigma, \tau, \upsilon) = \sum_{\substack{\sigma_0+\sigma_1+\cdots+\sigma_k=\sigma \\ \tau_0+\tau_1+\cdots+\tau_k=\tau \\ \upsilon_0+\upsilon_1+\cdots+\upsilon_k=\upsilon}} u_{t_k}^t(\sigma_k, \tau_k, \upsilon_k) G_k u_{t_{k-1}}^{t_k}(\sigma_{k-1}, \tau_{k-1}, \upsilon_{k-1}) G_{k-1} \cdots G_2 u_{t_1}^{t_2}(\sigma_1, \tau_1, \upsilon_1) G_1 u_s^{t_1}(\sigma_0, \tau_0, \upsilon_0).$$

Then for $f \in \mathscr{K}$

$$\mathscr{O}(F)g = U_{t_k}^t G_k U_{t_{k-1}}^{t_k} G_{k-1} \cdots G_2 U_{t_1}^{t_2} G_1 U_s^{t_1} g.$$

Proof The case $k = 0$ is clear. We prove by induction from $k-1$ to k. Put for short

$$u(i) = u_{t_i}^{t_{i+1}}(\sigma_i, \tau_i, \upsilon_i).$$

Then

$$F = \sum u(k) G_k \cdots u(1) G_1 u(0) = \sum_{\substack{\sigma_k+\sigma'=\sigma \\ \tau_k+\tau'=\tau \\ \upsilon_k+\upsilon'=\upsilon}} u(k) G_k F'(\sigma', \tau', \upsilon')$$

with

$$F'(\sigma', \tau', \upsilon') = \sum_{\substack{\sigma_0+\cdots+\sigma_{k-1}=\sigma' \\ \tau_0+\cdots+\tau_{k-1}=\tau' \\ \upsilon_0+\cdots+\upsilon_{k-1}=\upsilon'}} u(k-1) G_{k-1} \cdots u(1) G_1 u(0).$$

Go back to the proof of Proposition 4.4.1. Put

$$C = \max\bigl(\|A_i\|, i = 1, 0, -1; \|B\|; \|G_i\|, i = 1, \dots, k\bigr).$$

For $g \in \mathscr{K}$ we have

$$\bigl\|\bigl(\mathscr{O}(\|F\|)\|g\|\bigr)\bigr\|_r \le c\|g\|_r$$
$$\bigl\|\bigl(\mathscr{O}(\|F'\|)\|g\|\bigr)\bigr\|_r \le c'\|g\|_r.$$

For $h \in \mathscr{K}$

$$\int h^+(\omega)\bigl(\mathscr{O}(F)g\bigr)(\omega)\lambda_\omega = \int h^+(\pi) F(\sigma, \tau, \upsilon) g(\varrho) \langle a_\pi a_{\sigma+\tau}^+ a_{\tau+\upsilon} a_\varrho^+ \rangle \lambda_{\pi+\upsilon}.$$

Using the same argument as in the proof of Proposition 4.4.1 the last term equals

$$\int h^+(\pi)u(k)G_k F'(\sigma',\tau',\upsilon')g(\varrho)\langle a_\pi a^+_{\sigma_k+\tau_k} a_{\tau_k+\upsilon_k} a^+_{\sigma'+\tau'} a_{\tau'+\upsilon'} a^+_\varrho\rangle \lambda_{\pi+\upsilon_k+\upsilon'}.$$

Now following Theorem 5.5.1

$$\langle a_\pi a^+_{\sigma_k+\tau_k} a_{\tau_k+\upsilon_k} a^+_{\sigma'+\tau'} a_{\tau'+\upsilon'} a^+_\varrho\rangle = \int_\omega \langle a_\pi a^+_{\sigma_k+\tau_k} a_{\tau_k+\upsilon_k} a^+_\omega\rangle\langle a_\omega a^+_{\sigma'+\tau'} a_{\tau'+\upsilon'} a^+_\varrho\rangle.$$

As

$$\int F'(\sigma',\tau',\upsilon')g(\varrho)\langle a_\omega a^+_{\sigma'+\tau'} a_{\tau'+\upsilon'} a^+_\varrho\rangle \lambda_{\upsilon'} = (\mathscr{O}(F')g)(\omega) = f(\omega)$$

the integrability conditions are fulfilled and one obtains

$$\int h^+(\omega)(\mathscr{O}(F)g)(\omega)\lambda_\omega = \int h^+(\pi)u(k)G_k f(\omega)\langle a_\pi a^+_{\sigma_k+\tau_k} a_{\tau_k+\upsilon_k} a^+_\omega\rangle \lambda_{\pi+\upsilon_k}.$$

The conditions of the preceding lemma are fulfilled, the last expression equals

$$\langle h|U^t_{t_k} f\rangle = \langle h|U^t_{t_k} G_k U^{t_k}_{t_{k-1}} G_{k-1}\cdots G_2 U^{t_2}_{t_1} G_1 U^{t_1}_s g\rangle$$

using the hypothesis of induction. □

Theorem 8.7.1 *For any k there exists a polynomial P of degree $\leq k$ with coefficients ≥ 0, such that, for $g \in \Gamma_k$,*

$$\|U^t_s g\|^2_{\Gamma_k} \leq P(|t-s|)\|g\|^2_{\Gamma_k}.$$

Proof Following Lemma 8.4.1 and Proposition 8.4.1, we have for $f \in \mathscr{K}$

$$\langle U^t_s f|\binom{N}{k}|U^t_s f\rangle = \int \|(U^t_s f)(\omega+\xi)\|^2 \mathbf{1}\{\#\xi = k\}\lambda_{\omega+\xi} < \infty.$$

Hence $\int \|(U^t_s f)(\omega+\xi)\|^2\lambda_\omega < \infty$ for almost all ξ. We have

$$(\mathscr{O}(u^t_s))(\omega) = \sum_{\omega_1+\omega_2+\omega_3=\omega} \int \lambda_\upsilon u^t_s(\omega_1,\omega_2,\upsilon) f(\omega_2+\omega_3+\upsilon)$$

and

$$(\mathscr{O}(u^t_s))(\omega+\xi)$$
$$= \sum_{\substack{\omega_1+\omega_2+\omega_3=\omega\\ \xi_1+\xi_2+\xi_3=\xi}} \int \lambda_\upsilon u^t_s(\omega_1+\xi_1,\omega_2+\xi_2,\upsilon) f(\omega_2+\xi_2+\omega_3+\xi_3+\upsilon)$$

$$= \sum_{\xi_1+\xi_2+\xi_3=\xi} \big(\mathscr{O}\big((u_s^t)_{\xi_1,\xi_2}\big)\big) f_{\xi_2+\xi_3})(\omega)$$

with

$$\big(u_s^t\big)_{\xi_1,\xi_2}(\sigma,\tau,\upsilon) = u_s^t(\sigma+\xi_1,\tau+\xi_2,\upsilon),$$
$$f_{\xi_1+\xi_2}(\varrho) = f(\xi_1+\xi_2+\varrho).$$

Assume that the multiset $\{s,t,t_\xi,t_{\sigma+\tau+\upsilon}\}^\bullet$ has no multiple points and order the set

$$\big\{(t_i,1): i\in\xi_1\big\} + \{t_i,0): i\in\xi_0\} = \big\{(t_1,i_1),\dots,(t_l,i_l)\big\}$$

with $t_1 < \cdots < t_l$ and $i_j \in \{1,0\}$. Then

$$\begin{aligned}
&\big(u_s^t\big)_{\xi_1,\xi_2}(\sigma,\tau,\upsilon)\\
&\quad= \sum_{\substack{\sigma_0+\sigma_1+\cdots+\sigma_l=\sigma\\ \tau_0+\tau_1+\cdots+\tau_l=\tau\\ \upsilon_0+\upsilon_1+\cdots+\upsilon_l=\upsilon}} u_{t_l}^t(\sigma_l,\tau_l,\upsilon_l)A_{i_l}u_{t_{l-1}}^{t_l}(\sigma_{l-1},\tau_{l-1},\upsilon_{l-1})A_{i_{l-1}}\\
&\qquad\cdots A_{i_2}u_{t_1}^{t_2}(\sigma_1,\tau_1,\upsilon_1)A_{i_1}u_s^{t_1}(\sigma_0,\tau_0,\upsilon_0).
\end{aligned}$$

Using the last lemma we obtain, for $h\in\mathscr{K}$,

$$\mathscr{O}\big(\big(u_s^t\big)_{\xi_1,\xi_2}\big)h = U_{t_l}^t A_{i_l}\cdots A_{i_2}U_{t_1}^{t_2}A_{i_1}U_s^{t_1}h.$$

If $C=\max(\|A_i\|,\|B\|,1)$, then

$$\big\|\mathscr{O}\big(\big(u_s^t\big)_{\xi_1,\xi_2}\big)h\big\|_\Gamma \le C^{\#(\xi_1+\xi_2)}\mathbf{1}\big\{t_{\xi_1+\xi_2}\subset[s,t]\big\}\|h\|_\Gamma.$$

Finally

$$\begin{aligned}
\big\langle U_s^t f\big|\binom{N}{k}\big|U_s^t f\big\rangle &= \int \big\|\big(U_s^t f\big)(\omega+\xi)\big\|^2 \mathbf{1}\{\#\xi=k\}\\
&= \int\bigg\|\sum_{\xi_1+\xi_2+\xi_3=\xi}\mathscr{O}\big(u_s^t\big)_{\xi_1,\xi_2}f_{\xi_2,\xi_3}(\omega)\mathbf{1}\{\#\xi=k\}\bigg\|^2\lambda_{\xi+\omega}\\
&\le C^{2k}3^k\int\lambda_{\xi+\omega}\\
&\qquad\times\sum_{\xi_1+\xi_2+\xi_3=\xi}\mathbf{1}\{\#\xi=k\}\mathbf{1}\big\{t_{\xi_1+\xi_2}\subset[s,t]\big\}\big\|f(\xi_2+\xi_3+\omega)\big\|^2\\
&\le C^{2k}3^k\sum_{k_1=0}^{k}\int\lambda_{\xi_1}\mathbf{1}\{\#\xi_1=k_1\}\mathbf{1}\big\{t_{\xi_1}\subset[s,t]\big\}
\end{aligned}$$

$$\times \int \lambda_{\xi_0+\omega} \mathbf{1}\{\#\xi_0 = k - k_1\} \| f(\xi_0 + \omega) \|^2$$

$$= C^{2k} 3^k \sum_{k_1}^{k} \frac{(t-s)^{k_1}}{k_1!} \langle f | \binom{N}{k-k_1} | f \rangle. \qquad \square$$

The previous theorem covers the case $s < t$; a proof for $t < s$ can be carried out in the same way.

8.8 The Hamiltonian

8.8.1 Definition of the One-Parameter Group $W(t)$

Denote by $\Theta(t)$ the right shift on $\mathbb{R}$, and extend it to $\mathfrak{R}$,

$$\Theta(t)(t_1, \ldots, t_n) = (t_1 + t, \ldots, t_n + t).$$

If $\{t_1, \ldots, t_n\}^\bullet$ is a multiset, we define

$$\Theta(t)\{t_1, \ldots, t_n\}^\bullet = \{t_1 + t, \ldots, t_n + t\}^\bullet.$$

In the notation $\{t_1, \ldots, t_n\}^\bullet = t_\alpha$ we write $\Theta(t)t_\alpha = t_\alpha + t\, e_\alpha$ with $e_\alpha = \{1, \ldots, 1\}^\bullet$.

If f is a function on $\mathfrak{R}$, then $(\Theta(t)f)(w) = f(\Theta(t)w)$. If μ is a measure on $\mathfrak{R}$, then $\Theta(t)\mu$ is defined by the property

$$\int \bigl(\Theta(t)\mu(\mathrm{d}w)\bigr)\bigl(\Theta(t)f(w)\bigr) = \int \mu(\mathrm{d}w) f(w).$$

If $\mu(\mathrm{d}w) = g(w)\mathrm{d}w$ then $\Theta(t)\mu(\mathrm{d}w) = (\Theta(t)g)(w)\,\mathrm{d}w$. Similar notations hold for $\mathfrak{R}^k$.

Lemma 8.8.1 *We have*

$$\bigl(\Theta(t)\varepsilon_x\bigr)(\mathrm{d}y) = \varepsilon_{x-t}(\mathrm{d}y).$$

If φ is function on $\mathbb{R}$, ν a measure on $\mathbb{R}$, f a function on $\mathfrak{R}$, and μ is measure on $\mathfrak{R}$, one calculates

$$\Theta(t)\bigl(a^+(\varphi)f\bigr) = a^+\bigl(\Theta(t)\varphi\bigr)\bigl(\Theta(t)f\bigr),$$
$$\Theta(t)\bigl(a^+(\nu)\mu\bigr) = a^+\bigl(\Theta(t)\nu\bigr)\bigl(\Theta(t)\mu\bigr),$$
$$\Theta(t)\bigl(a(\nu)f\bigr) = a\bigl(\Theta(t)\nu\bigr)\bigl(\Theta(t)f\bigr),$$
$$\Theta(t)\bigl(a(\varphi)\mu\bigr) = a\bigl(\Theta(t)\varphi\bigr)\bigl(\Theta(t)\mu\bigr).$$

Proof The identities follow directly from the definitions. □

Lemma 8.8.2 *If W is an admissible sequence, then $\langle W\rangle\lambda_{\omega_-\backslash\omega_+}$ is a shift-invariant measure.*

Proof According to the considerations in Sect. 5.6, this measure is a sum of measures, each one a tensor product of measures of the form

$$\Lambda(\mathrm{d}t_1,\dots,\mathrm{d}t_n):\int \Lambda(\mathrm{d}t_1,\dots,\mathrm{d}t_n)f(t_1,\dots,t_n)=\int \mathrm{d}t f(t,\dots,t).$$

So it is clearly invariant due to the shift-invariance of Lebesgue measure $\mathrm{d}t$. □

Upon using the defining formulas, one obtains immediately

Lemma 8.8.3 *One has*

$$\Theta(r)u_s^t=u_{s-r}^{t-r}$$

for $r,s,t\in\mathbb{R}$.

Proposition 8.8.1 *Define a unitary operator $\Theta(t)$ on Γ by $f\mapsto\Theta(t)f$. The operators U_s^t, $s,t\in\mathbb{R}$, form a cocycle with respect to $\Theta(t)$, i.e.*

$$\Theta(r)U_s^t\Theta(-r)=U_{s-r}^{t-r}.$$

Proof Use the invariance of

$$\mathfrak{m}=\langle a_\pi a_{\sigma+\tau}^{+}a_{\tau+\upsilon}a_\varrho^{+}\rangle\lambda_{\pi+\upsilon}$$

and obtain

$$\begin{aligned}\int_\Gamma f^+(\pi)u_{s-r}^{t-r}(\sigma,\tau,\upsilon)g(\varrho)\mathfrak{m}&=\int f^+(\pi)\big(\Theta(r)u_s^t(\sigma,\tau,\upsilon)\big)g(\varrho)\mathfrak{m}\\&=\int\big(\Theta(-r)f^+\big)(\pi)u_s^t(\sigma,\tau,\upsilon)\big(\Theta(-r)g\big)(\varrho)\mathfrak{m}\\&=\langle\Theta(-r)f|U_s^t\Theta(-r)g\rangle=\langle f|\Theta(r)U_s^t\Theta(-r)g\rangle.\end{aligned}$$

□

Proposition 8.8.2 *Define, for $t\in\mathbb{R}$,*

$$W(t)=\Theta(t)\,U_0^t=U_{-t}^0\Theta(t);$$

then $W(t)$ is a unitary strongly continuous one-parameter group on Γ.

Proof We have $W(0) = 1$ and

$$W(s+t) = \Theta(t+s)U_0^{t+s} = \Theta(t)\Theta(s)U_s^{t+s}\Theta(-s)\Theta(s)U_0^s = W(s)W(t),$$

and also

$$W(t)^+ = U_t^0\Theta(-t) = \Theta(-t)U_0^{-t} = W(-t).$$ □

An immediate consequence of Proposition 8.4.1 and Theorem 8.7.1 is

Proposition 8.8.3 *The operators $W(t)$ map the space Γ_k into itself, they form a strongly continuous one-parameter group on Γ_k, and*

$$\|W(t)f\|^2_{\Gamma_k} \leq P(|t|)\|f\|^2_{\Gamma_K}$$

where P is a polynomial of degree $\leq k$.

8.8.2 *Definition of* $\hat{\mathfrak{a}}$, $\hat{\mathfrak{a}}^+$ *and* $\hat{\partial}$

If φ is an integrable function on the real line, we define

$$\Theta(\varphi) = \int \varphi(t)\Theta(t)\mathrm{d}t,$$

which is, for any k, an operator mapping Γ_k into Γ_k.

If ν is a measure on $\mathbb{R}$ and f a locally integrable function, symmetric on $\mathfrak{R}$, then

$$\big(a^+(\nu)f\lambda\big)(\omega) = \sum_{c\in\omega}\nu(c)f(\omega\setminus c)\lambda(\omega\setminus c).$$

We shall use again L. Schwartz's convention [37], and denote $f\lambda$ by f. So we write

$$\big(a^+(\nu)f\big)(\omega) = \sum_{c\in\omega}\nu(c)f(\omega\setminus c).$$

We set

$$\mathfrak{a} = a(\varepsilon_0) = a(0), \qquad \mathfrak{a}^+ = a^+(\varepsilon_0).$$

We use Gothic $\mathfrak{a}^+$ in order to distinguish it from the $a^+(\mathrm{d}x) = a^+(\varepsilon(\mathrm{d}x))$ used in the preceding text; $a(\varepsilon_0) = a(0) = \mathfrak{a}$ is the same as before. We have

$$\big(\mathfrak{a}^+f\big)(\omega) = \sum_{c\in\omega}\varepsilon_0(\mathrm{d}t_c)f(t_{\omega\setminus c}) = \sum_{c\in\omega}\varepsilon_0(\mathrm{d}t_c)f(t_{\omega\setminus c})\lambda(\mathrm{d}t_{\omega\setminus c}).$$

The duality relation (see Sect. 5.6) becomes

$$\int g(\omega)\big(\mathfrak{a}^+f\big)(\omega) = \int (\mathfrak{a}g)(\omega)f(\omega)\lambda(\omega).$$

Lemma 8.8.4 *Assume $f \in L^2(\mathbb{R}^n)$ and $\varphi \in (L^1 \cap L^2)(\mathbb{R})$. Then $\Theta(\varphi)$ maps the singular measure $\mathfrak{a}^+ f$ into an absolute continuous measure identified with its density, and we have*

$$\big(\Theta(\varphi)\mathfrak{a}^+ f\big)(t_\omega) = \sum_{c\in\omega} \varphi(-t_c)\big(\Theta(-t_c)f\big)(\omega \setminus c).$$

The map $\Theta(\varphi)\mathfrak{a}^+$ can be extended to a mapping $\Gamma_k \to \Gamma_{k-1}$, and we have

$$\big\|\Theta(\varphi)\mathfrak{a}^+ f\big\|_{\Gamma_{k-1}} \le \|\varphi\|_{L^2}\|f\|_{\Gamma_k}.$$

We have

$$\int g(\omega)^+\big(\Theta(\varphi)\mathfrak{a}^+ f\big)(t_\omega)\mathrm{d}\omega = \int \big(\mathfrak{a}\Theta(\varphi^+)g\big)^+(\omega)f(\omega)\mathrm{d}\omega$$

with $\varphi^+(t) = \overline{\varphi(-t)}$. One obtains

$$\big(\mathfrak{a}\Theta(\varphi)f\big)(t_1,\ldots,t_n) = \int \varphi(s)\mathrm{d}s f(s, t_1+s,\ldots,t_n+s).$$

This map can be extended to a mapping $\Gamma_k \to \Gamma_{k-1}$, and we have

$$\big\|\mathfrak{a}\Theta(\varphi)f\big\|_{\Gamma_{k-1}} \le 2^{k/2}\|\varphi\|_{L^2}\|f\|_{\Gamma_k}.$$

Proof One has

$$\begin{aligned}
&\int \mathrm{d}s\,\varphi(s)\Theta(s)\Big(\sum_{c\in\omega}\varepsilon_0(\mathrm{d}t_c)f(t_{\omega\setminus c})\lambda(\mathrm{d}t_{\omega\setminus c})\Big)\\
&\quad= \int \mathrm{d}s\,\varphi(s)\sum_{c\in\omega}\varepsilon_{-s}(\mathrm{d}t_c)\big(\Theta(s)f\big)(t_{\omega\setminus c})\lambda(\mathrm{d}t_{\omega\setminus c})\\
&\quad= \int \mathrm{d}s\,\varphi(-s)\sum_{c\in\omega}\varepsilon_{s}(\mathrm{d}t_c)\big(\Theta(-s)f\big)(t_{\omega\setminus c})\lambda(\mathrm{d}t_{\omega\setminus c})\\
&\quad= \sum_{c\in\omega}\varphi(-t_c)\big(\Theta(-t_c)f\big)(\omega\setminus c)\lambda(\omega\setminus c),
\end{aligned}$$

by changing the variable $s \mapsto -s$ to get the second equality, and using for the third

$$\int \mathrm{d}s\,\varphi(s)\varepsilon_s(t_c)\psi(t_c) = \int \varphi(t_c)\psi(t_c)\mathrm{d}t_c,$$

or more succinctly

$$\int \mathrm{d}s\,\varepsilon_s(\mathrm{d}t_c) = \mathrm{d}t_c.$$

So $\Theta(\varphi)$ works as a mollifier, as it is called in Schwartz's theory of distributions, making a function out of a singular measure. The other results follow by simple calculations. □

Lemma 8.8.5 *If $\varphi \in (L^1 \cap L^2)(\mathbb{R})$ and $f \in L^2(\mathbb{R}^{n+1})$, then*

$$x \in \mathbb{R}^{n+1} \mapsto g(x)$$
$$g(x)(t_1, \ldots, t_n) = \big(\Theta(\varphi) f\big)\big((0, t_1, \ldots, t_n) + x\big)$$

maps $\mathbb{R}^{n+1}$ into $L^2(\mathbb{R}^n)$, and furthermore $x \mapsto g(x)$ is a continuous function bounded by $\|\varphi\|_{L^2(\mathbb{R})} \|f\|_{L^2(\mathbb{R}^{n+1})}$.

Proof We have with $e = (1, 1, \ldots, 1)$

$$\begin{aligned}
&\|g(x) - g(y)\|^2_{L^2(\mathbb{R}^n)} \\
&\quad = \int \mathrm{d}t_1 \cdots \mathrm{d}t_n \left\| \int \mathrm{d}s\, \varphi(s) \big(f\big((0, t_1, \ldots, t_n) + se + x\big) \right. \\
&\qquad \left. - f\big((0, t_1, \ldots, t_n) + se + y\big) \right\|^2 \\
&\quad \le \|\varphi\|^2_{L^2(\mathbb{R})} \int \mathrm{d}t_1 \cdots \mathrm{d}t_n \int \mathrm{d}s \\
&\qquad \times \big\| f(s + x_0, t_1 + s + x_1, \ldots, t_n + s + x_n) \\
&\qquad - f(s + y_0, t_1 + s + y_1, \ldots, t_n + s + y_n) \big\|^2 \\
&\quad = \|\varphi\|^2_{L^2(\mathbb{R})} \int \mathrm{d}t_0 \cdots \mathrm{d}t_n \big\| f(t_0 + x_0, \ldots, t_n + x_n) - f(t_0 + y_0, \ldots, t_n + y_n) \big\|^2 \\
&\quad = \|\varphi\|^2_{L^2(\mathbb{R})} \big\| \big(T(x) - T(y)\big) f \big\|^2_{L^2(\mathbb{R}^{n+1})}
\end{aligned}$$

where $T(x)$ denotes translation by x. The bound for $\|g(x)\|$ can be shown in the same way. □

Lemma 8.8.6 *Assume $f \in L^2(\mathbb{R}^n_s)$, and that $\eta \in L^2(\mathbb{R})$ is a continuous bounded function on $\mathbb{R} \setminus \{0\}$ with right and left limits at 0, or, in other words, η is a continuous bounded function on $\mathbb{R}_0$. If*

$$x \in \mathbb{R}^{n+1} \mapsto g(x) \in L^2(\mathbb{R}^n)$$
$$g(x)(t_1, \ldots, t_n) = \big(\Theta(\eta) \mathrm{a}^+ f\big)\big((0, t_1, \ldots, t_n) + x\big)$$

then

$$\|g(x)\|_{L^2(\mathbb{R}^n)} \le (n + 1) \|\eta\|_\infty \|f\|_{L^2(\mathbb{R}^n)}$$

for all $x \in L^2(\mathbb{R}^{n+1})$, *and* $x \mapsto g(x)$ *is continuous on* $\{(x_0, x_1, \dots, x_n) : x_0 \neq 0\}$. *We have that the limits* $g(0\pm, x_1, \dots, x_n)$ *exist and*

$$g(0+, x_1, \dots, x_n) - g(0-, x_1, \dots, x_n) = -\big(\eta(0+) - \eta(0-)\big)T(x_1, \dots, x_n)f$$

where $T(x)$ *is the translation by* x.

Proof We have

$$\big(\Theta(\eta)\mathrm{a}^+ f\big)(t_0, t_1, \dots, t_n) = k_0(t_0, t_1, \dots, t_n) + \dots + k_n(t_0, t_1, \dots, t_n)$$

with

$$\begin{aligned}
k_0(t_0, t_1, \dots, t_n) &= \eta(-t_0) f(t_1 - t_0, t_2 - t_0, \dots, t_n - t_0),\\
k_1(t_0, t_1, \dots, t_n) &= \eta(-t_1) f(t_0 - t_1, t_2 - t_1, \dots, t_n - t_1),\\
&\vdots\\
k_n(t_0, t_1, \dots, t_n) &= \eta(-t_n) f(t_0 - t_n, t_1 - t_n, \dots, t_{n-1} - t_n).
\end{aligned}$$

Define

$$g_i(x)(t_1, \dots, t_n) = k_i\big((0, t_1, \dots, t_n) + x\big)$$

and first discuss g_i with $i \neq 0$, for example, g_n. We have

$$\begin{aligned}
g_n(x)(t_1, \dots, t_n) &= k_n\big((0, t_1, \dots, t_n) + x\big) = k_n(x_0, t_1 + x_1, \dots, t_n + x_n)\\
&= \eta(-x_n - t_n) f(x_0 - x_n - t_n, x_1 - x_n + t_1 - t_n, \dots,\\
&\qquad x_{n-1} - x_n + t_{n-1} - t_n)\\
&= \eta(-x_n - t_n)\big(T\big(x'\big)f\big)(-t_n, t_1 - t_n, \dots, t_{n-1} - t_{n-1})
\end{aligned}$$

with

$$x' = (x_0 - x_n, x_1 - x_n, \dots, x_{n-1} - x_n).$$

From there one obtains

$$\big\|g_n(x)\big\| \leq \|\eta\|_\infty \|f\|.$$

We have

$$\begin{aligned}
&\int \mathrm{d}t_1 \cdots \mathrm{d}t_n \big\|k_n\big((0, t_1, \dots, t_n) + x\big) - k_n\big((0, t_1, \dots, t_n) + y\big)\big\|^2\\
&\quad \leq 2 \int \mathrm{d}t_1 \cdots \mathrm{d}t_n \big|\eta(-x_n - t_n) - \eta(-y_n - t_n)\big|^2\\
&\qquad \times \big\|\big(T\big(x'\big)f\big)(-t_n, t_1 - t_n, \dots, t_{n-1} - t_{n-1})\big\|^2
\end{aligned}$$

$$+2\|\eta\|_\infty^2 \int \mathrm{d}t_1 \cdots \mathrm{d}t_n \left\| \left(T(x') - T(y')\right) f(-t_n, t_1 - t_n, \ldots, t_{n-1} - t_{n-1}) \right\|^2.$$

For $y \to x$, the first term goes to zero by the theorem of Lebesgue, and from the second term we observe that $z \mapsto T(z)f$ is norm continuous. So $g_n(x)$ and thus $g_i(x), i \neq 0$ are continuous for all x. We have

$$g_0(x) = \eta(-t_0) f(t_1 + x_1 - x_0, t_2 + x_2 - x_0, \ldots, t_n + x_n - x_0).$$

From there one obtains the result. □

We double the point 0 to $\{-0, +0\}$, and introduce

$$\mathbb{R}_0 =]-\infty, -0] + [+0, \infty[$$

with the usual topology, i.e., $\mathbb{R}_0 = \mathbb{R}_{\leq 0} + \mathbb{R}_{\geq 0}$. A function f on $\mathbb{R}_0$ is continuous, if its restriction to $\mathbb{R} \setminus \{0\}$ is continuous and if both limits $f(\pm 0)$ exist. We define

$$\mathfrak{R}_0 = \{\emptyset\} + \mathbb{R}_0 + \mathbb{R}_0^2 + \cdots .$$

We introduce on $\mathbb{R}_0$ and on $\mathfrak{R}_0$ the Lebesgue measure λ. A continuous function on $\mathbb{R} \setminus \{0\}$ which has left and right limits at 0 can be considered as a function on $\mathbb{R}_0$. We define the measures $\varepsilon_{\pm 0}$, and, for symmetric functions f on $\mathfrak{R}_0$, the operators $\mathfrak{a}_\pm = a(\varepsilon_{\pm 0})$ and $\mathfrak{a}_\pm^+ = a^+(\varepsilon_{\pm 0})$ and shall use similar conventions to those above. We put

$$\hat{\varepsilon}_0 = \frac{1}{2}(\varepsilon_{+0} + \varepsilon_{-0}), \qquad \hat{\mathfrak{a}} = \frac{1}{2}(\mathfrak{a}_+ + \mathfrak{a}_-), \qquad \hat{\mathfrak{a}}^+ = \frac{1}{2}\left(\mathfrak{a}_+^+ + \mathfrak{a}_-^+\right).$$

A δ-sequence is a sequence of functions $\varphi_n \in C_c^\infty$ such that

$$\int \varphi_n(t)\mathrm{d}t = 1, \qquad \int \left|\varphi_n(t)\right| \mathrm{d}t \leq C < \infty, \qquad \operatorname{supp}(\varphi_n) \subset]-\varepsilon_n, \varepsilon_n[$$

and $\varepsilon_n \downarrow 0$.

Definition 8.8.1 We term a δ-sequence φ_n a *symmetric* δ-sequence, if the φ_n are real and $\varphi_n(t) = \varphi_n(-t)$ for all n and t.

Proposition 8.8.4 *Assume we have two functions f and η fulfilling the conditions of Lemma* 8.8.6, *then $\hat{\mathfrak{a}}\Theta(\eta)\mathfrak{a}^+ f$ exists, and*

$$\left\|\hat{\mathfrak{a}}\Theta(\eta)\mathfrak{a}^+ f\right\|_\Gamma \leq \|\eta\|_\infty \|f\|_{\Gamma_2};$$

if φ_n is a symmetric δ-sequence, then $\mathfrak{a}\Theta(\varphi_n)\Theta(\eta)\mathfrak{a}^+ f$ exists, and

$$\mathfrak{a}\Theta(\varphi_n)\Theta(\eta)\mathfrak{a}^+ f \to \hat{\mathfrak{a}}\Theta(\eta)\mathfrak{a}^+ f.$$

Proof We apply Lemma 8.8.6. We have

$$\big(\Theta(\eta)\mathfrak{a}^+ f\big)(s, t_1, \ldots, t_n) = g(s, 0, \ldots, 0)(t_1, \ldots, t_n)$$

and $g(s, 0, \ldots, 0) \to g(0+, 0, \ldots, 0)$ for $s \downarrow 0$. From there one concludes the existence of $\hat{\mathfrak{a}}\Theta(\eta)\mathfrak{a}^+ f$. We have

$$\big(\mathfrak{a}\Theta(s)\Theta(\eta)\mathfrak{a}^+ f\big)(t_1, \ldots, t_n) = g(s, \ldots, s)(t_1, \ldots, t_n)$$

and $g(s, \ldots, s) \to g(0+, 0, \ldots, 0)$ for $s \downarrow 0$. From there one obtains the rest of the proposition. □

Assume φ_n to be a symmetric δ-sequence and $f \in D$, then $\mathfrak{a}\Theta(\varphi_n) f \to \hat{\mathfrak{a}} f$ in the norm of Γ. If $f \in C^1(\mathbb{R}^n)$ and φ_n is a δ-sequence, then

$$\begin{aligned}\Theta(\varphi_n') f(t) &= \int \varphi_n'(s) f(t + se)\mathrm{d}s \\ &= -\int \varphi_n(s) f'(t + se)\mathrm{d}s \to -\sum \frac{\partial f}{\partial t_i}(t) = -(\partial f)(t).\end{aligned}$$

This motivates

Definition 8.8.2 We define

$$\hat{\partial} = -\lim \Theta(\varphi_n'),$$

where φ_n is a *symmetric* δ-sequence.

Proposition 8.8.5 *Assume we are given a function $\eta \in L^2(\mathbb{R})$, which is bounded and C^1 on $\mathbb{R} \setminus \{0\}$ and has left and right limits at 0, so the Schwartz derivative of η equals*

$$\partial\eta = \big(\eta(+0) - \eta(-0)\big)\delta + \partial^{\mathrm{c}}\eta,$$

where $\partial^{\mathrm{c}}\eta$ is the continuous part of the derivative. Assume, furthermore, that $\partial^{\mathrm{c}}\eta \in L^2(\mathbb{R})$. Put

$$\mathscr{L}^n(\eta) = \big\{\Theta(\eta)\mathfrak{a}^+ f,\ f \in L_s^2(\mathbb{R}^n)\big\}$$

and let $\mathscr{L}^n(\eta)^\dagger$ denote the space of all semilinear functionals $\mathscr{L}^n(\eta) \to \mathbb{C}$. Then $\hat{\partial}$ defines a linear mapping

$$\mathscr{L}^n(\eta) \to \mathscr{L}^n(\eta)^\dagger$$

given by the sesquilinear form on $\mathscr{L}^n(\eta)$

$$\langle u|\hat{\partial}|v\rangle = -\lim\langle u|\Theta(\varphi_n')|v\rangle.$$

This sesquilinear form is antisymmetric. One has

$$\hat{\partial}\Theta(\eta) f = \big(\eta(+0) - \eta(-0)\big)\hat{\mathfrak{a}}^+ f + \Theta(\partial^{\mathrm{c}}\eta)\mathfrak{a}^+ f.$$

Proof We have

$$\langle \Theta(\eta)\mathfrak{a}^+ g|\Theta(\varphi_n')\Theta(\eta)\mathfrak{a}^+ f\rangle = \langle \Theta(\eta)\mathfrak{a}^+ g|\Theta(\varphi_n' \star \eta)\mathfrak{a}^+ f\rangle,$$

where $\star$ denotes the usual convolution. As

$$\varphi_n' \star \eta = \varphi_n \star \eta' = \big(\eta(+0) - \eta(-0)\big)\varphi_n + \varphi_n \star \partial_c \eta$$

we continue

$$= \big(\eta(+0) - \eta(-0)\big)\langle \Theta(\eta)\mathfrak{a}^+ g|\Theta(\varphi_n)\mathfrak{a}^+ f\rangle + \langle \Theta(\eta)\mathfrak{a}^+ g|\Theta(\varphi_n \star \partial_c\eta)\mathfrak{a}^+ f\rangle.$$

The second term converges to

$$\langle \Theta(\eta)\mathfrak{a}^+ g|\Theta(\partial_c\eta)\mathfrak{a}^+ f\rangle.$$

For the first term observe that

$$\langle \Theta(\eta)\mathfrak{a}^+ g|\Theta(\varphi_n)\mathfrak{a}^+ f\rangle = \langle \mathfrak{a}\Theta(\varphi_n)\Theta(\eta)\mathfrak{a}^+ g|f\rangle$$

using $\varphi^+ = \varphi$. By Proposition 8.8.4 this expression converges to

$$\langle \hat{\mathfrak{a}}\Theta(\eta)\mathfrak{a}^+ g|f\rangle = \langle \Theta(\eta)\mathfrak{a}^+ g|\hat{\mathfrak{a}}^+ f\rangle.$$

In order to show that $\hat{\partial}$ is antisymmetric, observe that φ_n' is antisymmetric,

$$(\varphi_n')(-t) = -\varphi_n'(t),$$

and apply Proposition 8.8.4 again. □

8.8.3 Characterization of the Hamiltonian

We recall the resolvent (from Sect. 3.1) associated to the group $W(t)$.

Definition 8.8.3 For $z \in \mathbb{C}$, $\operatorname{Im} z \neq 0$, the resolvent $R(z)$ is defined by

$$R(z) = \begin{cases} -\mathrm{i}\int_0^\infty \mathrm{e}^{\mathrm{i}zt} W(t)\mathrm{d}t & \text{for } \operatorname{Im} z > 0, \\ +\mathrm{i}\int_{-\infty}^0 \mathrm{e}^{\mathrm{i}zt} W(t)\mathrm{d}t & \text{for } \operatorname{Im} z < 0. \end{cases}$$

The Hamiltonian H of $W(t)$ is so defined that $-\mathrm{i}H$ is the generator of the group $W(t)$ (see Sect. 3.1). Its domain is the set

$$D = R(z)\Gamma,$$

where $z \in \mathbb{C}$, $\operatorname{Im} z \neq 0$. The set D is independent of the z chosen. The Hamiltonian is a selfadjoint operator given by the equation

$$HR(z)f = -f + zR(z)f.$$

Definition 8.8.4 Furthermore, we set

$$S(z) = \begin{cases} -\mathrm{i}\int_0^\infty \mathrm{e}^{\mathrm{i}zt} W(t)a(t) & \text{for } \operatorname{Im} z > 0, \\ +\mathrm{i}\int_{-\infty}^0 \mathrm{e}^{\mathrm{i}zt} W(t)a(t) & \text{for } \operatorname{Im} z < 0 \end{cases}$$

and

$$\kappa(z) = \begin{cases} -\mathrm{i}\mathbf{1}\{t > 0\}\mathrm{e}^{\mathrm{i}zt+Bt} & \text{for } \operatorname{Im} z > 0, \\ +\mathrm{i}\mathbf{1}\{t < 0\}\mathrm{e}^{\mathrm{i}zt-B^+t} & \text{for } \operatorname{Im} z < 0 \end{cases}$$

and

$$\tilde{R}(z) = \Theta\big(\kappa(z)\big) = \begin{cases} -\mathrm{i}\int_0^\infty \mathrm{e}^{\mathrm{i}zt}\mathrm{e}^{Bt}\Theta(t)\mathrm{d}t & \text{for } \operatorname{Im} z > 0, \\ +\mathrm{i}\int_{-\infty}^0 \mathrm{e}^{\mathrm{i}zt}\mathrm{e}^{-B^+t}\Theta(t)\mathrm{d}t & \text{for } \operatorname{Im} z < 0. \end{cases}$$

Proposition 8.8.6 *We have, for* $f \in \mathscr{K}$,

$$\begin{aligned} R(z)f &= \tilde{R}(z)f \\ &\quad + \begin{cases} \mathrm{i}\tilde{R}(z)\mathfrak{a}^+ A_1 R(z)f + \mathrm{i}\tilde{R}(z)\mathfrak{a}^+ A_0 S(z)f + \mathrm{i}\tilde{R}(z)A_{-1}S(z)f \\ -\mathrm{i}\tilde{R}(z)\mathfrak{a}^+ A_{-1}^+ R(z)f - \mathrm{i}\tilde{R}(z)\mathfrak{a}^+ A_0^+ S(z)f - \mathrm{i}\tilde{R}(z)A_1^+ S(z)f. \end{cases} \end{aligned}$$

The upper line holds for $\operatorname{Im} z > 0$, *the lower one for* $\operatorname{Im} z < 0$.

Proof Directly from the definition for $t > s$, considering first the variation in t and then in s, we have

$$\begin{aligned} u_s^t(\sigma, \tau, \upsilon) &= \mathrm{e}^{B(t-s)}\mathbf{e}(\sigma, \tau, \upsilon) \\ &\quad + \sum_{c\in\sigma} \mathrm{e}^{B(t-t_c)}\mathbf{1}\{t_c \in [s,t]\} A_1 u_s^{t_c}(\sigma \setminus c, \tau, \upsilon) \\ &\quad + \sum_{c\in\tau} \mathrm{e}^{B(t-t_c)}\mathbf{1}\{t_c \in [s,t]\} A_0 u_s^{t_c}(\sigma, \tau \setminus c, \upsilon) \\ &\quad + \sum_{c\in\upsilon} \mathrm{e}^{B(t-t_c)}\mathbf{1}\{t_c \in [s,t]\} A_{-1} u_s^{t_c}(\sigma, \tau, \upsilon \setminus c) \\ &= \mathrm{e}^{B(t-s)}\mathbf{e}(\sigma, \tau, \upsilon) \\ &\quad + \sum_{c\in\sigma} u_{t_c}^t(\sigma \setminus c, \tau, \upsilon) A_1 \mathrm{e}^{B(t_c-s)}\mathbf{1}\{t_c \in [s,t]\} \\ &\quad + \sum_{c\in\tau} u_{t_c}^t(\sigma, \tau \setminus c, \upsilon) A_0 \mathrm{e}^{B(t_c-s)}\mathbf{1}\{t_c \in [s,t]\} \\ &\quad + \sum_{c\in\upsilon} u_{t_c}^t(\sigma, \tau, \upsilon \setminus c) A_{-1} \mathrm{e}^{B(t_c-s)}\mathbf{1}\{t_c \in [s,t]\}. \end{aligned}$$

Hence, working with the second formula above since adjoint will reverse the order of things in a product,

$$\begin{aligned}(u_s^t)^+(\sigma,\tau,\upsilon) &= \big(u_s^t(\upsilon,\tau,\sigma)\big)^+\\ &= e^{B^+(t-s)}\mathbf{e}(\sigma,\tau,\upsilon)\\ &\quad+\sum_{c\in\sigma} e^{B^+(t_c-s)}\mathbf{1}\{t_c\in[s,t]\}A_{-1}^+\big(u_s^{t_c}\big)^+(\sigma\setminus c,\tau,\upsilon)\\ &\quad+\sum_{c\in\tau} e^{B^+(t_c-s)}\mathbf{1}\{t_c\in[s,t]\}A_0^+\big(u_s^{t_c}\big)^+(\sigma,\tau\setminus c,\upsilon)\\ &\quad+\sum_{c\in\upsilon} e^{B^+(t_c-s)}\mathbf{1}\{t_c\in[s,t]\}A_1^+\big(u_s^{t_c}\big)^+(\sigma,\tau,\upsilon\setminus c).\end{aligned}$$

Assume $f, g \in \mathscr{K}$ and using the same arguments as in the proof of Theorem 8.2.1 we obtain

$$\begin{aligned}\langle f|U_s^t g\rangle &= \langle f|e^{B(t-s)}g\rangle + \int_s^t dr\big(\langle a(r)f|e^{B(t-r)}A_1U_s^r g\rangle\\ &\quad+\langle a(t)f|e^{B(t-r)}A_0U_s^r a(r)g\rangle + \langle f|e^{B(t-r)}A_{-1}U_s^r a(r)g\rangle\big)\end{aligned}$$

and

$$\begin{aligned}\langle f|(U_s^t)^+ g\rangle &= \langle f|e^{B^+(t-s)}g\rangle + \int_s^t dr\big(\langle a(r)f|e^{B^+(r-s)}A_{-1}^+(U_r^t)^+ g\rangle\\ &\quad+\langle a(r)f|e^{B^+(r-s)}A_0^+(U_r^t)^+ a(r)g\rangle + \langle f|e^{B^+(r-s)}A_1^+(U_r^t)^+ a(r)g\rangle\big).\end{aligned}$$

Finally, for $t > 0$,

$$\begin{aligned}\langle f|U_0^t g\rangle &= \langle f|e^{Bt}g\rangle + \int_0^t dr\big(\langle a(r)f|e^{B(t-r)}A_1U_0^r g\rangle\\ &\quad+\langle a(r)f|e^{B(t-r)}A_0U_0^r a(r)g\rangle + \langle f|e^{B(t-r)}A_{-1}U_0^r a(r)g\rangle\big)\end{aligned}$$

and, for $t < 0$,

$$\begin{aligned}\langle f|U_0^t g\rangle &= \langle f|(U_t^0)^+ g\rangle\\ &= \langle f|e^{-B^+t}g\rangle + \int_t^0 dr\big(\langle a(r)f|e^{B^+(r-t)}A_{-1}^+U_0^r g\rangle\\ &\quad+\langle a(r)f|e^{B^+(r-t)}A_0^+U_0^r a(r)g\rangle + \langle f|e^{B^+(r-t)}A_1^+U_0^r a(r)g\rangle\big).\end{aligned}$$

We want now to calculate the resolvent for $\operatorname{Im} z > 0$

$$\langle f|R(z)g\rangle = -i\int_0^\infty dt\, e^{izt}\langle f|\Theta(t)U_0^t g\rangle = -i\int_0^\infty dt\, e^{izt}\langle \Theta(-t)f|U_0^t g\rangle$$

and consider, for example, the term

$$-\mathrm{i}\int_0^\infty \mathrm{d}t\mathrm{e}^{\mathrm{i}zt}\int_0^t \mathrm{d}r\langle a(r)\Theta(-t)f|\mathrm{e}^{B(t-r)}A_0U_0^r a(r)g\rangle$$
$$= -\mathrm{i}\int_0^\infty \mathrm{d}r\int_r^\infty \mathrm{e}^{\mathrm{i}zt}\langle a(r)\Theta(-t)f|\mathrm{e}^{B(t-r)}A_0U_0^r a(r)g\rangle$$

Introduce $t' = t - r$ and call it again t and continue

$$= -\mathrm{i}\int_0^\infty \mathrm{d}r\int_0^\infty \mathrm{d}t\mathrm{e}^{\mathrm{i}z(t+r)}\langle a(r)\Theta(-t-r)f|\mathrm{e}^{Bt}A_0U_0^r a(r)g\rangle$$
$$= -\mathrm{i}\int_0^\infty \mathrm{d}r\int_0^\infty \mathrm{d}t\mathrm{e}^{\mathrm{i}z(t+r)}\langle a(0)\mathrm{e}^{B^+t}\Theta(-t)f|A_0\Theta(r)U_0^r a(r)g\rangle$$
$$= \langle \mathfrak{a}\tilde{R}(z)^+g|\mathrm{i}A_0S(z)g\rangle = \mathrm{i}\langle f|\tilde{R}(z)\mathfrak{a}^+A_0S(z)g\rangle.$$

By similar calculations one finishes the proof. □

Corollary 8.8.1 *If $f \in \mathscr{K}$, we may write*

$$R(z)f = \tilde{R}(z)\big(f_0(z) + \mathfrak{a}^+ f_1(z)\big)$$

with

$$f_0(z) = f + \begin{cases} +\mathrm{i}A_{-1}S(z)f & \textit{for } \operatorname{Im} z > 0, \\ -\mathrm{i}A_1^+S(z)f & \textit{for } \operatorname{Im} z < 0 \end{cases}$$

and

$$f_1(z) = \begin{cases} +\mathrm{i}A_1R(z)f + \mathrm{i}A_0S(z)f & \textit{for } \operatorname{Im} z > 0, \\ -\mathrm{i}A_{-1}^+R(z)f - \mathrm{i}A_0 + S(z)f & \textit{for } \operatorname{Im} z < 0. \end{cases}$$

Definition 8.8.5 The vector space $\hat{D} \subset \Gamma$ is defined by

$$\hat{D} = \big\{f = \tilde{R}(z)\big(f_0 + \mathfrak{a}^+ f_1\big) : f_0 \in \Gamma_1, f_1 \in \Gamma_2\big\}.$$

Proposition 8.8.7 *The resolvent maps $\mathscr{K}$ to $\hat{D}$:*

$$R(z) : \mathscr{K} \to \hat{D}.$$

Proof By Proposition 8.8.3

$$\|W(t)f\|^2_{\Gamma_k} \le P(|t|)\|f\|^2_{\Gamma_k},$$

where $P(|t|)$ is a polynomial in t of degree $\leq k$ with coefficients ≥ 0. So, for example, for $\operatorname{Im} z > 0$

$$\|R(z)f\|_{\Gamma_k} \leq \int_0^\infty dt \exp(-t \operatorname{Im} z)\sqrt{P(|t|)}\|f\|_{\Gamma_k}.$$

If $f \in \mathscr{K}$, the function $f \in \Gamma_k$ for all k. The functions $a(t)f, t \in \mathbb{R}$ are uniformly bounded in any Γ_k-norm. Hence $R(z)f$ and $S(z)f$ are in Γ_k for all k. □

Proposition 8.8.8 *For $f \in \mathscr{K}$, we have*

$$\mathring{a}R(z)f = S(z)f + \begin{cases} \frac{1}{2}A_1 R(z)f + \frac{1}{2}A_0 S(z)f & \text{for } \operatorname{Im} z > 0, \\ \frac{1}{2}A_{-1}^+ R(z)f + \frac{1}{2}A_0^+ S(z)f & \text{for } \operatorname{Im} z < 0. \end{cases}$$

Proof We first prove the case $\operatorname{Im} z > 0$. We have

$$\begin{aligned}(U_{-t}^0 f)(\omega + c) &= (U_{-t}^0 a(t_c) f)(\omega) \\ &\quad + \mathbf{1}\{t_c \in [-t, 0]\}\big((U_{t_c}^0 A_1 U_{-t}^{t_c} f)(\omega) + (U_{t_c}^0 A_0 U_{-t}^{t_c} a(t_c) f)(\omega)\big)\end{aligned}$$

and

$$\begin{aligned}&(R(z)f)(\omega + c) \\ &= -\mathrm{i}\int_0^\infty dt e^{\mathrm{i}zt}(\Theta(t)U_0^t f)(\omega + c) = -\mathrm{i}\int_0^\infty dt e^{\mathrm{i}zt}(U_{-t}^0 \Theta(t) f)(\omega + c) \\ &= -\mathrm{i}\int_0^\infty dt (U_{-t}^0 \Theta(t) a(t_c + t) f)(\omega) - \mathrm{i}\mathbf{1}\{t_c < 0\} U_{t_c}^0 \\ &\quad \times \left(A_1 \int_{-t_c}^\infty dt e^{\mathrm{i}zt}(U_{-t}^{t_c}\Theta(t) f)(\omega) + A_0 \int_{-t_c}^\infty dt e^{\mathrm{i}zt}(U_{-t}^{t_c}\Theta(t) a(t_c + t) f)(\omega)\right).\end{aligned}$$

One concludes

$$\begin{aligned}(R(z)f)(0+, t_1, \ldots, t_n) &= (S(z)f)(t_1, \ldots, t_n), \\ (R(z)f)(0-, t_1, \ldots, t_n) &= (S(z)f)(t_1, \ldots, t_n) + A_1(R(z)f)(t_1, \ldots, t_n) \\ &\quad + A_0(S(z)f)(t_1, \ldots, t_n).\end{aligned}$$

Similarly, for $\operatorname{Im} z < 0$,

$$\begin{aligned}(R(z)f)(0+, t_1, \ldots, t_n) &= (S(z)f)(t_1, \ldots, t_n) + A_{-1}^+(R(z)f)(t_1, \ldots, t_n) \\ &\quad + A_0^+(S(z)f)(t_1, \ldots, t_n), \\ (R(z)f)(0-, t_1, \ldots, t_n) &= (S(z)f)(t_1, \ldots, t_n).\end{aligned}$$

□

We start with an *Ansatz* $\hat{H}$ for H.

Definition 8.8.6 Assume we have four operators $M_0, M_{\pm 1}, G \in B(\mathfrak{k})$ such that

$$M_0^+ = M_0, \qquad M_1^+ = M_{-1}, \qquad G^+ = G.$$

Define a mapping $\hat{D} \to \hat{D}^\dagger$ ($\hat{D}^\dagger$ is the space of all semilinear functionals $\hat{D} \to \mathbb{C}$) by

$$\hat{H} = \mathrm{i}\hat{\partial} + M_1\hat{\mathfrak{a}}^+ + M_0\hat{\mathfrak{a}}^+\hat{\mathfrak{a}} + M_{-1}\hat{\mathfrak{a}} + G.$$

The following lemma is a direct consequence of Proposition 8.8.5 and the assumptions about the coefficients.

Lemma 8.8.7 *The sesquilinear form*

$$f, g \in D \mapsto \langle f|\hat{H}g\rangle = \big\langle f|(\mathrm{i}\hat{\partial} + G)g\big\rangle + \langle\hat{\mathfrak{a}}f|M_1 g\rangle + \langle\hat{\mathfrak{a}}f|\hat{\mathfrak{a}}M_0 g\rangle + \langle f|\hat{\mathfrak{a}}M_{-1}g\rangle$$

exists and is symmetric.

As already stated in Sect. 4.2.2, we may embed Γ into $\hat{D}^\dagger$ by the mapping

$$f \in \Gamma \mapsto \big(g \in \hat{D} \mapsto \langle g|f\rangle\big).$$

As $\hat{D}$ is dense in Γ, we can embed

$$\hat{D} \subset \Gamma \subset \hat{D}^\dagger.$$

Proposition 8.8.9 *Assume* $f = \tilde{R}(z)(f_0 + \mathfrak{a}^+ f_1) \in \hat{D}$. *Then*

$$\hat{H}f = -\big(f_0 + \hat{\mathfrak{a}}^+ f_1\big) + \big(z - \mathrm{i}C(z)\big)f + M_1\hat{\mathfrak{a}}^+ f + M_0\hat{\mathfrak{a}}^+\hat{\mathfrak{a}}f + M_{-1}\hat{\mathfrak{a}}f$$

with

$$C(z) = \begin{cases} +B & \textit{for } \operatorname{Im} z > 0, \\ -B^+ & \textit{for } \operatorname{Im} z < 0. \end{cases}$$

Then $\hat{H}f \in \Gamma$ *if and only if*

$$-f_1 + M_1 f + M_0\hat{\mathfrak{a}}f = 0.$$

Proof We calculate the Schwartz derivative

$$\kappa(z)' = -\mathrm{i}\delta + \partial^c\kappa(z) = -\mathrm{i}\delta + \big(\mathrm{i}z + C(z)\big)\kappa(z)$$

and obtain (see Definition 8.8.4)

$$\begin{aligned}\hat{\partial}\tilde{R}(z)\big(f_0 + \mathfrak{a}^+ f_1\big) &= -\lim \Theta\big(\varphi_n'\big)\Theta\big(\kappa(z)\big)\big(f_0 + \mathfrak{a}^+ f_1\big) \\ &= -\lim \Theta\big(\varphi_n * \kappa(z)'\big)\big(f_0 + \mathfrak{a}^+ f_1\big)\end{aligned}$$

$$= -\lim \Theta\big(-\mathrm{i}\varphi_n + \varphi_n \star \partial^{\mathrm{c}}\kappa(z)\big)\big(f_0 + \mathfrak{a}^+ f_1\big)$$
$$= \mathrm{i}\big(f_0 + \hat{\mathfrak{a}}^+ f_1\big) - \Theta\big(\partial^{\mathrm{c}}\kappa(z)\big)\big(f_0 + \mathfrak{a}^+ f_1\big)$$
$$= \mathrm{i}\big(f_0 + \hat{\mathfrak{a}}^+ f_1\big) - \big(\mathrm{i}z + C(z)\big) f.$$

Finally

$$\hat{H} f = -\big(f_0 + \hat{\mathfrak{a}}^+ f_1\big) + \big(z - \mathrm{i}C(z)\big) f + M_1 \hat{\mathfrak{a}}^+ f + M_0 \hat{\mathfrak{a}}^+ \hat{\mathfrak{a}} f + M_{-1} \hat{\mathfrak{a}} f.$$

This formula shows that the singular part of $\hat{H} f$ vanishes if and only if the corresponding equation in the proposition is fulfilled. □

Definition 8.8.7 Define D_0 as the subspace of those functions $f = \tilde{R}(z)(f_0 + \mathfrak{a}^+ f_1) \in \hat{D}$ which obey the condition of Proposition 8.8.9, i.e., $f = \tilde{R}(z)(f_0 + \mathfrak{a}^+ f_1) \in \Gamma$, and denote by H_0 the restriction of $\hat{H}$ to $D_0 f$.

Lemma 8.8.8 *As $\hat{H}$ is symmetric on $\hat{D}$, it is symmetric on D_0 too.*

The conditions for the unitarity of the operators $\mathscr{O}(u_s^t)(A_i, B)$ were (Theorem 8.6.1) that the operators A_i, $i = 1, 0, -1$ and B fulfill the following conditions: There exists a unitary operator Υ such that

$$A_0 = \Upsilon - 1,$$
$$A_1 = -\Upsilon A_{-1}^+,$$
$$B + B^+ = -A_1^+ A_1 = -A_{-1} A_{-1}^+.$$

Theorem 8.8.1 *The operator $\hat{H}$ fulfills the equation*

$$\hat{H} R(z) f = -f + z R(z) f \qquad (*)$$

for all $f \in \mathscr{K}$ if and only if

$$\begin{aligned} A_1 &= \frac{1}{\mathrm{i} - M_0/2} M_1, \\ A_0 &= \frac{M_0}{\mathrm{i} - M_0/2}, \\ A_{-1} &= M_{-1} \frac{1}{\mathrm{i} - M_0/2}, \\ B &= -\mathrm{i}G - \frac{\mathrm{i}}{2} M_{-1} \frac{1}{\mathrm{i} - M_0/2} M_1. \end{aligned} \qquad (**)$$

As a consequence

$$\Upsilon = \frac{\mathrm{i} + M_0/2}{\mathrm{i} - M_0/2}.$$

If equation (∗∗) is fulfilled, then $R(z)$ maps $\mathscr{K}$ into D_0. The domain D of the Hamiltonian H of $W(t)$ contains D_0 and the restriction of H to D_0 coincides with the restriction H_0 of $\hat{H}$ to D_0, and furthermore D_0 is dense in Γ and H is the closure of H_0.

Proof Assume at first $\operatorname{Im} z > 0$. By Propositions 8.8.8 and 8.8.9,

$$\mathrm{i}\hat{\partial} Rf = -f - \mathrm{i}A_1 SF - \hat{\mathfrak{a}}^+(\mathrm{i}A_1 Rf + \mathrm{i}A_0 Sf) + (z - \mathrm{i}B)Rf$$

$$\hat{\mathfrak{a}} Rf = Sf + \frac{1}{2}A_1 Rf + \frac{1}{2}A_0 Sf.$$

Then

$$\hat{H} Rf = -f + zRf + C_1 \hat{\mathfrak{a}}^+ Rf + C_2 \hat{\mathfrak{a}}^+ Sf + C_3 Sf + C_4 Rf$$

with

$$C_1 = -\mathrm{i}A_1 + M_1 + \frac{1}{2}M_0 A_1,$$

$$C_2 = -\mathrm{i}A_0 + M_0 + \frac{1}{2}M_0 A_0,$$

$$C_3 = -\mathrm{i}A_{-1} + M_{-1} + \frac{1}{2}M_{-1} A_0,$$

$$C_4 = -\mathrm{i}B + G + \frac{1}{2}M_{-1} A_1.$$

The equations $C_1 = C_2 = C_3 = C_4 = 0$ are equivalent to (∗∗).

For $\operatorname{Im} z < 0$, we obtain

$$\mathrm{i}\partial R = -\left(f - \mathrm{i}A_1^+ Sf\right) + \hat{\mathfrak{a}}^+\left(\mathrm{i}A_{-1} Rf + A_0^+ Sf\right) + \left(z + \mathrm{i}B^+\right)Rf,$$

$$\hat{\mathfrak{a}} Rf = Sf + \frac{1}{2}A_{-1}^+ Rf + \frac{1}{2}A_0^+ Sf.$$

Again

$$\hat{H} Rf = -f + zRf + C_1' \hat{\mathfrak{a}}^+ Rf + C_2' \hat{\mathfrak{a}}^+ Sf + C_3' Sf + C_4' Rf$$

with

$$C_1' = \mathrm{i}A_{-1}^+ + M_1 + \frac{1}{2}M_0 A_{-1}^+,$$

$$C_2' = \mathrm{i}A_0^+ + M_0 + \frac{1}{2}M_0 A_0^+,$$

$$C_3' = \mathrm{i}A_1^+ + M_{-1} + \frac{1}{2}M_{-1} A_0^+,$$

$$C_4' = \mathrm{i}B^+ + G + \frac{1}{2}M_{-1} A_{-1}^+.$$

Equations $C_1' = C_2' = C_3' = C_4' = 0$ are equivalent to $(**)$ as well, as it should be. We know already that $R(z)$ maps $\mathcal{K}$ into D for $\operatorname{Im} z \neq 0$. Formula $(*)$ shows that $R(z)$ maps $\mathcal{K}$ into D_0.

We studied in Sect. 3.1 the following situation. Assume a unitary group $U(t)$ and a dense subspace $V_0 \subset V$. Assume given a subspace $D_0 \subset V$ and that z and $\overline{z}$ are in the resolvent set of the Hamiltonian, and, furthermore, that $R(z)V_0$ and $R(\overline{z})V_0$ are contained in D_0. Let there be given a *symmetric* operator $H_0 : D_0 \to V$, i.e.

$$(f|H_0 g) = (H_0 g|f)$$

for $f, g \in D_0$, and assume that

$$H_0 R(z)\xi = -\xi + zR(z)\xi,$$
$$H_0 R(\overline{z})\xi = -\xi + \overline{z}R(\overline{z})\xi$$

for $\xi \in V_0$.

Then the subspace D_0 is dense in V and $D_0 \subset D$, the domain of H; also

$$H_0 = H \upharpoonright D_0,$$

and H is the closure of H_0.

We apply this result to $U(t) \to W(t)$, $V \to \Gamma$, $V_0 \to \mathcal{K}$ and finish the proof. □

Remark 8.8.1 L. Accardi [2, 4] and J. Gough [20] studied the so-called Hamiltonian form of quantum stochastic differential equations, and arrived at similar formulae. In particular, a Cayley transform, like that in equation $(**)$, shows up. Writing a Hamiltonian form for the equations is different from finding a Hamiltonian.

Another representation of the Hamiltonian prior to our representation was found by Gregoratti [22], who used the ideas of Chebotarev [14]. Chebotarev had obtained a characterization of the Hamiltonian for the Hudson-Parthasarathy equation with commuting coefficients.

Chapter 9
The Amplified Oscillator

Abstract We study the quantum stochastic differential equation of the amplified oscillator. The solution can be given as a series of normal ordered monomials. The series can be summed with the help of Wick's theorem. From there one gets an a priori estimate. As the solution is a $\mathscr{C}^1$-process, we can prove that it is a unitary cocycle. We obtain the Heisenberg equation studied in Chap. 4, and from there an a posteriori estimate strong enough to calculate the explicit form of the Hamiltonian. We show how amplification works and how the classical Yule process is a part of the quantum stochastic process.

9.1 The Quantum Stochastic Differential Equation

A quantum oscillator has the energy levels $\{nh\nu : n = 0, 1, 2, \ldots\}$. A *damped* oscillator has the property, if the oscillator is in level n, then it emits a photon and jumps to level $n-1$, then it emits a second photon and jumps to level $n-2$, and so on. After some approximations and normalizations it can be described by the QSDE

$$\frac{\mathrm{d}U_0^t}{\mathrm{d}t} = -\mathrm{i}ba^\dagger(t)U_0^t - \mathrm{i}b^+U_0^t a(t) - \frac{1}{2}b^+b\, U_0^t.$$

Here b and b^+ are the usual oscillator operators, but we have carefully distinguished $a^\dagger$ which only can act to the left, in contrast to an ordinary adjoint such as b^+, and is defined by (cf. Sect. 2.4)

$$\langle f | a^\dagger(t) = \langle a(t) f |.$$

Its restriction to the one-excitation case has been studied in Sect. 4.2 and in Sect. 8.3.1.

An *amplified* oscillator has the property, if the oscillator is in level n, then it emits a photon and jumps to level $n+1$, then it emits a second photon and jumps to level $n+2$, and so on. The number of emissions per time is proportional to the number of photons. So an avalanche is created. It can be described by the QSDE

$$\frac{\mathrm{d}U_0^t}{\mathrm{d}t} = -\mathrm{i}b^+a^\dagger(t)U_0^t - \mathrm{i}bU_0^t a(t) - \frac{1}{2}bb^+U_0^t.$$

W. von Waldenfels, *A Measure Theoretical Approach to Quantum Stochastic Processes*, Lecture Notes in Physics 878, DOI 10.1007/978-3-642-45082-2_9,

The physical background has been explained in Sect. 4.2. The differential equation studied here differs from that in Sect. 4.4 and in Sect. 8.3.3 by a scaling factor $\sqrt{2\pi}$.

The quantum stochastic differential equation has been studied by Hudson and Ion [25]. They used another method and obtained the solution as a Bogolyubov transform of the Heisenberg equation. More explicit is Berezin's treatment [8]. The amplified oscillator has a quadratic Hamiltonian and the time evolution can be calculated. There is, however, the inversion of a complicated operator involved. Mandel and Wolf [32] treat the problem with the help of a master equation. It would be nice, to compare the different approaches.

Define

$$\Gamma^* = \Gamma \otimes l^2(\mathbb{N})$$

and, for $f \in \Gamma^*$,

$$|f\rangle = \sum_{m,k=0}^{\infty} 1/(m!k!) \int f_{m,k}(x_1, \dots x_m) a^+(\mathrm{d}x_1) \cdots a^+(\mathrm{d}x_m)|\emptyset\rangle \otimes b^{+k}|0\rangle$$

with

$$f_{m,k} \in L(m) = L^2_{\mathrm{s}}(\mathbb{R}^m)$$

and

$$\|f\|^2 = \sum_{m,k=0}^{\infty} 1/(m!k!) \int \mathrm{d}x_1 \cdots \mathrm{d}x_m \left| f_{m,k}(x_1, \dots, x_m) \right|^2 = \langle f | f \rangle.$$

The functions in Γ^* can be considered as functions on $\mathfrak{R} \times \mathbb{N}$. Denote by Γ^*_{f} the subspace consisting of finite sums in m and k. We denote by $\mathscr{K}^*$ the subspace of those functions f, where all $f_{m,k}$ are continuous with compact support and where the sum over m and k has finitely many terms.

9.2 Closed Solution

The solution can be represented by the series

$$U_s^t = \sum_{n=0}^{\infty} (-\mathrm{i})^n U_{n,s}^t$$

with

$$U_{n,s}^t = \int \cdots \int_{s<t_1<\cdots<t_n<t} \mathrm{d}t_1 \cdots \mathrm{d}t_n \sum_{\vartheta_1,\dots,\vartheta_n} \mathbb{O}_a\left(\mathrm{e}^{-bb^+(t-t_n)/2} b^{\vartheta_n} a^{\vartheta_n}(t_n)\right.$$

$$\times \mathrm{e}^{-bb^+(t_n-t_{n-1})/2} b^{\vartheta_{n-1}} a^{\vartheta_{n-1}}(t_{n-1}) \cdots \mathrm{e}^{-bb^+(t_2-t_1)/2} b^{\vartheta_1} a^{\vartheta_1}(t_1)$$

$$\times \mathrm{e}^{-bb^+(t_1-s)/2}\Big),$$

where, $\vartheta = \pm 1$,

$$b^{+1} = b^+, \qquad b^{-1} = b,$$

$$a^{+1} = a^\dagger, \qquad a^{-1} = a$$

and $\mathbb{O}_a$ denotes normal ordering with respect to $a^\dagger, a$.

We introduce ordering with respect to t, and denote it again by an ordering symbol $\mathbb{O}_t$. As a result of $\mathbb{O}_t$ a function of $t_1, \ldots, t_n$ becomes symmetric in $t_1, \ldots, t_n$ and

$$\int \cdots \int_{s<t_1<\cdots<t_n<t} \mathrm{d}t_1 \cdots \mathrm{d}t_n = \mathbb{O}_t \frac{1}{n!} \int_s^t \cdots \int_s^t \mathrm{d}t_1 \cdots \mathrm{d}t_n.$$

Use the formula

$$\mathrm{e}^{bb^+t/2} b^\vartheta \mathrm{e}^{-bb^+t/2} = \mathrm{e}^{\vartheta t/2} b^\vartheta$$

and the time-ordering operator $\mathbb{O}_t$ to arrive at

$$U_{n,s}^t = \mathrm{e}^{-bb^+(t-s)/2}$$

$$\mathbb{O}_t \mathbb{O}_a \frac{1}{n!} \int_s^t \cdots \int_s^t \mathrm{d}t_1 \cdots \mathrm{d}t_n \sum_{\vartheta_1,\ldots,\vartheta_n} \mathrm{e}^{\vartheta_n t/2} a^{\vartheta_n}(t_n) b^{\vartheta_n} \ldots \mathrm{e}^{\vartheta_1 t/2} a^{\vartheta_1}(t_1) b^{\vartheta_1}.$$

Consider the expression

$$f(t, \vartheta) = \mathrm{e}^{\vartheta t/2} a^\vartheta(t) b^\vartheta$$

and

$$F = \mathbb{O}_a \mathbb{O}_t \sum_{\vartheta_1,\ldots,\vartheta_n} \mathrm{e}^{\vartheta_n t/2} b^{\vartheta_n} \ldots \mathrm{e}^{\vartheta_1 t/2} b^{\vartheta_1} = \mathbb{O}_a \mathbb{O}_t \sum_{\vartheta_1,\ldots,\vartheta_n} f(t_n, \vartheta_n) \cdots f(t_1, \vartheta_1).$$

The operator $\mathbb{O}_a$ has as a consequence that the order of $a, a^\dagger$ in expressions to the right of it does not matter; effectively in such expressions the quantities $a, a^\dagger$ commute. We apply Wick's theorem (Sect. 1.3) for the orderings with respect to t and to ϑ. Ordering with respect to ϑ means normal ordering with respect to b^+, b. We define

$$C(t, \vartheta; t', \vartheta') = \big[f(t, \vartheta), f(t', \vartheta')\big]\big(\mathbf{1}\{t > t'\} - \mathbf{1}\{\vartheta > \vartheta'\}\big)$$

and consider the fact that in this context $a, a^\dagger$ are commuting quantities, so

$$C(t, \vartheta; t', \vartheta') = \mathrm{e}^{\vartheta t/2 + \vartheta' t'/2} a^\vartheta(t) a^{\vartheta'}(t') \begin{cases} \mathbf{1}\{t' > t\} & \text{for } \vartheta = 1, \vartheta' = -1, \\ \mathbf{1}\{t > t'\} & \text{for } \vartheta = -1, \vartheta' = 1. \end{cases}$$

Denote by $\mathfrak{P}(n)$ the set of partitions of $[1,n]$ into singletons and pairs. So $\mathfrak{p} \in \mathfrak{P}(n)$ is of the form

$$\mathfrak{p} = \big\{\{u_1\}, \dots, \{u_l\}, \{r_1, s_1\}, \dots, \{r_m, s_m\}\big\}.$$

Define

$$F_\mathfrak{p} = \mathbb{O}_a \mathbb{O}_t \big(\mathbb{O}_\vartheta f(t_{u_1}, \vartheta_{u_1}) \cdots f(t_{u_l}, \vartheta_{u_l})\big) C_{r_1,s_1} \cdots C_{r_m,s_m}$$

in which we note that

$$C_{r,s} = C(t_r, \vartheta_r; t_s, \vartheta_s) = C_{s,r}.$$

Then

$$F = \mathbb{O}_a \mathbb{O}_t \sum_{\vartheta_1,\dots,\vartheta_n} \sum_{\mathfrak{p} \in \mathfrak{P}(n)} F_\mathfrak{p}.$$

Now $F_\mathfrak{p}$ is a function of the pairs $(t_1, \vartheta_1), \dots, (t_n, \vartheta_n)$ symmetric in its variables under those permutations of $1, \dots, n$, which leave $\mathfrak{p}$ invariant. So $\sum_\mathfrak{p} F_\mathfrak{p}$ is invariant under all permutations of $(t_1, \vartheta_1), \dots, (t_n, \vartheta_n)$, and $\mathbb{O}_a \sum_{\vartheta_1,\dots,\vartheta_n} \sum_{\mathfrak{p}\in\mathfrak{P}(n)} F_\mathfrak{p}$ is a symmetric function in $t_1, \dots, t_n$; we may forget about $\mathbb{O}_t$. We calculate

$$\begin{aligned}
U^t_{n,s} &= \mathrm{e}^{-bb^+(t-s)/2} \mathbb{O}_a \frac{1}{n!} \int_s^t \cdots \int_s^t \mathrm{d}t_1 \cdots \mathrm{d}t_n \sum_{\vartheta_1,\dots,\vartheta_n} \sum_{\mathfrak{p}\in\mathfrak{P}(n)} F_\mathfrak{p} \\
&= \mathrm{e}^{-bb^+(t-s)/2} \mathbb{O}_a \sum_{l+2m=n} \frac{1}{l!2^m m!} \int_s^t \cdots \int_s^t \mathrm{d}t_1 \cdots \mathrm{d}t_l \\
&\quad \times \sum_{\vartheta_1,\dots,\vartheta_l} \mathbb{O}_\vartheta \big(f(t_1,\vartheta_1) \cdots f(t_l, \vartheta_l)\big) \bigg(\int_s^t \int_s^t \mathrm{d}t_1 \mathrm{d}t_2 \sum_{\vartheta_1,\vartheta_2} C_{12}\bigg)^m \\
&= \mathrm{e}^{-bb^+(t-s)/2} \mathbb{O}_a \sum_{l_1+l_2+2m=n} \frac{1}{l_1! l_2! m!} g(1)^{l_1} g(-1)^{l_2} D^m
\end{aligned}$$

with

$$\begin{aligned}
g(1) &= \int_s^t \mathrm{d}t_1 f(t_1 - s, 1) = \int_s^t \mathrm{d}t_1 \mathrm{e}^{(t_1-s)/2} a^\dagger(t_1) b^+, \\
g(-1) &= \int_s^t \mathrm{d}t_1 f(t_1, -1) = \int_s^t \mathrm{d}t_1 \mathrm{e}^{-(t_1-s)/2} a(t_1) b, \\
D &= \frac{1}{2} \mathbb{O}_a \int_s^t \int_s^t \mathrm{d}t_1 \mathrm{d}t_2 \sum_{\vartheta_1,\vartheta_2} C_{12} = \iint_{s<t_1<t_2<t} \mathrm{d}t_1 \mathrm{d}t_2\, \mathrm{e}^{t_1/2 - t_2/2} a^\dagger(t_1) a(t_2).
\end{aligned}$$

Explicitly

$$U^t_{n,s} = \mathrm{e}^{-bb^+(t-s)/2} \sum_{l_1+l_2+2m=n} \frac{1}{l_1! l_2! m!} \bigg(\int_s^t \mathrm{d}t_1 \mathrm{e}^{(t_1-s)/2} a^\dagger(t_1) b^+\bigg)^{l_1}$$

$$\times \mathbb{O}_a\left(\left(\iint_{s<t_1<t_2<t} dt_1 dt_2 e^{t_1/2-t_2/2} a^\dagger(t_1)a(t_2)\right)^m\right)$$

$$\times \left(\int_s^t dt_1 e^{-(t_1-s)/2} a(t_1) b\right)^{l_2}.$$

Using again the formula

$$e^{bb^+t/2} b^\vartheta e^{-bb^+t/2} = e^{\vartheta t/2} b^\vartheta$$

we obtain

$$U_{n,s}^t = \sum_{l_1+l_2+2m=n} \frac{1}{l_1! l_2! m!} \left(\int_s^t dt_1 e^{-(t-t_1)/2} a^\dagger(t_1) b^+\right)^{l_1} e^{-bb^+(t-s)/2}$$

$$\times \mathbb{O}_a\left(\left(\iint_{s<t_1<t_2<t} dt_1 dt_2 e^{t_1/2-t_2/2} a^\dagger(t_1)a(t_2)\right)^m\right)$$

$$\times \left(\int_s^t dt_1 e^{-(t_1-s)/2} a(t_1) b\right)^{l_2}.$$

If $f \in \Gamma_{\mathrm{f}}^*$, $f \geq 0$ then $b^{+l} U_{n,s}^t b^{+k} |f\rangle$ and $b^{+l} (U_{n,s}^t)^+ b^{+k} |f\rangle$ can be considered as Borel functions ≥ 0 on $\mathfrak{R} \times \mathbb{N}$ which are symmetric on $\mathfrak{R}$. We set

$$Y_s^t = \sum_{n=0}^\infty U_{n,s}^t,$$

$$(Y_s^t)^+ = \sum_{n=0}^\infty (U_{n,s}^t)^+.$$

The functions $b^{+l} Y_s^t b^{+k} |f\rangle$ and $b^{+l} (U_s^t)^+ b^{+k} |f\rangle$ are defined, are symmetric and ≥ 0, and they have possibly the value ∞. We obtain

Proposition 9.2.1

$$Y_s^t = \exp\left(\int_s^t dt_1 e^{-(t-t_1)/2} a^\dagger(t_1) b^+\right) e^{-bb^+(t-s)/2}$$

$$\times \mathbb{O}_a\left(\exp\left(\iint_{s<t_1<t_2<t} dt_1 dt_2 e^{t_1/2-t_2/2} a^\dagger(t_1)a(t_2)\right)\right)$$

$$\times \exp\left(\int_s^t dt_1 e^{-(t_1-s)/2} a(t_1) b\right)$$

and

$$(Y_s^t)^+ = \exp\left(\int_s^t dt_1 e^{-(t_1-s)/2} a^\dagger(t_1) b^+\right) e^{-bb^+(t-s)/2}$$

$$\times \mathbb{O}_a\left(\exp\left(\iint_{s<t_1<t_2<t} dt_1 dt_2 e^{t_1/2-t_2/2} a^\dagger(t_2) a(t_1)\right)\right)$$

$$\times \exp\left(\int_s^t dt_1 e^{-(t-t_1)/2} a(t_1) b\right).$$

We want to show that $b^{+l} Y_s^t b^{+k}|f\rangle$ and $b^{+l} (Y_s^t)^+ b^{+k}|f\rangle$ are in Γ^*, and to give estimates for their norms. We start with some lemmata.

Lemma 9.2.1 *Consider two pairs of quantum oscillators with the usual operators a, a^+ and b, b^+ and*

$$T = \exp(sa^+ b^+),$$

with $s \in \mathbb{C}$ and $|s|^2 < 1$.

Then

$$\langle 0|a^m b^n T^+ b^k b^{+k} T a^{+m} b^{+n}|0\rangle = m!(n+k)!{}_2F_1(m+1, n+k+1, 1, |s|^2)$$

where ${}_2F_1$ is the Gauss hypergeometric function (see [5]).

Proof We calculate

$$\langle 0|a^m b^n e^{\bar{s}ab} b^k b^{+k} e^{sa^+b^+} a^{+m} b^{+n}|0\rangle$$

$$= \sum_{l_1,l_2=0}^{\infty} \frac{1}{l_1! l_2!} \bar{s}^{l_1} s^{l_2} \langle 0|a^{m+l_1} (a^+)^{l_2+m}|0\rangle \langle 0|b^{n+l_1+k} (b^+)^{l_2+n+k}|0\rangle$$

$$= \sum_l \frac{1}{(l!)^2} |s|^{2l} (m+l)!(n+l+k)!.$$

We use Pochhammer's symbol

$$(a)_0 = 1, \qquad (a)_p = a(a+1)\cdots(a+p-1)$$

and obtain

$$\langle 0|a^m b^n e^{\bar{s}ab} b^k b^{+k} e^{sa^+b^+} a^{+m} b^{+n}|0\rangle = m!(m+k)! \sum_{l=0}^{\infty} \frac{(m+1)_l (n+k+1)_l}{(l!)^2} |s|^{2l}.$$

□

Lemma 9.2.2 *We have*

$$\langle 0|a^m b^n T^+ b^k b^{+k} T a^{+m} b^{+n}|0\rangle \le (k+m+n)!(1-|s|^2)^{-(k+m+n+1)}.$$

This estimate is optimal.

Proof We have

$$\frac{(m+l)!(n+l+k)!}{(l!)^2} = (l+1)\cdots(m+l)(l+1)\cdots(n+k+l)$$
$$\leq (l+1)\cdots(l+m)(l+m+1)\cdots(l+m+n+k)$$
$$= \frac{(l+m+n+k)!}{l!}.$$

For $0 \leq x < 1$, we have

$$\sum_{l=0}^{\infty} \frac{(k+m+n+1)_l}{l!} x^l = (1-x)^{-(k+m+n+1)}.$$

This estimate is optimal, as using Theorem 2.1.3 in Askey's book [5] we have

$$\lim_{x\to 1-0} {}_2F_1(m+1, n+k+1, 1; x)(1-x)^{2m+k+1} = \frac{\Gamma(1)\Gamma(m+n+k+1)}{\Gamma(m+1)\Gamma(n+k+1)}$$
$$= \frac{(m+n+k)!}{m!(n+k)!}.$$ □

Lemma 9.2.3 *Assume given a Lebesgue square-integrable $K : \mathbb{R}^2 \to \mathbb{C}$ and consider the operator*

$$L = \mathbb{O}_a \exp\left(\int K(s,t)a^+(\mathrm{d}s)a(t)\mathrm{d}t\right)$$
$$= \sum_{l=0}^{\infty}(1/l!)\int\cdots\int K(s_1,t_1)\cdots K(s_l,t_l)a^+(\mathrm{d}s_1)a^+(\mathrm{d}s_l)a(t_1)\cdots a(t_l)\mathrm{d}t_1\cdots\mathrm{d}t_l.$$

Then L maps $L_s^2(\mathbb{R}^n)$ into itself, and, for $f \in L_s^2(\mathbb{R}^n)$, we have

$$\|Lf\| \leq (1+\|K\|_{HS})^l\|f\|,$$

where

$$\|K\|_{HS} = \left(\iint \mathrm{d}s\,\mathrm{d}t\left|K(s,t)\right|^2\right)^{1/2}$$

is the Hilbert-Schmidt norm of the operator defined by K.

Proof We have for $f, g \in L_s^2(\mathbb{R}^n)$ in easily understandable notation

$$\langle f|L|g\rangle = \sum_{l=0}^{n}\frac{1}{l!}\frac{1}{n!^2}\int \overline{f}(x_{[1,n]})K(s_{[1,l]},t_{[1,l]})g(y_{[1,l]})\mathrm{m}$$

with

$$\mathfrak{m} = \big\langle a(x_{[1,n]})a^+(\mathrm{d}s_{[1,l]})a(t_{[1,l]})a^+(\mathrm{d}y_{[1,n]})\big\rangle \mathrm{d}x_{[1,n]}\mathrm{d}s_{[1,l]}.$$

Using Wick's theorem

$$\mathfrak{m} = \sum_{I\subset[1,n],\#I=l}\ \sum_{J\subset[1,n],\#J=l}\ \sum_{\varphi:[1,l]\twoheadrightarrow I}\ \sum_{\psi:[1,l]\twoheadrightarrow J}\ \sum_{\varphi:I^c\twoheadrightarrow J^c} \mathfrak{m}(\varphi,\psi,\chi)$$

(where $\twoheadrightarrow$ denotes a bijective mapping) and

$$\begin{aligned}\mathfrak{m}(\varphi,\psi,\chi) = {} & \prod_{i\in[1,l]}\big(\varepsilon(x_{\varphi(i)},\mathrm{d}s_i)\big)\mathrm{d}x_{\varphi(i)} \prod_{i\in[1,l]}\big(\varepsilon(t_i,\mathrm{d}y_{\psi(i)})\big)\mathrm{d}t_i \\ & \times \prod_{i\in I^c}\big(\varepsilon(x_i,\mathrm{d}y_{\chi(i)})\big)\mathrm{d}x_i.\end{aligned}$$

Hence

$$\begin{aligned}F(\varphi,\psi,\chi) &= \int \overline{f}(x_{[1,n]})K(s_{[1,l]},t_{[1,l]})g(y_{[1,l]})\mathfrak{m}(\varphi,\psi,\chi) \\ &= \iint \mathrm{d}s_{[1,l]}\mathrm{d}t_{[1,l]}K(s_{[1,l]},t_{[1,l]}) \\ &\quad \times \int \mathrm{d}x_{[1,n]\setminus I}\overline{f}\big((s_{\varphi^{-1}(i)})_{i\in I},x_{[1,n]\setminus I}\big)g\big((t_{\psi^{-1}(i)})_{i\in J},x_{\chi^{-1}(i)}\big)_{i\in[1,n]\setminus I}.\end{aligned}$$

By the Cauchy-Schwarz inequality

$$\begin{aligned}&\big|F(\varphi,\psi,\chi)\big|^2 \\ &\quad \le \iint \mathrm{d}s_{[1,l]}\mathrm{d}t_{[1,l]}\big|K(s_{[1,l]},t_{[1,l]})\big|^2 \\ &\qquad \times \int \mathrm{d}s_{[1,l]}\mathrm{d}t_{[1,l]}\bigg|\int \mathrm{d}x_{[1,n]\setminus I}\overline{f}\big((s_{\varphi^{-1}(i)})_{i\in I},x_{[1,n]\setminus I}\big) \\ &\qquad \times g\big((t_{\psi^{-1}(i)})_{i\in J},x_{\chi^{-1}(i)}\big)_{i\in[1,n]\setminus I}\bigg|^2 \\ &\quad \le \|K\|_{HS}^{2l}(n!)^2\|f\|_\Gamma^2\|g\|_\Gamma^2.\end{aligned}$$

Finally

$$\begin{aligned}\big|\langle f|L|g\rangle\big| &\le \sum_{l=0}^{n}\frac{1}{l!}\frac{1}{n!}\big(n(n-1)\cdots(n-l+1)\big)^2(n-l)!\|K\|_{HS}^l\|f\|_\Gamma\|g\|_\Gamma \\ &= \sum_{l=0}^{n}\binom{n}{l}\|K\|_{HS}^l\|f\|_\Gamma\|g\|_\Gamma.\end{aligned}$$

From there one obtains the result. □

Use the notation, for $f \in L(m)$,

$$a^+(f) = \frac{1}{m!} \int a^+(\mathrm{d}x_1) \cdots a^+(\mathrm{d}x_m) f(x_1, \dots, x_m)$$

and

$$|f\rangle = a^+(f)|\emptyset\rangle \otimes |0\rangle$$

where $|\emptyset\rangle$ is the vacuum of the heat bath and $|0\rangle$ is the ground state of the oscillator.

Lemma 9.2.4 *If $\varphi \in L^1(\mathbb{R})$ and $f \in L^2_s(\mathbb{R}^m)$, then*

$$\mathrm{e}^{a(\varphi)b} b^{+n} |f\rangle \in \bigoplus_l \big(L^2_s(\mathbb{R}^{m-l}) \otimes b^{+(n-l)}|0\rangle\big)$$

and

$$\big\|\mathrm{e}^{a(\varphi)b} b^{+n}|f\rangle\big\|^2 \le n!\|f\|^2\, {}_2F_1\big(-m, -n, 1; \|\varphi\|^2\big),$$

where the Gauss hypergeometric function ${}_2F_1(-m, -n, 1; \|\varphi\|^2)$ is a finite polynomial in $\|\varphi\|^2$ with coefficients ≥ 0.

Proof We calculate

$$\begin{aligned}
\big\|\mathrm{e}^{a(\varphi)b} b^{+n}|f\rangle\big\|^2 &= \langle f|b^n \mathrm{e}^{a^+(\varphi)b} \mathrm{e}^{a(\varphi)b} b^{+n}|f\rangle \\
&= \sum_l (1/l!)^2 \langle f|a(\varphi)^l a^+(\varphi)^l|f\rangle \langle 0|b^n b^{+l} b^l b^{+n}|0\rangle \\
&\le \sum_l (1/l!)^2 m(m-1)\cdots(m-l+1)\|\varphi\|^{2l}\|f\|^2 \\
&\quad \times \big(n(n-1)\cdots(n-l+1)\big)^2 (n-l)! \\
&= \sum_l (1/l!)^2 (-m)_l (-n)_l \|\varphi\|^{2l} n! \|f\|^2 \\
&= n!\|f\|^2\, {}_2F_1\big(-m, -n, 1; \|\varphi\|^2\big).
\end{aligned}$$

□

Proposition 9.2.2 *We have for $f \in L(m)$ with $f \ge 0$ the estimates*

$$\begin{aligned}
\big\|b^{+l} Y_s^t b^{+k}|f\rangle\big\| &\le C(t-s; m, l, k)\|f\|, \\
\big\|b^{+l} \big(Y_s^t\big)^+ b^{+k}|f\rangle\big\| &\le C(t-s; m, l, k)\|f\|
\end{aligned}$$

with

$$C(t-s;m,l,k) = \mathrm{e}^{(l+m+k+1)(t-s)/2}\left(\sum_{j=0}^{m}(j+k+l)!/j!\right)^{1/2}$$
$$\times (1+\sqrt{t-s})^m\sqrt{{}_2F_1(-m,-k,1;1)}\sqrt{k!}).$$

Proof Use the notation

$$\varphi(t_1) = \mathbf{1}\{s<t_1<t\}\mathrm{e}^{-(t-t_1)/2},$$
$$\psi(t_1) = \mathbf{1}\{s<t_1<t\}\mathrm{e}^{-(t_1-s)/2},$$
$$K(t_1,t_2) = \mathbf{1}\{s<t_1<t_2<t\}\mathrm{e}^{-(t_2-t_1)/2}.$$

Then

$$Y_s^t = \mathrm{e}^{a^+(\varphi)b^+}\mathrm{e}^{-bb^+(t-s)/2}\mathbb{O}_a\mathrm{e}^{\iint \mathrm{d}t_1\mathrm{d}t_2 K(t_1,t_2)a^\dagger(t_1)a(t_2)}\mathrm{e}^{a(\psi)b},$$
$$\left(Y_s^t\right)^+ = \mathrm{e}^{a^+(\psi)b^+}\mathrm{e}^{-bb^+(t-s)/2}\mathbb{O}_a\mathrm{e}^{\iint \mathrm{d}t_1\mathrm{d}t_2 K(t_1,t_2)a^\dagger(t_2)a(t_1)}\mathrm{e}^{a(\varphi)b}.$$

We have

$$\|\varphi\| = \|\psi\| = \sqrt{1-\mathrm{e}^{-(t-s)}}.$$

Putting $a = a(\varphi)/\|\varphi\|$ or $a(\varphi) = \|\varphi\|a$, Lemma 9.2.2 yields

$$\langle 0|a(\varphi)^m b^n \mathrm{e}^{a(\varphi)b} b^k b^{+k} e^{a^+(\varphi)b^+} a^+(\varphi)^m b^{+n}|0\rangle$$
$$\leq (k+m+n)!\left(1-\mathrm{e}^{-(t-s)}\right)^m \mathrm{e}^{(k+m+n+1)(t-s)}. \tag{i}$$

We recall the equation, holding for two Hilbert spaces V_1, V_2 and a bounded linear mapping $A: V_1 \to V_2$,

$$\left(\ker(A)\right)^\perp = \mathrm{image}\left(A^+\right).$$

Here $\ker(A) = \{v_1 \in V_1 : Av_1 = 0\}$, $\perp$ denotes the orthogonal complement of A, and A^+ is the adjoint of A.

Consider the annihilation operator $a(\varphi): L(n) \to L(n-1)$ with $\varphi \in L(1)$. The adjoint of $a(\varphi)$ is $a^+(\varphi): L(n-1) \to L(n)$. We split $L(n)$ into the orthogonal sum

$$L(n) = a^+(\varphi)L(n-1) \oplus \ker\left(a(\varphi)\right), \qquad f = a^+(\varphi)g_n + f_n$$

with $g_n \in L(n-1)$ and $af_n = 0$. We continue and see

$$a^+(\varphi)g_n = \left(a^+(\varphi)\right)^2 g_{n-1} + a^+(\varphi)f_{n-1}$$

and finally obtain

$$f = f_n + a^+(\varphi)f_{n-1} + \cdots + \left(a^+(\varphi)\right)^n f_0$$

with $f_i \in L(i)$ and $a(\varphi) f_i = 0$.

We have

$$\langle (a^+(\varphi))^i f_{n-i} | (a^+(\varphi))^j f_{n-j} \rangle = \delta_{i,j} j! \|\varphi\|^{2j} \|f_{n-j}\|^2$$

and

$$\|f\|^2 = \sum_{j=0}^{n} j! \|\varphi\|^{2j} \|f_{n-j}\|^2.$$

We calculate

$$\begin{aligned} &\| (b^+)^k \mathrm{e}^{a^+(\varphi) b^+} a^+(\varphi)^j b^{+l} f_{n-j} \|^2 \\ &= \langle f_{n-j} | b^l a(\varphi)^j \mathrm{e}^{a(\varphi) b} b^k (b^+)^k \mathrm{e}^{a^+(\varphi) b^+} a^+(\varphi)^j b^{+l} | f_{n-j} \rangle. \end{aligned}$$

As $a^+(f_{n-j})$ commutes with $a(\varphi)$ we obtain

$$= \|f_{n-j}\|^2 \langle 0 | b^l a(\varphi)^j \mathrm{e}^{a(\varphi) b} b^k (b^+)^k \mathrm{e}^{a^+(\varphi) b^+} a^+(\varphi)^j b^{+l} | 0 \rangle = \|f_{n-j}\|^2 c_j^2$$

and by equation (i)

$$c_j^2 = \| (b^+)^k \mathrm{e}^{a^+(\varphi) b^+} a^+(\varphi)^j b^{+l} | 0 \rangle \|^2 \leq \|\varphi\|^{2j} (j + l + k)! \mathrm{e}^{(k+j+l+1)(t-s)}$$

and, for $f \in L(n)$,

$$\begin{aligned} &\| (b^+)^k \mathrm{e}^{a^+(\varphi) b^+} b^{+l} | f \rangle \| \\ &\leq \sum_{j=0}^{n} \| (b^+)^k \mathrm{e}^{a^+(\varphi) b^+} a^+(\varphi)^j b^{+l} | f_{n-j} \rangle \| \\ &= \sum_{j=0}^{n} c_j \|f_{n-j}\| \leq \left(\sum \frac{c_j^2}{j! \|\varphi\|^{2j}} \right)^{1/2} \left(\sum j! \|\varphi\|^{2j} \|f_{n-j}\|^2 \right)^{1/2} \end{aligned}$$

and finally

$$\| (b^+)^k \mathrm{e}^{a^+(\varphi) b^+} b^{+l} | f \rangle \| \leq \left(\sum_{j=0}^{n} \frac{(j+l+k)!}{j!} \mathrm{e}^{(k+j+l+1)(t-s)} \right)^{1/2} \|f\|. \qquad \text{(ii)}$$

The expression

$$\mathbb{O}_a \left(\exp \left(\iint_{s < r_1 < r_2 < t} a^+(r_1) e^{r_1/2} a(r_2) e^{-r_2/2} \mathrm{d}r_1 \mathrm{d}r_2 \right) \right)$$

is of the form considered in Lemma 9.2.3 with

$$K(r_1, r_2) = \mathbf{1}\{s < r_1 < r_2 < t\} \mathrm{e}^{-(r_2 - r_1)/2}$$

and

$$\|K\|_{HS} = \left(\iint \mathrm{d}s\,\mathrm{d}t\,K(s,t)^2\right)^{1/2} \le \sqrt{t-s}.$$

It defines a mapping $L(m) \to L(m)$ with the operator norm

$$\left\| \mathbb{O}_a\left(\exp\left(\iint_{s<r_1<r_2<t} a^+(r_1)\mathrm{e}^{r_1/2}a(r_2)\mathrm{e}^{-r_2/2}\mathrm{d}r_1\mathrm{d}r_2\right)\right)\right\| \le (1+\sqrt{t-s})^m. \tag{iii}$$

The operator norm

$$\left\|\mathrm{e}^{-bb^+(t-s)/2}\right\|_{\Gamma^*} \le 1. \tag{iv}$$

For $f \in L(m)$

$$\mathrm{e}^{a(\psi)b)}b^{+n}|f\rangle \in \bigoplus_{l=0}^{m} L(m)\otimes b^{n-l}|0\rangle$$

and by Lemma 9.2.4

$$\left\|\mathrm{e}^{a(\psi)b)}b^{+n}|f\rangle\right\| \le \sqrt{{}_2F_1(-m,-n,1;1)n!}\,\|f\|. \tag{v}$$

By combining equations (i) to (v) we obtain the result for Y_s^t. For $(Y_s^t)^+$ all goes the same way. □

A consequence of the last proposition is the following theorem.

Theorem 9.2.1 *We have the explicit formulae*

$$\begin{aligned} U_s^t &= \sum_{n=0}^{\infty}(-\mathrm{i})^n U_{n,s}^t \\ &= \exp\left(-\mathrm{i}\int_s^t \mathrm{e}^{-(t-t_1)/2}a^+(t_1)\mathrm{d}t_1 b^+\right)\exp\left(-bb^+(t-s)/2\right) \\ &\quad\times \mathbb{O}_a\left(\exp\left(-\iint_{s<r_1<r_2<t} a^+(r_1)\mathrm{e}^{r_1/2}a(r_2)\mathrm{e}^{-r_2/2}\mathrm{d}r_1\mathrm{d}r_2\right)\right) \\ &\quad\times \exp\left(-\mathrm{i}\int_s^t a(t_1)\mathrm{e}^{-(t_1-s)/2}\mathrm{d}t_1 b\right), \\ (U_s^t)^+ &= \sum_{n=0}^{\infty}\mathrm{i}^n (U_{n,s}^t)^+ \\ &= \exp\left(\mathrm{i}\int_s^t \mathrm{d}t_1 \mathrm{e}^{-(t_1-s)/2}a^+(t_1)b^+\right)\mathrm{e}^{-bb^+(t-s)/2} \end{aligned}$$

$$\times \mathbb{O}_a\left(\exp\left(-\iint_{s<t_1<t_2<t} dt_1 dt_2 e^{t_1/2-t_2/2} a^+(t_2)a(t_1)\right)\right)$$
$$\times \exp\left(i\int_s^t dt_1 e^{-(t-t_1)/2} a(t_1)b\right).$$

The sums $\sum_{n=0}^{\infty}(-i)^n U_{n,s}^t b^{+k}|f\rangle$ *and* $\sum_{n=0}^{\infty} i^n (U_{n,s}^t)^+ b^{+k}|f\rangle$ *converge in norm for fixed* $f \in L_s^2(\mathbb{R}^m)$ *and* k.

Lemma 9.2.5 *For* $f \in L_s^2(\mathbb{R}^m)$, *as* $t \downarrow s$

$$\sum_{n=1}^{\infty} \| U_{n,s}^t b^{+k}|f\rangle \| \downarrow 0.$$

Proof Recall

$$U_{n,s}^t = \int\cdots\int_{s<t_1<\cdots<t_n<t} dt_1\cdots dt_n \sum_{\vartheta_1,\ldots,\vartheta_n} \mathbb{O}_a\big(e^{-bb^+(t-t_n/2)} b^{\vartheta_n} a^{\vartheta_n}(t_n)$$
$$\times e^{-bb^+(t_n-t_{n-1})/2} b^{\vartheta_{n-1}} a^{\vartheta_{n-1}}(t_{n-1})\cdots e^{-bb^+(t_2-t_1)/2} b^{\vartheta_1} a^{\vartheta_1}(t_1)$$
$$\times e^{-bb^+(t_1-s)/2}\big).$$

Hence, for $f \in L_s^2(\mathbb{R}^m)$,

$$\begin{aligned}
\| U_{n,s}^t b^{+k}|f\rangle \| &\le \sum_{\vartheta_1,\ldots,\vartheta_n} \| b^{\vartheta_n}\cdots b^{\vartheta_1} b^{+k}|0\rangle \| \\
&\quad \times \int\cdots\int_{s<t_1<\cdots<t_n<t} dt_1\cdots dt_n \| \mathbb{O}_a a^{\vartheta_n}(t_n)\cdots a^{\vartheta_1}(t_1)|f\rangle \| \\
&\le \sum_{\vartheta_1,\ldots,\vartheta_n} \sqrt{(k+1)\cdots(k+n)} \frac{(t-s)^n}{n!} \sqrt{(m+1)\cdots(m+n)} \|f\| \\
&\le 2^n \frac{(l+1)_n}{n!} (t-s)^n \|f\|
\end{aligned}$$

for $l = \max(k, m)$. Hence, if $t - s < 1/2$, for $t \downarrow s$,

$$\sum_{n=1}^{\infty} \| U_{n,s}^t b^{+k}|f\rangle \| \le \left(\left(1-2(t-s)\right)^{-l-1} - 1\right)\|f\| \downarrow 0.$$

□

9.3 The Unitary Evolution

Recall

$$U_{n,s}^t = \int \cdots \int_{s<t_1<\cdots<t_n<t} dt_1 \cdots dt_n \sum_{\vartheta_1,\ldots,\vartheta_n} \mathbb{O}_a\big(e^{-bb^+(t-t_n/2)} b^{\vartheta_n} a^{\vartheta_n}(t_n)$$

$$e^{-bb^+(t_n-t_{n-1})/2} b^{\vartheta_{n-1}} a^{\vartheta_{n-1}}(t_{n-1}) \cdots e^{-bb^+(t_2-t_1)/2} b^{\vartheta_1} a^{\vartheta_1}(t_1) e^{-bb^+(t_1-s)/2}\big)$$

$$= \int \cdots \int_{s<t_1<\cdots<t_n<t} dt_1 \cdots dt_n \mathbb{O}_a\big(e^{-bb^+(t-t_n/2)}\big(b^+a^\dagger(t_n) + ba(t_n)\big)$$

$$\times e^{-bb^+(t_n-t_{n-1})/2}$$

$$\big(b^+a^\dagger(t_{n-1}) + ba(t_{n-1})\big) \cdots e^{-bb^+(t_2-t_1)/2}\big(b^+a^\dagger(t_1) + ba(t_1)\big)$$

$$\times e^{-bb^+(t_1-s)/2}\big).$$

Then

$$\big(U_{n,s}^t\big)^+ = \int \cdots \int_{s<t_1<\cdots<t_n<t} dt_1 \cdots dt_n \mathbb{O}_a\big(e^{-bb^+(t_1-s)}\big(b^+a^\dagger(t_1) + ba(t_1)\big)$$

$$\times e^{-bb^+(t_2-t_1)/2}$$

$$\big(b^+a^\dagger(t_2) + ba(t_2)\big) \cdots e^{-bb^+(t_n-t_{n-1})/2}\big(b^+a^\dagger(t_n) + ba(t_n)\big)$$

$$\times e^{-bb^+(t-t_n)/2}\big).$$

We go back to the measure-theoretic formulation and write

$$U_{n,s}^t = \int \big(u_{n,s}^t\big)(\sigma,\tau) a_\sigma^+ a_\tau \lambda_\tau,$$

$$\big(U_{n,s}^t\big)^+ = \int \big(\tilde{u}_{n,s}^t\big)(\sigma,\tau) a_\sigma^+ a_\tau \lambda_\tau$$

with

$$\big(u_{n,s}^t\big)(\sigma,\tau) = \mathbf{1}\big\{t_\sigma + t_\tau \subset]s,t[\big\}\mathbf{1}\{\#\sigma + \#\tau = n\}$$

$$e^{-bb^+/2(t-t_n)} b^{\vartheta_n} e^{-bb^+/2(t_n-t_{n-1})} b^{\vartheta_{n-1}} \ldots b^{\vartheta_2} e^{-bb^+/2(t_2-t_1)} b^{\vartheta_1}$$

$$e^{-bb^+/2(t_1-s)}$$

and

$$\big(\tilde{u}_{n,s}^t\big)(\sigma,\tau) = \big(u_{n,s}^t\big)^+(\sigma,\tau) = \mathbf{1}\big\{t_\sigma + t_\tau \subset]s,t[\big\}\mathbf{1}\{\#\sigma + \#\tau = n\}$$

$$e^{-bb^+/2(t_1-s)}b^{\vartheta_1}e^{-bb^+/2(t_2-t_1)}b^{\vartheta_2}\dots b^{\vartheta_{n-1}}e^{-bb^+/2(t_{n-1}-t_n)}b^{\vartheta_n}$$
$$e^{-bb^+/2(t-t_n)}$$

under the assumptions, that $\{s,t,t_\sigma,t_\tau\}^\bullet$ contains no multiple points and

$$\{s,t,t_\sigma,t_\tau\}=\{s<t_1<\cdots<t_{n-1}<t_n<t\}$$

and $\vartheta_i=1$ if $t_i\in t_\sigma$ and $\vartheta_i=-1$ if $t_i\in t_\tau$.

Make the two definitions

$$u_s^t=\sum_{n=0}^{\infty}(-\mathrm{i})^n u_{n,s}^t,$$
$$\tilde{u}_s^t=\sum_{n=0}^{\infty}\mathrm{i}^n\tilde{u}_{n,s}^t.$$

We want to apply Ito's theorem, and observe that $\langle 0|b^l u_s^t b^{+k}|0\rangle$ and $\langle 0|b^l(u_s^t)^+b^{+k}|0\rangle$ are $\mathscr{C}^1$ functions with values in $\mathbb{R}$.

For our purposes we have to adapt Ito's theorem. Assume we have two matrix-valued functions

$$F=(F_{kl}),\qquad G=(G_{kl}),\quad k,l=0,1,2,\ldots:\mathfrak{R}^2\to\mathbb{R},$$

where all the matrix elements are Lebesgue measurable. Define the measure

$$\mathfrak{m}(\pi,\sigma_1,\tau_1,\sigma_2,\tau_2,\rho)=\langle a_\pi a_{\sigma_1}^+a_{\tau_1}a_{\sigma_2}^+a_{\tau_2}a_\rho^+\rangle\lambda_{\pi+\tau_1+\tau_2}$$

and the matrix-valued sesquilinear form

$$f,g\in\mathscr{K}_s(\mathfrak{R})\to\langle f|\mathscr{B}(F,G)|g\rangle=\langle f|\big(\mathscr{B}(F,G)\big)_{kl}|g\rangle,$$
$$\langle f|\mathscr{B}(F,G)_{kl}|g\rangle=\int\mathfrak{m}\sum_m 1/m!\overline{f}(\pi)F_{km}(\sigma_1,\tau_1)G_{ml}(\sigma_2,\tau_2)g(\rho),$$

provided the integral combined with the sum converges absolutely.

Theorem 9.3.1 *Assume x_t and y_t to be matrix-valued functions, where all their matrix elements of class $\mathscr{C}^1$, and that, for $f,g\in\mathscr{K}_s(\mathfrak{R},\mathbb{R})$, all the sesquilinear forms $\langle f|\mathscr{B}(F_t,G_t)|g\rangle$ exist so that $t\in\mathbb{R}\mapsto\langle f|\mathscr{B}(F_t,G_t)|g\rangle$ is locally integrable, where F_t is any function drawn from $\{x_t,\partial^c x_t,R_\pm^1 x_t,R_\mp^{-1}x_t\}$, and similarly G_t is any function drawn from $\{y_t,\partial^c y_t,R_\pm^1 y_t,R_\mp^{-1}y_t\}$. Then $t\mapsto\langle f|(\mathscr{B}(x_t,y_t))_{kl}|g\rangle$ is continuous and has as Schwartz derivative a locally integrable function. The Schwartz derivative is*

$$\partial\langle f|\mathscr{B}(x_t,y_t)|g\rangle=\langle f|\mathscr{B}\big(\partial^c x_t,y_t\big)+\mathscr{B}\big(f,\partial^c y_t\big)+I_{-1,+1,t}|g\rangle$$

$$+ \langle a(t)f | \mathscr{B}(D^1 x_t, y_t) + \mathscr{B}(f, D^1 y_t) | g \rangle$$
$$+ \langle f | \mathscr{B}(D^{-1} x_t, y_t) + \mathscr{B}(f, D^{-1} y_t) | a(t)g \rangle$$

with

$$I_{-1,1,t} = \mathscr{B}(R_+^{-1} x_t, R_+^1 y_t) - \mathscr{B}(R_-^{-1} x_t, R_-^1 y_t).$$

We recall the operators ∂^c and $R_\pm^i$, from Sect. 6.3, and the following properties of them that will be useful in our calculations:

Lemma 9.3.1 *With ∂^c and $R_\pm^i$ acting on the upper index t of u_s^t we have*

$$\partial_t^c u_s^t(\sigma, \tau) = -bb^+/2u_s^t(\sigma, \tau),$$
$$(R_+^1 u_s^{\cdot})_t(\sigma, \tau) = u_s^{t+0}(t_\sigma + t, t_\tau) = -ib^+ u_s^t(\sigma, \tau),$$
$$(R_-^1 u_s^{\cdot})_t(\sigma, \tau) = u_s^{t-0}(t_\sigma + t, t_\tau) = 0,$$
$$(R_+^{-1} u_s^{\cdot})_t(\sigma, \tau) = u_s^{t+0}(t_\sigma, t_\tau + t) = -ib u_s^t(\sigma, \tau),$$
$$(R_-^{-1} u_s^{\cdot})_t(\sigma, \tau) = u_s^{t-0}(t_\sigma, t_\tau + t) = 0;$$

and acting on the lower index s of u_s^t

$$\partial_s^c u_s^t(\sigma, \tau) = u_s^t(\sigma, \tau)(bb^+/2),$$
$$(R_+^1 u_{\cdot}^t)_s(\sigma, \tau) = u_{s+0}^t(t_\sigma + s, t_\tau) = 0,$$
$$(R_-^1 u_{\cdot}^t)_s(\sigma, \tau) = u_{s-0}^t(t_\sigma + s, t_\tau) = u_s^t(-ib^+),$$
$$(R_+^{-1} u_{\cdot}^t)_s(\sigma, \tau) = u_{s+0}^t(t_\sigma, t_\tau + s) = 0,$$
$$(R_-^{-1} u_{\cdot}^t)_s(\sigma, \tau) = u_{s-0}^t(t_\sigma, t_\tau + s) = u_s^t(\sigma, \tau)(-ib);$$

then acting on the upper t index of $\tilde{u}_s^t$, we have

$$\partial_t^c(\tilde{u}_s^t)(\sigma, \tau) = \tilde{u}_s^t(\sigma, \tau)(-bb^+/2),$$
$$(R_+^1 \tilde{u}_s^{\cdot})_t(\sigma, \tau) = \tilde{u}_s^{t+0}(t_\sigma + t, t_\tau) = \tilde{u}_s^t(\sigma, \tau)ib^+,$$
$$(R_-^1 \tilde{u}_s^{\cdot})_t(\sigma, \tau) = \tilde{u}_s^{t-0}(t_\sigma + t, t_\tau) = 0,$$
$$(R_+^{-1} \tilde{u}_s^{\cdot})_t(\sigma, \tau) = \tilde{u}_s^{t+0}(t_\sigma, t_\tau + t) = \tilde{u}_s^t(\sigma, \tau)ib,$$
$$(R_-^{-1} \tilde{u}_s^{\cdot})_t(\sigma, \tau) = \tilde{u}_s^{t-0}(t_\sigma, t_\tau + t) = 0;$$

and, finally, acting on the index s of $\tilde{u}_s^t$

$$\partial_s^c \tilde{u}_s^t(\sigma, \tau) = (bb^+/2)\tilde{u}_s^t(\sigma, \tau)(bb^+/2),$$

$$\left(R^1_+\tilde{u}^t_\cdot\right)_s(\sigma,\tau) = \tilde{u}^t_{s+0}(t_\sigma + s, t_\tau) = 0,$$

$$\left(R^1_-\tilde{u}^t_\cdot\right)_s(\sigma,\tau) = \tilde{u}^t_{s-0}(t_\sigma + s, t_\tau) = ib^+u^t_s,$$

$$\left(R^{-1}_+\tilde{u}^t_\cdot\right)_s(\sigma,\tau) = \tilde{u}^t_{s+0}(t_\sigma, t_\tau + s) = 0,$$

$$\left(R^{-1}_-\tilde{u}^t_\cdot\right)_s(\sigma,\tau) = \tilde{u}^t_{s-0}(t_\sigma, t_\tau + s) = ib\tilde{u}^t_s(\sigma,\tau).$$

Theorem 9.3.2 *There exist uniquely determined operators $\hat{U}^t_s$ on Γ^*, whose restrictions to $\mathscr{K}^*$ coincide with U^t_s. We shall write U^t_s instead of $\hat{U}^t_s$ and use the notation*

$$U^t_s = \left(U^s_t\right)^+ \quad \textit{for } t < s.$$

We have

$$U^t_r U^r_s = U^t_s \quad \textit{for } s,t,r \in \mathbb{R}.$$

The U^t_s form a strongly continuous unitary evolution on Γ^*.

Proof We want to apply Ito's theorem to

$$\left(u^t_s\right)_{kl} = \langle 0|b^k u^t_s b^{+l}|0\rangle,$$

$$\left(u^t_s\right)^+_{kl} = \langle 0|b^l \left(u^t_s\right)^+ b^{+k}|0\rangle.$$

We have to show, e.g., that we have a well-defined

$$\mathscr{B}\left(\left(u^t_s\right)^+, u^t_s\right).$$

But this relation, and the other relations needed for the application of Ito's theorem, follow directly from Proposition 9.2.2 and Theorem 9.3.1.

We obtain with the help of Lemma 9.3.1

$$\partial_r \mathscr{B}\left(u^t_r, u^r_s\right) = 0,$$

hence

$$\left(\mathscr{B}\left(u^t_r, u^r_s\right)\right)_{kl} = \left(\mathscr{B}\left(u^t_s, u^s_s\right)\right)_{kl} = \left(\mathscr{B}\left(u^t_s, \mathbf{e}(\emptyset,\emptyset)\right)\right)_{kl}$$

$$= \int \left(u^t_s\right)_{kl}(\sigma,\tau) a^+_\sigma a_\tau \lambda_\tau = \mathscr{B}\left(u^t_s\right)_{kl},$$

and so

$$U^t_s = \left(U^s_t\right)^+ \quad \text{for } t < s.$$

In the same way,

$$\partial_t \mathscr{B}\left(\left(u^t_s\right)^+, u^t_s\right) = 0$$

and

$$\partial_s \mathscr{B}(u_s^t, (u_s^t)^+) = 0,$$

giving

$$(\mathscr{B}((u_s^t)^+, u_s^t))_{kl} = (\mathscr{B}(u_s^t, (u_s^t)^+))_{kl} = \delta_{kl}.$$

Therefore the mappings U_s^t and $(U_s^t)^+$ have the property that, for $f, g \in \Gamma_f^*$,

$$\langle f|(U_s^t)^+ U_s^t|g\rangle = \langle f|U_s^t (U_s^t)^+|g\rangle = \langle f|g\rangle,$$

so U_s^t and $(U_s^t)^+$ are the restrictions of unitary operators from Γ_{f}^* to Γ^*. These unitary operators we denote again by U_s^t and $(U_s^t)^+$. We have, for $s < r < t$,

$$U_r^t U_s^r = U_s^t.$$

If we put for $s > t$

$$U_s^t = (U_t^s)^+,$$

the relation $U_r^t U_s^r = U_s^t$ holds for all $s, t, r \in \mathbb{R}$.

The strong continuity follows from Lemma 9.2.5. □

9.4 Heisenberg Equation

Lemma 9.4.1 *We have, for $s < t$,*

$$\partial_t\big((U_s^t)^+ b^+ U_s^t\big) = \frac{1}{2}(U_s^t)^+ b^+ U_s^t + \mathrm{i}a(t),$$

and

$$\partial_s\big(U_s^t b^+ (U_s^t)^+\big) = -\frac{1}{2} U_s^t b^+ (U_s^t)^+ + \mathrm{i}a(s).$$

Hence

$$(U_0^t)^+ b^+ U_0^t = \mathrm{e}^{t/2} b^+ + \mathrm{i}\int_0^t \mathrm{e}^{(t-s)/2} a(s)\mathrm{d}s$$

and

$$\lim_{t\to\infty} \mathrm{e}^{-t/2}(U_0^t)^+ b^+ U_0^t = b^+ + \mathrm{i}\int_0^\infty \mathrm{e}^{-s/2} a(s)\mathrm{d}s.$$

Proof We want to calculate $(U_s^t)b^+U_s^t$. Define

$$\mathfrak{m} = \mathfrak{m}(\pi, \sigma_1, \tau_1, \sigma_2, \tau_2, \rho) = \langle a_\pi a_{\sigma_1}^+ a_{\tau_1} a_{\sigma_2}^+ a_{\tau_2} a_\rho^+ \rangle \lambda_{\pi+\tau_1+\tau_2}.$$

For $f, g \in \mathscr{K}_s(\mathbb{R})$, we have

$$\langle f|((U_s^t)b^+U_s^t)_{kl}|g\rangle = \int \mathrm{m}\overline{f}(\pi)\langle 0|b^k(u_s^t)^+(\sigma_1,\tau_1)b^+u_s^t(\sigma_2,\tau_2)b^{+l}|0\rangle g(\rho).$$

With the help of Ito's theorem, we obtain

$$\begin{aligned}
\partial_t & \int \mathrm{m}\overline{f}(\pi)\big((u_s^t)^+(\sigma_1,\tau_1)b^+u_s^t(\sigma_2,\tau_2)\mathrm{i}bb^+u_s^t(\sigma_2,\tau_2)b^+\big)_{kl}g(\rho) \\
&= \int \overline{f}(\pi)\langle 0|b^k(u_s^t)^+(\sigma_1,\tau_1) \\
&\quad \times \big(\mathrm{m}(-b(b^+)^2/2 - b^+bb^+/2 + b(b^+)^2) \\
&\quad + a^\dagger(t)\mathrm{m}(\mathrm{i}(b^+)^2 - \mathrm{i}(b^+)^2) + \mathrm{m}a(t)(\mathrm{i}bb^+ - \mathrm{i}b^+b)\big)u_s^t(\sigma_2,\tau_2)b^{+l}|0\rangle g(\rho) \\
&= \int \overline{f}(\pi)\langle 0|b^k(u_s^t)^+(\sigma_1,\tau_1)\big(\mathrm{m}b^+/2 + \mathrm{i}\mathrm{m}a(t)\big)u_s^t(\sigma_2,\tau_2)b^{+l}|0\rangle g(\rho)
\end{aligned}$$

or

$$\partial_t(U_s^t)^+b^+U_s^t = (U_s^t)^+b^+U_s^t/2 + \mathrm{i}a(t).$$

Integrate the differential equation and obtain

$$(U_s^t)^+b^+U_s^t = \mathrm{e}^{(t-s)/2}b^+ + \mathrm{i}\int_s^t \mathrm{e}^{(r-s)/2}a(r)\mathrm{d}r.$$

The second equation of the lemma is obtained in the same way.

One calculates

$$\lim_{t\to\infty} \mathrm{e}^{-t/2}(U_0^t)^+b^+U_0^t = b^+ + \mathrm{i}\int_0^\infty \mathrm{e}^{-s/2}a(s)\mathrm{d}s. \qquad \square$$

Lemma 9.4.2 *For* $s, t \in \mathbb{R}$

$$(U_s^t)^+b^+U_s^t = \mathrm{e}^{|t-s|/2}\left(b^+ + \mathrm{i}\int_s^t \mathrm{d}s'\mathrm{e}^{-|s'-s|/2}a(s')\right).$$

Proof Integrate the differential equations, and obtain for $s < t$

$$\begin{aligned}
(U_s^t)^+b^+U_s^t &= \mathrm{e}^{(t-s)/2}b^+ + \mathrm{i}\int_s^t \mathrm{e}^{(t-s')/2}a(s')\mathrm{d}s' \\
&= \mathrm{e}^{(t-s)/2}\left(b^+ + \mathrm{i}\int_s^t \mathrm{e}^{(s-s')/2}a(s')\mathrm{d}s'\right)
\end{aligned}$$

and

$$U_s^t b^+ U_s^{t+} = e^{(t-s)/2} b^+ - i \int_s^t ds' e^{(s'-s)/2} a(s')$$
$$= e^{(t-s)/2} \left(b^+ - i \int_s^t ds' e^{(s'-t)/2} a(s') ds' \right),$$

and for $s > t$, upon interchanging the roles of s and t in the last equation,

$$(U_s^t)^+ b^+ U_s^t = e^{(s-t)/2} \left(b^+ - i \int_t^s e^{(s'-s)/2} a(s') ds' \right)$$
$$= e^{|t-s|/2} \left(b^+ + i \int_s^t ds' e^{-|s'-s|/2} a(s') \right).$$

□

Lemma 9.4.3 *For $r \neq s, t$*

$$[a_r, U_s^t] = \mathbf{1}_{[s,t]}(r) U_r^t (-ib^+) U_s^r,$$

and

$$[U_s^t, a^+(dr)] = \mathbf{1}_{[s,t]}(r) U_r^t (-ib) U_s^r \, dr.$$

Proof Recall from Sect. 9.3 that

$$U_s^t = \sum (-i)^n U_{n,s}^t,$$

and

$$U_{n,s}^t = \int_{\sigma,\tau} u_{n,s}^t(\sigma,\tau) a_\sigma^+ a_\tau \lambda_\tau$$

and also that $u_{n,s}^t(\sigma,\tau) = 0$ for $\#\sigma + \#\tau \neq n$. Calculate, for $s \neq 0, t$,

$$[a_r, U_{n,s}^t] = \int_{\sigma,\tau} [a_r, a_\sigma^+ a_\tau] u_{n,s}^t(\sigma,\tau)\lambda_\tau = \int_{\sigma,\tau} \sum_{c\in\sigma} a_{\sigma\setminus c}^+ a_\tau \varepsilon(r, t_c) u_{n,s}^t(\sigma,\tau)\lambda_\tau$$
$$= \int_{\sigma,\tau,c} \mathbf{1}_{[s,t]}(t_c) u_{n,s}^t(\sigma + c, \tau) \varepsilon(r, t_c) a_\sigma^+ a_\tau \lambda_\tau.$$

Assume $s < r < t$, and introduce

$$N(t_{\sigma+\tau}) = \begin{cases} 1 & \text{if } \{t_{\sigma+\tau}, s, r, t\}^\bullet \text{ has a multiple point,} \\ 0 & \text{otherwise.} \end{cases}$$

As $N(t_{\sigma+\tau})$ is a null function, we have, for $f, g \in \mathscr{K}^*$,

$$
\begin{aligned}
&\langle f|[a_r, U^t_{n,s}]|g\rangle \\
&= \int (1 - N(t_{\sigma+\tau}))\overline{f}(\pi) u^t_{n,s}(\sigma + c, \tau)\varepsilon(r, t_c) g(\rho)\langle a_\pi a^+_\sigma a_\tau a^+_\rho\rangle \lambda_{\pi+\tau} \\
&= \int (1 - N(t_{\sigma+\tau}))\overline{f}(\pi) u^t_{n,s}(t_\sigma + \{r\}, t_\tau) g(\rho)\langle a_\pi a^+_\sigma a_\tau a^+_\rho\rangle \lambda_{\pi+\tau}.
\end{aligned}
$$

Since

$$
u^t_{n,s}(t_\sigma + \{r\}, t_\tau) = \sum_{\substack{n_1+n_2=n \\ \sigma_1+\sigma_2=\sigma \\ \tau_1+\tau_2=\tau}} u^t_{n_2,r}(\sigma_2, \tau_2)(-ib^+) u^r_{n_1,s}(\sigma_1, \tau_1)
$$

we can continue the reckoning with

$$
\begin{aligned}
&= \sum_{n_1+n_2=n} \int (1 - N(t_{\sigma_1+\tau_1+\sigma_2+\tau_2}))\overline{f}(\pi) u^t_{n_2,r}(\sigma_2, \tau_2)(-ib^+) u^r_{n_1,s}(\sigma_1, \tau_1) g(\rho) \\
&\quad \times \langle a_\pi a^+_{\sigma_2} a^+_{\sigma_1} a_{\tau_2} a_{\tau_1} a^+_\rho\rangle \lambda_{\pi+\tau_1+\tau_2} \\
&= \sum_{n_1+n_2=n} \int \overline{f}(\pi) u^t_{n_2,r}(\sigma_2, \tau_2)(-ib^+) u^r_{n_1,s}(\sigma_1, \tau_1) g(\rho) \\
&\quad \times \langle a_\pi a^+_{\sigma_2} a_{\tau_2} a^+_{\sigma_1} a_{\tau_1} a^+_\rho\rangle \lambda_{\pi+\tau_1+\tau_2}
\end{aligned}
$$

as the integrals over all commutators of a_{τ_2} and $a^+_{\sigma_1}$ vanish (see Lemma 8.5.1) and N is again a null function. Finally we have

$$
\langle f|[a_r, U^t_{n,s}]|g\rangle = \sum_{n_1+n_2=n} \langle f|U^t_{n,r}(-ib^+) U^r_{n_2,s}|g\rangle.
$$

By the results of Sect. 9.2 the sum over n converges, and we obtain the first equation of the lemma. The second equation is obtained in a similar way. □

We can use the commutator identities to give formulas for the adjoint action of U^t_s on a and a^+ like those above for b and b^+.

Lemma 9.4.4 *For $r \neq s, t$ with $t > s$*

$$
(U^t_s)^+ a(r) U^t_s = a(r) + \mathbf{1}_{[s,t]}(r)(U^r_s)^+(-ib^+)U^r_s
$$

and

$$
U^t_s a^+(\mathrm{d}r)(U^t_s)^+ = a^+(\mathrm{d}r) + \mathbf{1}_{[s,t]}(r) U^t_r(-ib)(U^t_r)\mathrm{d}r.
$$

With a type of matrix notation we now put all the equations for the adjoint action together in a succinct form. The index s' in the proposition below carries with it an implicit integration over s'.

Proposition 9.4.1 *We have*

$$(U_0^t)^+\begin{pmatrix} b^+ \\ a(s) \end{pmatrix} U_0^t = \begin{pmatrix} V_{00} & (V_{01})_{s'} \\ V_{10} & (V_{11})_{ss'} \end{pmatrix} \begin{pmatrix} b^+ \\ a(s') \end{pmatrix}$$

with

$$\begin{aligned} V_{00} &= \mathrm{e}^{t/2}, \\ (V_{01})_{s'} &= \mathrm{i}\mathbf{1}\{0 < s' < t\}\mathrm{e}^{(t-s')/2}\delta(s-s'), \\ V_{10} &= -\mathrm{i}\mathbf{1}\{0 < s < t\}\mathrm{e}^{s/2}, \\ (V_{11})_{ss'} &= \delta(s-s') + \mathbf{1}\{0 < s' < s < t\}\mathrm{e}^{(s-s')/2}. \end{aligned}$$

Furthermore

$$(U_0^t)^+\begin{pmatrix} b \\ a^+(\mathrm{d}s) \end{pmatrix} U_0^t = \begin{pmatrix} \tilde{V}_{00} & (\tilde{V}_{01})_{s'} \\ \tilde{V}_{10} & (\tilde{V}_{11})_{ss'} \end{pmatrix} \begin{pmatrix} b \\ a^+(\mathrm{d}s') \end{pmatrix}$$

with

$$\begin{aligned} \tilde{V}_{00} &= \mathrm{e}^{t/2}, \\ (\tilde{V}_{01})_{s'} &= -\mathrm{i}\mathbf{1}\{0 < s' < t\}\mathrm{e}^{(t-s')/2}\delta(s-s'), \\ \tilde{V}_{10} &= \mathrm{i}\mathbf{1}\{0 < s < t\}\mathrm{e}^{s/2}, \\ (\tilde{V}_{11})_{ss'} &= \delta(s-s') + \mathbf{1}\{0 < s' < s < t\}\mathrm{e}^{(s-s')/2}. \end{aligned}$$

$\tilde{V}$ can be represented as the solution of a quantum stochastic differential equation. This equation differs from that in Sect. 4.3 and Sect. 8.3.3 by a scaling factor of $\sqrt{2\pi}$.

Proposition 9.4.2 *Define*

$$N = \int a^+(\mathrm{d}s)a(s) = \int \mathrm{d}s a^\dagger(s)a(s);$$

then $N - bb^+$ is an integral of motion, i.e.,

$$(U_0^t)^+(N - bb^+)U_0^t = N - bb^+.$$

Proof For an operator A use the abbreviation

$$A^t = (U_0^t)^+ A U_0^t.$$

Then

$$(bb^+)^t = bb^+ + \int_0^t \left(\mathrm{d}s\,\frac{1}{2}b^s - \mathrm{i}a^+(\mathrm{d}s)\right)(b^+)^s + \int_0^t \mathrm{d}s\, b^s\left(\frac{1}{2}(b^+)^s + \mathrm{i}a(s)\right)$$

and

$$\begin{aligned}N^t &= \int \left(a^+(\mathrm{d}s) + \mathrm{i}b^s \mathbf{1}\{0 < s < t\}\mathrm{d}s\right)\left(a(s) - \mathrm{i}(b^+)^s \mathbf{1}\{0 < s < t\}\right) \\ &= N + \mathrm{i}\int_0^t \mathrm{d}s\, b^s a(s) - \mathrm{i}\int_0^t a^+(\mathrm{d}s)(b^+)^s + \int_0^t \mathrm{d}s\,(bb^+)^s. \qquad \square\end{aligned}$$

Remark 9.4.1 One obtains in the same way, for $s < t$,

$$(U_s^t)^+ b^+ U_s^t = \mathrm{e}^{(t-s)/2} b^+ + \mathrm{i}\int_s^t \mathrm{e}^{(t-s')/2} a(s')\mathrm{d}s'$$

and

$$(U_s^t)^+ (N - bb^+) U_s^t = N - bb^+.$$

Definition 9.4.1 Denote by Γ_k^* the subspace of Γ^* consisting of those functions $f \in \Gamma^*$ for which

$$\|f\|_{\Gamma_k^*} = \langle f | (N + bb^+)^k | f\rangle < \infty.$$

In the definition, we use $(N + bb^+)$, the total number of excitations, because we want an upper bound on functions, and $(N - bb^+)$ leaves things invariant. We need the following theorem only for even k, and formulate it for simplicity just for that case.

Theorem 9.4.1 *The operators U_s^t, for $s, t \in \mathbb{R}$, map each Γ_k^* for $k = 0, 2, 4, \ldots$ into itself, and there exist constants C_k such that for $f \in \Gamma_k^*$ and $s, t \in \mathbb{R}$,*

$$\|U_s^t f\|_{\Gamma_k^*} \le C_k \mathrm{e}^{k|t-s|} \|f\|_{\Gamma_k^*}.$$

Proof With the notation

$$M = N + bb^+,$$

we have

$$\begin{aligned}(U_s^t)^+ M U_s^t &= (U_s^t)^+ (2bb^+ + N - bb^+) U_s^t \\ &= (U_s^t)^+ (2bb^+) U_t^s + N - bb^+ = \mathrm{e}^{|t-s|} A(s,t).\end{aligned}$$

Define

$$f(s,t)(s') = \operatorname{sign}(t-s)\, \mathrm{i}\mathbf{1}_{[s,t]}(s')\mathrm{e}^{-|s-s'|/2}\left(1 - \mathrm{e}^{-|t-s|}\right)^{-1/2}.$$

Then

$$\int \mathrm{d}s' |f(s,t)(s')|^2 = 1,$$

and we consider

$$A(s,t) = \big(b + (1 - \mathrm{e}^{-|s-t|})^{1/2}\big)a^+\big(f(s,t)\big)\big(b^+ + (1 - \mathrm{e}^{-|s-t|})^{1/2}\big)a\big(f(s,t)\big)$$
$$+ \mathrm{e}^{-|s-t|}\big(N - bb^+\big).$$

The operator $A(s,t)$ maps $\mathscr{K}^*$ into itself, and we calculate for $g \in \mathscr{K}^*$,

$$g = \sum b^{+k} |g_{k,m}\rangle,$$

the norm

$$\begin{aligned}
&\big\|M^{l-1}a^+(f)b^+M^{-l}g\big\|^2_{\Gamma^*} \\
&\quad = \sum_{k,m} \big\|M^{l-1}a^+(f)b^+M^{-l}b^k|g_{k,m}\rangle\big\|^2_{\Gamma^*} \\
&\quad \le \sum (k+m+3)^{2(l-1)}(k+1)(m+1)(k+m+1)^{-2l}\big\|b^k|g_{k,m}\rangle\big\|^2_{\Gamma^*} \\
&\quad \le \sum \big((k+m+3)/(k+m+1)\big)^{2l}\big\|b^k|g_{k,m}\rangle\big\|^2_{\Gamma^*} \le 3^{2l}\|g\|^2_{\Gamma^*}.
\end{aligned}$$

We have

$$\begin{aligned}
&\big\|M^{l-1}a(f)b^+M^{-l}g\big\|^2_{\Gamma^*} \\
&\quad = \sum_{k,m} \big\|M^{l-1}a^+(f)b^+M^{-l}b^k|g_{k,m}\rangle\big\|^2_{\Gamma^*} \\
&\quad \le \sum (k+m+1)^{2(l-1)}k(m+1)(k+m+1)^{-2l}\big\|b^k|g_{k,m}\rangle\big\|^2_{\Gamma^*} \\
&\quad \le \sum \big\|b^k|g_{k,m}\rangle\big\|^2_{\Gamma^*} \le \|g\|^2_{\Gamma^*}.
\end{aligned}$$

Similar inequalities hold if one replaces $a(f)b^+$ by $a^+(f)b$, or by $a(f)b$, or by $N - bb^+$. Hence

$$\big\|M^{l-1}A(s,t)M^{-l}g\big\|_{\Gamma^*} \le \big(4 + 3^l\big)\|g\|^2_{\Gamma^*}.$$

One obtains

$$A^kM^{-k}g = AM^{-1}M^1AM^{-2}M^2AM^{-3}\cdots M^{k-1}AM^{-k}g,$$

so

$$\big\|A^kM^{-k}g\big\|_{\Gamma^*} \le (4+3)\big(4+3^2\big)\cdots\big(4+3^k\big)\|g\|_{\Gamma^*} = C_k\|g\|_{\Gamma^*}.$$

Hence

$$M^{-k}A^{2k}M^{-k} \le C_k^2 1_{\Gamma^+}$$

or

$$A^{2k} \le M^{2k} C_k^2.$$

From there follows the result as $\mathscr{K}^*$ is dense in Γ^*. □

9.5 The Hamiltonian

We use the same notation and results as in Sects. 8.8.1 and 8.8.2. Define for $t \in \mathbb{R}$

$$W(t) = \Theta(t) U_0^t = U_{-t}^0 \Theta(t);$$

then $W(t)$ is a unitary strongly continuous one-parameter group on Γ^*. An immediate consequence of Theorem 9.4.1 is

Proposition 9.5.1 *The operators $W(t)$ map the space Γ_k^* into itself and there exist constants C_k such that*

$$\|W(t)f\|_{\Gamma_k^*}^2 \le C_k e^{k|t|} \|f\|_{\Gamma_K}^2.$$

We shall use the notations and results of Sect. 8.8.3.

Definition 9.5.1 For $z \in \mathbb{C}$, $\operatorname{Im} z \neq 0$, we define the resolvent $R(z)$ by

$$R(z) = \begin{cases} -\mathrm{i} \int_0^\infty e^{\mathrm{i}zt} W(t) \mathrm{d}t & \text{for } \operatorname{Im} z > 0, \\ \mathrm{i} \int_{-\infty}^0 e^{\mathrm{i}zt} W(t) \mathrm{d}t & \text{for } \operatorname{Im} z < 0. \end{cases}$$

Furthermore we set

$$S(z) = \begin{cases} -\mathrm{i} \int_0^\infty e^{\mathrm{i}zt} W(t) a(t) & \text{for } \operatorname{Im} z > 0, \\ \mathrm{i} \int_{-\infty}^0 e^{\mathrm{i}zt} W(t) a(t) & \text{for } \operatorname{Im} z < 0 \end{cases}$$

and

$$\kappa(z)(t) = \begin{cases} -\mathrm{i}1\{t > 0\} e^{\mathrm{i}zt - tbb^+/2} & \text{for } \operatorname{Im} z > 0, \\ \mathrm{i}1\{t < 0\} e^{\mathrm{i}zt + tbb^+/2} & \text{for } \operatorname{Im} z < 0 \end{cases}$$

and

$$\tilde{R}(z) = \Theta\big(\kappa(z)\big) = \begin{cases} -\mathrm{i} \int_0^\infty e^{\mathrm{i}zt} e^{-tbb^+/2} \Theta(t) \mathrm{d}t & \text{for } \operatorname{Im} z > 0, \\ +\mathrm{i} \int_{-\infty}^0 e^{\mathrm{i}zt} e^{tbb^+/2} \Theta(t) \mathrm{d}t & \text{for } \operatorname{Im} z < 0. \end{cases}$$

Taking into account the a priori estimate Proposition 9.2.2, one proves, as in Sect. 8.8.3, with $\mathfrak{a}$ and $\mathfrak{a}^+$ defined as in Sect. 8.8.2,

Proposition 9.5.2 *We have for $f \in \mathscr{K}^*$*

$$R(z)f = \tilde{R}(z)\big(f + bS(z)f + \mathfrak{a}^+ b^+ R(z)f\big).$$

Corollary 9.5.1 *If $f \in \mathscr{K}^*$, then*

$$R(z)f = \tilde{R}(z)\big(f_0 + \mathfrak{a}^+ b^+ f_1\big)$$

with

$$f_0 = f + S(z)f$$

and

$$f_1 = R(z)f.$$

Again using the a priori estimate Proposition 9.2.2, the same arguments as in Proposition 8.8.8 establish the following proposition:

Proposition 9.5.3 *For $f \in \mathscr{K}^*$ we have*

$$\hat{\mathfrak{a}} R(z) = S(z) - \mathrm{i}(1/2) b^+ R(z).$$

Definition 9.5.2 The vector space $\hat{D} \subset \Gamma^*$ is defined by

$$\hat{D} = \big\{f = \tilde{R}(z)\big(f_0 + \mathfrak{a}^+ b^+ f_1\big) : f_0 \in \Gamma_2^*, f_1 \in \Gamma_4^*\big\}.$$

We have the following consequence of Corollary 9.5.1:

Proposition 9.5.4 *Recall the constants C_k of Proposition* 9.5.1; *then for* $|\operatorname{Im} z| > C_4$ *the operator $R(z)$ maps $\mathscr{K}^*$ into $\hat{D}$.*

We make at first an *Ansatz* for the Hamiltonian H. Recall Sect. 8.8.3 and the space $\hat{D}^\dagger$ of all semilinear functionals $\hat{D} \to \mathbb{C}$. We have in an analogous way to 8.8.3

$$\hat{D} \subset \Gamma^* \subset \hat{D}^\dagger.$$

Definition 9.5.3 Define an operator $\hat{D} \to \hat{D}^\dagger$ by

$$\hat{H} = \mathrm{i}\hat{\partial} + \mathfrak{a}^+ b^+ + \mathfrak{a} b,$$

or, equivalently, the sesquilinear form $\hat{H}$ on $\hat{D}$ given by

$$\langle f | \hat{H} | g\rangle = \langle f | \mathrm{i}\hat{\partial} | g\rangle + \langle \hat{\mathfrak{a}} b f | g\rangle + \langle f | \hat{\mathfrak{a}} b g\rangle.$$

Proposition 9.5.5 *The operator $\hat{H}$ exists and is symmetric. One obtains for $f = \tilde{R}(z)(f_0 + \mathfrak{a}^+ b^+ f_1)$*

$$\hat{H} f = (\mathrm{i}\hat{\partial} + \hat{\mathfrak{a}}^+ b^+ + \hat{\mathfrak{a}} b) f = -(f_0 + b^+ \hat{\mathfrak{a}}^+ f_1) + \left(z + \frac{\mathrm{i}}{2} b b^+\right) f + \hat{\mathfrak{a}}^+ b^+ f + \hat{\mathfrak{a}} b f.$$

So $\hat{H} f \in \Gamma^*$ if and only if $-f_1 + f = 0$.

Proof An element in Γ^* can be represented in the form

$$f = \sum_{k,m=0}^{\infty} 1/(l!m!) b^{+l} a^+(f_{l,m})|0\rangle = \sum_{l,m=0}^{\infty} 1/(l!m!) b^{+l} |f_{l,m}\rangle$$

with $f_{l,m} \in L(m) = L^2_s(\mathbb{R}^m)$,

$$\|f\|^2_{\Gamma^*} = \sum 1/(l!m!) \|f_{l,m}\|^2_{L(m)}$$

and

$$\|f_{l,m}\|^2_{L(m)} = \int \mathrm{d}t_1 \cdots \mathrm{d}t_m \left| f_{l,m}(t_1, \ldots, t_m) \right|^2.$$

Fix an element $z \in \mathbb{C}$ with $\operatorname{Im} z \neq 0$, and write κ for $\kappa(z)$. One has

$$\tilde{R}(z) f = \Theta(\kappa) f = \sum 1/(l!m!) b^{+l} \Theta(\kappa_l) |f_{k,m}\rangle$$

with

$$\kappa_l(t) = \begin{cases} -\mathrm{i}\mathbf{1}\{t > 0\} \mathrm{e}^{\mathrm{i}zt - (l+1)/2} & \text{for } \operatorname{Im} z > 0, \\ \mathrm{i}\mathbf{1}\{t < 0\} \mathrm{e}^{\mathrm{i}zt + (l+1)/2} & \text{for } \operatorname{Im} z < 0. \end{cases}$$

As κ_l fulfills all conditions for φ and η in the lemmata of Sect. 8.8.2, and $\|\kappa_l\|_{L^2} \leq 1$, we can sum up and obtain that $\Theta(\kappa)\mathfrak{a}^+$ defines a mapping $\Gamma^*_{k-1} \to \Gamma^*_k$ with

$$\left\| \Theta(\kappa) \mathfrak{a}^+ f \right\|_k \leq 2^{k/2} \|f\|_{k-1},$$

and $\mathfrak{a}\Theta(\kappa)$ also defines a mapping $\Gamma^*_{k-1} \to \Gamma^*_k$ with

$$\left\| \Theta(\kappa) \mathfrak{a}^+ f \right\|_k \leq \|f\|_{k-1}.$$

One establishes by arguments similar to those of Sect. 8.8, that for $f \in \hat{D}$ the element $\hat{\mathfrak{a}} f$ is well defined. We calculate for $f \in \hat{D}$

$$\begin{aligned} \mathrm{i}\hat{\partial} f &= -\mathrm{i} \lim \Theta(\varphi'_n) f \\ &= -\mathrm{i} \lim \Theta(\varphi'_n * \kappa)(f_0 + b^+ \mathfrak{a}^+ f_1) = -\mathrm{i} \lim \Theta(\varphi_n * \kappa')(f_0 + b^+ \mathfrak{a}^+ f_1). \end{aligned}$$

Now

$$-\mathrm{i}\varphi_n * \kappa' = -\mathrm{i}\varphi_n * \left(-\mathrm{i}\delta + \left(\mathrm{i}z - \frac{1}{2}bb^+\right)\kappa\right) = -\varphi_n + \varphi_n * \left(z + \frac{\mathrm{i}}{2}bb^+\right)\kappa$$

and

$$\mathrm{i}\hat{\partial} f = -\left(f_0 + \hat{\mathfrak{a}}^+ f_1\right) + \left(z + \frac{\mathrm{i}}{2}bb^+\right) f. \qquad \square$$

Definition 9.5.4 Define

$$D_0 = \{f \in \hat{D} : f = f_1\};$$

denote by H_0 the restriction of $\hat{H}$ to D_0.

Proposition 9.5.6 *For* $f \in \mathscr{K}^*$

$$\hat{H} R(z) f = -f + z R(z) f$$

and $R(z) f \in D_0$.

Proof By Corollary 9.5.1 we have, for $f \in \mathscr{K}^*$ and $\operatorname{Im} z > 4$,

$$R(z) f = \tilde{R}(z)\left(f_0 + \mathfrak{a}^+ b^+ f_1\right),$$
$$f_0 = f + S(z) f,$$
$$f_1 = R(z) f.$$

With the help of Proposition 9.5.5 we obtain

$$\begin{aligned}
\hat{H} f &= \left(\mathrm{i}\hat{\partial} + \hat{\mathfrak{a}}^+ b^+ + \hat{\mathfrak{a}} b\right) f \\
&= -\left(f_0 + b^+ \hat{\mathfrak{a}}^+ f_1\right) + \left(z + \frac{\mathrm{i}}{2}bb^+\right) f + \hat{\mathfrak{a}}^+ b^+ f + \hat{\mathfrak{a}} b f \\
&= -\left(f + bS(z) f + \hat{\mathfrak{a}}^+ b^+ R(z) f\right) + \left(z + \frac{\mathrm{i}}{2}bb^+\right) R(z) f \\
&\quad + \hat{\mathfrak{a}}^+ b^+ R(z) f + bS(z) f - \frac{\mathrm{i}}{2} bb^+ R(z) f \\
&= -f + z R(z) f.
\end{aligned} \qquad \square$$

Just as for Theorem 8.8.1, we obtain with the help of Proposition 3.1.9,

Theorem 9.5.1 *The domain D of the Hamiltonian H of $W(t)$ contains D_0 and the restriction of H to D_0 coincides with H_0, the restriction of*

$$\hat{H} = \mathrm{i}\hat{\partial} + \hat{\mathfrak{a}}^+ b^+ + \hat{\mathfrak{a}} b$$

to D_0; furthermore, D_0 is dense in Γ and H is the closure of H_0.

9.6 Amplification

The amplified oscillator yields a model for a photo multiplier. Recall the Fourier-Weyl transform. If ρ is a density operator on Γ, then the Fourier-Weyl transform is given by

$$\mathscr{W}(\rho)(\varphi, z) = \text{Trace } \rho e^{i(a(\varphi)+a^+(\varphi)+zb+\bar{z}b^+)}$$

for $\varphi \in \mathscr{K}(\mathbb{R})$ and $z \in \mathbb{C}$. The time development of ρ is given by

$$\rho(t) = U_0^t \rho \left(U_0^t\right)^+.$$

Hence

$$\mathscr{W}\big(\rho(t)\big)(z, \varphi) = \text{Trace } \rho \exp\big(i(U_0^t)^+\big(a(\varphi) + a^+(\varphi) + zb + \bar{z}b^+\big)U_0^t\big).$$

According to Lemma 9.4.1 and Lemma 9.4.4, we have

$$\begin{aligned}
(U_0^t)^+ b^+ U_0^t &= e^{t/2}\left(b^+ + i\int_0^t ds\, e^{-s/2} a(s)\right),\\
(U_0^t)^+ a(s) U_0^t &= a(s) + \mathbf{1}\{0 < s < t\}(U_0^s)^+ b^+ U_0^s\\
&= a(s) + \mathbf{1}\{0 < s < t\} e^{s/2}\left(b^+ + i\int_0^s ds' e^{-s'/2}\right) a(s'),\\
(U_0^t)^+ a(\varphi) U_0^t &= \int ds\, \varphi(s) (U_0^t)^+ a(s) U_0^t\\
&= a(\varphi) + \int_0^t ds\, \varphi(s) e^{s/2}\left(b^+ + i\int_0^s ds' e^{-s'/2} a(s')\right).
\end{aligned}$$

For $t \to \infty$

$$e^{-t/2}(U_0^t)^+ b^+ U_0^t \to b^+ + i\int_0^\infty ds\, e^{-s/2} a(s) = b^+ + a(\psi)$$

with

$$\psi(t) = -i\mathbf{1}\{t > 0\} e^{-t/2}$$

and

$$e^{-t/2}(U_0^t)^+ a(\varphi) U_0^t \to 0,$$

since

$$e^{-t/2}\int_0^t ds\, e^{s/2}\varphi(s) \to 0.$$

So

$$\mathscr{W}\big(\rho(t)\big)\big(e^{-t/2}z, e^{-t/2}\varphi\big) \to \text{Trace } \rho e^{iz(b+a^+(\psi))+\bar{z}(b^+ + a(\psi))}.$$

As

$$[(b + a^+(\psi)), b^+ + a(\psi)] = 0,$$

the last expression can be understood as the Fourier transform of a classical probability measure p on the complex plane given by

$$\int p(\mathrm{d}\xi)\mathrm{e}^{\mathrm{i}z\xi + \mathrm{i}\bar{z}\bar{\xi}} = \hat{p}(z) = \text{Trace } \rho \mathrm{e}^{\mathrm{i}z(b+a^+(\psi)) + \bar{z}(b^+ + a(\psi))}.$$

So p may be understood as an amplification of ρ.

Examples

- Assume $\rho = \rho_0 \otimes |\emptyset\rangle\langle\emptyset|$, where ρ_0 is the initial density matrix of the oscillator and $|\emptyset\rangle$ is the ground state of the heat bath,

$$\hat{p}(z) = \mathrm{e}^{-|z|^2/2}\text{Trace } \rho_0 \mathrm{e}^{\mathrm{i}(zb + \bar{z}b^+)} = \mathrm{e}^{-|z|^2/2}\hat{\rho}_0(z),$$

$$p(\mathrm{d}\xi) = (2/\pi)\int \text{Wigner}(\rho_0)(\eta)\mathrm{e}^{-2|\xi-\eta|^2}\mathrm{d}\eta\mathrm{d}\xi,$$

  where Wigner(ρ_0) is the Wigner transform of ρ_0. So p is the Wigner transform of ρ_0 smeared out with a Gaussian distribution.
- Assume $\rho = |0\rangle\langle 0| \otimes |\emptyset\rangle\langle\emptyset|$, the ground state of both oscillator and heat bath; then

$$p(\mathrm{d}\xi) = (1/\pi)\mathrm{e}^{-|\xi|^2}\mathrm{d}\xi.$$

- Assume a coherent state $\rho_0 = |\psi\rangle\langle\psi|$

$$\psi = \mathrm{e}^{-|\beta|^2/2}\mathrm{e}^{\beta b^+}|0\rangle,$$

  then

$$p(\mathrm{d}\xi) = (1/\pi)\mathrm{e}^{-|\xi-\beta|^2}\mathrm{d}\xi,$$

  the translated probability measure for the vacuum. So we recover β with an additional uncertainty.

9.7 The Classical Yule-Markov Process

The Yule process is a pure birth process. Individuals live forever. For each individual living at time t, during the period from t to $t + \mathrm{d}t$ there is a chance equal $\mathrm{d}t$ of having a child. Thus each individual gives birth at rate 1. The state space is $\mathbb{N}_0 = \{0, 1, 2, \ldots\}$. If $Z(t)$ is the random number of individuals at time t, then the conditional probability

$$p_{m,l}(t - s) = \mathbb{P}\{Z(t) = m | Z(s) = l\}$$

obeys the differential equation

$$\frac{\mathrm{d}p_{m+1,l}(t-s)}{\mathrm{d}t} = -(m+1)p_{m+1,l}(t-s) + mp_{m,l}(t-s).$$

Hence

$$\begin{aligned} p_{m+1,l}(t-s) = \int \cdots \int_{s<s_1<\cdots<s_{m-l}<t} & \mathrm{d}s_1 \cdots \mathrm{d}s_{m-l} \\ & \times \mathrm{e}^{-(t-s_{m-l})(m+1)} m \mathrm{e}^{-(s_{m-l}-s_{m-l-1})m}(m-1)\cdots \\ & \times \mathrm{e}^{(s_2-s_1)(l+1)}(l+1)\mathrm{e}^{-(l+1)(s_1-s)}. \end{aligned}$$

For the following discussion it is convenient to introduce the vectors

$$\eta_m = \sqrt{1/m!}\, b^{+m}|0\rangle.$$

They form an orthonormal basis of $L^2(\mathbb{R})$ and

$$\begin{aligned} \langle \eta_l | \eta_m \rangle &= \delta_{lm}, \\ b^+ \eta_m &= \sqrt{m+1}\eta_{m+1}, \\ b\eta_m &= \sqrt{m}\eta_{m-1}. \end{aligned}$$

Consider

$$U_s^t \eta_l \Phi = \sum_{j,m} (1/j!)\eta_m \int f_{jm}(s_1,\ldots,s_j)a^+(\mathrm{d}s_1)\cdots a^+(\mathrm{d}s_j)\Phi,$$

where Φ is the vacuum state of the heat bath, and note that $m = l + j$, so we have

$$\begin{aligned} f_{jm}(s_1,\ldots,s_j) &= \langle \eta_m | u_s^t(\{s_1,\ldots,s_j\},\emptyset)|\eta_l\rangle \\ &= (-\mathrm{i})^j \langle \eta_m | \mathrm{e}^{-bb^+(t-s_j)/2} b^+ \cdots b^+ \mathrm{e}^{-bb^+(s_2-s_1)/2} b^+ \mathrm{e}^{-(s_1-s)}\eta_l \rangle \\ &= (-\mathrm{i})^j \mathrm{e}^{-(m+1)(t-s_j)/2}\sqrt{m}\cdots\sqrt{l+2}\mathrm{e}^{-(l+2)(s_2-s_1)/2} \\ &\quad \times \sqrt{l+1}\mathrm{e}^{-(l+1)(s_1-s)/2}. \end{aligned}$$

Now look at the coefficient of η_m. We have

$$\begin{aligned} &\left\| (1/j!) \int f_{jm}(s_1,\ldots,s_j)a^+(\mathrm{d}s_1)\cdots a^+(\mathrm{d}s_j)\Phi \right\|_\Gamma^2 \\ &\quad = (1/j!) \int \mathrm{d}s_1 \cdots \mathrm{d}s_j \left| f_{jm}(s_1,\ldots,s_j) \right|^2 \\ &\quad = \int_{s<s_1<\cdots<s_j<t} \mathrm{d}s_1 \cdots \mathrm{d}s_j \left| \langle \eta_m | u_s^t(\{s_1,\ldots,s_j\},\emptyset)\eta_l \rangle \right|^2 = p_{l,m}(t-s). \end{aligned}$$

Consider $\eta_n = \eta_n \otimes \mathrm{id}$ and $\eta_n^+ = \eta_n^+ \otimes \mathrm{id}$ as operators, and define

$$X_n = \eta_n \eta_n^+,$$
$$X_n(t) = \left(U_0^t\right)^+ X_n U_0^t = U_t^0 X_n U_0^t.$$

Lemma 9.7.1 *Write $\Phi = \mathrm{id} \otimes \Phi$ for short, then for $s < t$*

$$\langle \Phi | U_t^s X_m U_s^t | \Phi \rangle = \sum_l p_{m,l}(t-s) X_l$$

where $\langle \Phi |$ stands for $\langle \Phi | \otimes \mathrm{id}$, and $| \Phi \rangle = \mathrm{id} \otimes | \Phi \rangle$.

Proof We have

$$\langle \Phi \otimes \eta_n | U_t^s X_m U_s^t | \eta_l \otimes \Phi \rangle = \langle \Phi \otimes \eta_n | \left(U_s^t\right)^+ \eta_m^+ \eta_m U_s^t | \eta_l \otimes \Phi \rangle$$

and

$$\eta_m U_s^t | \eta_l \otimes \Phi \rangle = \int_\sigma \langle \eta_m | u_s^t(\sigma, \emptyset) | \eta_l \rangle a_\sigma^+ \Phi.$$

Hence

$$\langle \Phi \otimes \eta_n | U_t^s X_m U_s^t | \eta_l \otimes \Phi \rangle = \int_{\sigma,\tau} \overline{\langle \eta_m | u_s^t(\tau, \emptyset) | \eta_n \rangle} \langle \eta_m | u_s^t(\sigma, \emptyset) | \eta_l \rangle \langle \Phi | a_\tau a_\sigma^+ | \Phi \rangle \lambda_\tau.$$

We have $\#\sigma = \#\tau = m - n = m - l$, hence $l = n$, unless the expression vanishes. We continue with

$$= \int_\sigma \lambda_\sigma \left| \langle \eta_m | u_s^t(\sigma, \emptyset) | \eta_l \rangle \right|^2 | \eta_l \rangle \langle \eta_l | \delta_{ln} = p_{m,l}(t-s) \delta_{ln}.$$

□

Theorem 9.7.1 *If ρ_0 is a density matrix on l^2, and $0 < t_1 < \cdots < t_p$, then*

$$\begin{aligned}
&\mathrm{Tr}\left(\left(\rho_0 \otimes |\Phi\rangle\langle\Phi| X_{m_1'}(t_1) \cdots X_{m_{p-1}'}(t_{p-1})\right) X_{m_p}(t_p) X_{m_{p-1}}(t_{p-1}) \cdots X_{m_1}(t_1)\right) \\
&= \delta_{m_1,m_1'} \cdots \delta_{m_{p-1},m_{p-1}'} \mathbb{P}_\pi \left\{ Z(t_1) = m_1, \ldots, Z(t_p) = m_p \right\} \\
&= \sum_l \delta_{m_1,m_1'} \cdots \delta_{m_{p-1},m_{p-1}'} p_{m_p,m_{p-1}}(t_p - t_{p-1}) \cdots p_{m_2,m_1}(t_2 - t_1) p_{m_1,l}(t_1) \pi_l
\end{aligned}$$

where π is the initial distribution of the Yule process,

$$\pi_l = \mathbb{P}\left\{ Z(0) = l \right\} = \langle \eta_l | \rho_0 | \eta_l \rangle.$$

Proof We carry out the proof by induction over p. For $p = 1$, we have

$$\mathrm{Tr}\left(\rho_0 \otimes |\Phi\rangle\langle\Phi| X_m(t)\right) = \mathrm{Tr}\left(\rho_0 \langle \Phi | U_t^0 X_m U_0^t | \Phi \rangle\right)$$

$$= \mathrm{Tr}\rho_0\Big(\sum p_{m,l}(t)X_l\Big)$$
$$= \sum p_{m,l}(t)\mathrm{Tr}\rho_0 X_l = \sum \pi_l p_{m,l}(t).$$

We calculate

$$\mathrm{Tr}\big((\rho_0 \otimes |\Phi\rangle\langle\Phi| X_{m'_1}(t_1)\cdots X_{m'_{p-1}}(t_{p-1})\big)X_{m_p}(t_p)X_{m_{p-1}}(t_{p-1})\cdots X_{m_1}(t_1)\big)$$
$$= \mathrm{Tr}\ \rho_0\langle\Phi|U_{t_1}^0 X_{m_1} U_{t_2}^{t_1}\cdots U_{t_{p-1}}^{t_{p-2}} X_{m'_{p-1}} U_{t_p}^{t_{p-1}} X_{m_p} U_{t_{p-1}}^{t_p}\cdots U_{t_1}^{t_2} X_{m_1} U_0^{t_1}|\Phi\rangle$$
$$= \int \mathrm{Tr}\ \rho_0\big(u_{t_1}^0(\emptyset,\tau_1)X_{m_1}u_{t_2}^{t_1}(\emptyset,\tau_2)\cdots u_{t_{p-1}}^{t_{p-2}}(\emptyset,\tau_{p-1})X_{m'_{p-1}}u_{t_p}^{t_{p-1}}(\emptyset,\tau_p)$$
$$\times X_{m_p}u_{t_{p-1}}^{t_p}(\sigma_p,\emptyset)\cdots u_{t_1}^{t_2}(\sigma_2,\emptyset)X_{m_1}u_0^{t_1}(\sigma_1,\emptyset)\big)$$
$$\times \langle\Phi|a_{\tau_1+\cdots+\tau_p}a^+_{\sigma_1+\cdots+\sigma_p}|\Phi\rangle\lambda_{\tau_1+\cdots+\tau_p}.$$

Here, for $t < s$, we have

$$U_t^s = \int u_t^s(\sigma,\tau)a_\sigma^+ a_\tau \lambda_\tau$$

and

$$u_t^s(\sigma,\tau) = \big(u_s^t\big)^+(\sigma,\tau) = \big(u_s^t(\tau,\sigma)\big)^+.$$

As $t_{\tau_1+\cdots+\tau_{p-1}} \subset [0,t_{p-1}]$ and $t_{\tau_p} \subset [t_{p-1},t_p]$, and $t_{\sigma_1+\cdots+\sigma_{p-1}} \subset [0,t_{p-1}]$ and $t_{\sigma_p} \subset [t_{p-1},t_p]$, we may, under the integral, replace (see Lemma 8.5.1)

$$\langle\Phi|a_{\tau_1+\cdots+\tau_p}a^+_{\sigma_1+\cdots+\sigma_p}|\Phi\rangle\lambda_{\tau_1+\cdots+\tau_p}$$

by

$$\langle\Phi|a_{\tau_1+\cdots+\tau_{p-1}}a^+_{\sigma_1+\cdots+\sigma_{p-1}}|\Phi\rangle\lambda_{\tau_1+\cdots+\tau_{p-1}}\langle\Phi|a_{\tau_p}a^+_{\sigma_p}|\Phi\rangle\lambda_{\tau_p}.$$

We split the integral, and perform first the integral over the second factor to obtain

$$\int u_{t_p}^{t_{p-1}}(\emptyset,\tau_p)X_{m_p}u_{t_{p-1}}^{t_p}(\sigma_p,\emptyset)\langle\Phi|a_{\tau_p}a^+_{\sigma_p}|\Phi\rangle\lambda_{\tau_p}$$
$$= \langle\Phi|U_{t_p}^{t_{p-1}}X_{m_p}U_{t_{p-1}}^{t_p}|\Phi\rangle = \sum_l X_l p_{m_p,l}(t_p - t_{p-1}).$$

We insert this result into the integral and obtain

$$= \sum_l p_{m_p,l}(t_p - t_{p-1})\int \mathrm{Tr}\ \rho_0\big(u_{t_1}^0(\emptyset,\tau_1)X_{m_1}u_{t_2}^{t_1}(\emptyset,\tau_2)\cdots u_{t_{p-1}}^{t_{p-2}}(\emptyset,\tau_{p-1})$$
$$X_{m'_{p-1}}X_l X_{m_{p-1}}u_{t_{p-2}}^{t_{p-1}}(\sigma_{p-1},\emptyset)\cdots u_{t_1}^{t_2}(\sigma_2,\emptyset)X_{m_1}u_0^{t_1}(\sigma_1,\emptyset)\big)$$

$$\langle \Phi | a_{\tau_1+\cdots+\tau_{p-1}} a^+_{\sigma_1+\cdots+\sigma_{p-1}} | \Phi \rangle \lambda_{\tau_1+\cdots+\tau_{p-1}}.$$

Using

$$X_{m'_{p-1}} X_l X_{m_{p-1}} = \delta_{m_p, m'_{p-1}} \delta_{m_{p-1}, l} X_{m_{p-1}}$$

one finishes the proof. □

Corollary 9.7.1 *We have by the result above, that for* $t_1 < \cdots < t_p$

$$\left\| X_{m_p}(t_p) \cdots X_{m_1}(t_1) | \eta_l \otimes \Phi \rangle \right\|_\Gamma^2 = \mathbb{P}_l \{ Z(t_p) = m_p, \ldots, Z(t_1) = m_1 \},$$

where $\mathbb{P}_l$ *is the probability distribution of the Yule process starting at* l *at time* 0.

Chapter 10
Approximation by Coloured Noise

Abstract We show that the Hudson-Parthasarathy equation can be approximated by coloured noise using the singular coupling limit.

10.1 Definition of the Singular Coupling Limit

We recall the Hudson-Parthasarathy quantum stochastic differential equation (QSDE)

$$\partial_t U(t) = A_1 a^\dagger(t) U(t) + A_0 a^\dagger(t) U(t) a(t) + A_{-1} U(t) a(t) + B U(t),$$
$$U(0) = 1$$

where A_i $(i = -1, 0, 1)$ and B are in $B(\mathfrak{k})$. Assuming that $U(t)$ is a power series in a and a^+, we write

$$\partial_t U(t) = {:}K(t)U(t){:}$$
$$K(t) = A_1 a^\dagger(t) + A_0 a^\dagger(t) a(t) + A_{-1} a(t) + B,$$

where $:\cdots:$ stands for normal ordering, and is also denoted by $\mathbb{O}_a \cdots$. This is Accardi's *normal ordered* form of the QSDE. The solution was given by an infinite series in Sect. 8.2:

$$U(t) = 1 + \sum_{n=1}^{\infty} \int \cdots \int_{0<t_1<\cdots<t_n<t} \mathrm{d}t_1 \cdots \mathrm{d}t_n {:}K(t_n) \cdots K(t_n){:}.$$

Recall

$$\mathcal{K} = \mathcal{K}_s(\mathfrak{R}, \mathfrak{k}).$$

If $f, g \in \mathcal{K}$, then $\langle f | U(t) | g \rangle$ is well defined, as the infinite sum on the right-hand side contains only finitely many terms.

Quantum white noise is called "white", because the correlation function

$$\langle \emptyset | a(s) a^+(\mathrm{d}t) | \emptyset \rangle = \varepsilon_s(\mathrm{d}t),$$

W. von Waldenfels, *A Measure Theoretical Approach to Quantum Stochastic Processes*, Lecture Notes in Physics 878, DOI 10.1007/978-3-642-45082-2_10,

or, again introducing $a^+(\mathrm{d}t) = a^\dagger(t)\mathrm{d}t$,

$$\langle\emptyset|a(s)a^\dagger(t)|\emptyset\rangle = \delta(t-s)$$

and the δ-function has a white spectrum, i.e., its Fourier transform is constant. "Coloured" noise means, that the spectrum of the correlation function is not white. We will understand by coloured noise that we make the replacements

$$\begin{aligned}
a(t) &\Rightarrow a(\varphi^t),\\
a^\dagger(t) &\Rightarrow a^+(\varphi^t),\\
a^+(\mathrm{d}t) &\Rightarrow a^+(\varphi^t)\mathrm{d}t,
\end{aligned}$$

where φ is a complex-valued continuous function on the real line, and

$$\varphi^t(s) = \varphi(s-t).$$

Then we define

$$\begin{aligned}
a(\varphi^t) &= \int \mathrm{d}s\overline{\varphi}^t(s)a(s),\\
a^+(\varphi_t) &= \int \mathrm{d}s\varphi^t(s)a^\dagger(s) = \int \mathrm{d}s\varphi^t(s)a^+(\mathrm{d}s).
\end{aligned}$$

So

$$\begin{aligned}
&a(\varphi^t) : \mathscr{K} \to \mathscr{K}, &&\big(a(\varphi^t)f\big)(t_1,\ldots,t_n) = \int \mathrm{d}t\,\overline{\varphi}^t(t_0)f(t_0,t_1,\ldots,t_n)\\
&a^+(\varphi^t) : \mathscr{K} \to \mathscr{K}, &&\big(a^+(\varphi^t)f\big)(t_1,\ldots,t_n) = \varphi^t(t_1)f(t_2,\ldots,t_n) + \cdots\\
& && \qquad + \varphi^t(t_n)f(t_1,\ldots,t_{n-1}).
\end{aligned}$$

The quantities $a(\varphi^t)$ and $a^+(\varphi^t)$ are called *coloured noise operators*. The correlation function is

$$\langle\emptyset|a(\varphi^s)a^+(\varphi^t)|\emptyset\rangle = \int \mathrm{d}r\,\overline{\varphi}(r-t)\varphi(r-s) = k(t-s)$$

and its Fourier transform is

$$\int \mathrm{d}t\,\mathrm{e}^{\mathrm{i}\omega t}k(t) = \left|\int \mathrm{d}t\,\varphi(t)\mathrm{e}^{\mathrm{i}\omega t}\right|^2,$$

a function vanishing at ∞.

We want to perform the *singular coupling limit* limit already used in Chap. 4, and put

$$\varphi_\varepsilon^t(s) = \frac{1}{\varepsilon}\varphi\left(\frac{s-t}{\varepsilon}\right).$$

For $\varepsilon \downarrow 0$, one obtains

$$\begin{aligned}\varphi_\varepsilon^t(s) &\to \gamma\delta_t(s) = \gamma\delta(s-t),\\ a(\varphi_\varepsilon^t) &\to \overline{\gamma}a(\delta_t) = \overline{\gamma}a(t),\\ a^+(\varphi_\varepsilon^t) &\to \gamma a^+(\delta_t) = \gamma a^+(t),\\ \gamma &= \int \varphi(s)\,\mathrm{d}s.\end{aligned}$$

Their correlation functions are

$$\begin{aligned}&\langle\emptyset|a(\varphi_\varepsilon^s)a^+(\varphi_\varepsilon^t)|\emptyset\rangle = k_\varepsilon(t-s),\\ &k_\varepsilon(t-s) = \int \mathrm{d}r\,\overline{\varphi}_\varepsilon^t(r)\varphi_\varepsilon^s(r)\mathrm{d}r = k_\varepsilon(t-s),\\ &k_\varepsilon(t-s) = \frac{1}{\varepsilon}k\left(\frac{t-s}{\varepsilon}\right) \to \left(\int \mathrm{d}r\,k(r)\right)\delta(t-s),\\ &\int \mathrm{d}r\,k(r) = |\gamma|^2.\end{aligned}$$

10.2 Approximation of the Hudson-Parthasarathy Equation

We investigate for $\varepsilon \downarrow 0$ the solution of the differential equation

$$\begin{aligned}&\partial_t U_\varepsilon(t) = H_\varepsilon(t)U_\varepsilon(t),\\ &U_\varepsilon(0) = 1,\\ &H_\varepsilon(t) = M_1 a^+(\varphi_\varepsilon^t) + M_0 a^+(\varphi_\varepsilon^t)a(\varphi_\varepsilon^t) + M_{-1}a(\varphi_\varepsilon^t).\end{aligned}$$

Theorem 10.2.1 *Assume*

$$\|M_0\| \int \mathrm{d}t\,|\varphi(t)|^2/2 < 1.$$

Then for $\varepsilon \downarrow 0$ and $f, g \in \mathscr{K}$,

$$\langle f|U_\varepsilon(t)|g\rangle \to \langle f|U(t)|g\rangle$$

such that $U(t)$ satisfies a QSDE with right-hand side $K(t)$ of standard form, which can be explicitly give as

$$\partial_t U(t) = :K(t)U(t):, \qquad K(t) = A_1 a^\dagger(t) + A_0 a^\dagger(t)a(t) + A_{-1}a(t) + B$$

with

$$A_1 = \frac{\gamma}{1-\kappa M_0} M_1, \qquad A_0 = \frac{|\gamma|^2 M_0}{1-\kappa M_0}, \qquad A_{-1} = M_{-1}\frac{\overline{\gamma}}{1-\kappa M_0},$$

$$B = M_{-1}\frac{\kappa}{1-\kappa M_0} M_1$$

and

$$\gamma = \int \mathrm{d}t\, \varphi(t), \qquad \kappa = \int_0^\infty k(t)\mathrm{d}t = \int_0^\infty \mathrm{d}t \int \mathrm{d}s\, \overline{\varphi}(s-t)\varphi(s).$$

We use a trick common in quantum field theory and introduce artificial time dependence in the M_i, and so we write $M_i(t)$ instead of M_i. Then we define

$$H_\varepsilon(t) = M_1(t) a^+\big(\varphi_\varepsilon^t\big) + M_0(t) a^+\big(\varphi_\varepsilon^t\big) a\big(\varphi_\varepsilon^t\big) + M_{-1}(t) a\big(\varphi_\varepsilon^t\big).$$

Lemma 10.2.1 *Assume $t_n > \cdots > t_1$, then*

$$H_\varepsilon(t_n)\cdots H_\varepsilon(t_1) = \sum_{\{I_1,\ldots,I_m\}\in\mathfrak{P}_n} \mathbb{O}_t{:}L(I_1)\cdots L(I_m){:}.$$

We denote by $\mathfrak{P}_n$ the set of all partitions of $[1,n]$. We put

$$L\big(\{t_1\}\big) = H_\varepsilon(t_1),$$

and, for $l \geq 2$,

$$\begin{aligned} &L\big(\{t_1,\ldots,t_l\}\big) \\ &\quad = \big(M_0(t_l)\cdots M_0(t_2) M_1(t_1) a^+\big(\varphi_\varepsilon^{t_l}\big) + M_0(t_l)\cdots M_0(t_1) a^+\big(\varphi_\varepsilon^{t_l}\big) a\big(\varphi_\varepsilon^{t_1}\big) \\ &\qquad + M_{-1}(t_l) M_0(t_{l-1})\cdots M_0(t_1) a\big(\varphi_\varepsilon^{t_1}\big) + M_{-1}(t_l) M_0(t_{l-1})\cdots M_0(t_2) M_1(t_1)\big) \\ &\qquad \times k_\varepsilon(t_l - t_{l-1})\cdots k_\varepsilon(t_2 - t_1). \end{aligned}$$

$\mathbb{O}_t$ is the time-ordering operator for the M_i and the $k(t_j - t_i)$, i.e., all monomials in M_i are ordered in such a way that the first factor is dependent on t_n and so on down to the last one on t_1, and similarly $k(t_i - t_j)$ becomes $k(t_{\max(i,j)} - t_{\min(i,j)})$.

Proof We perform the proof by induction. The case $n = 1$ is clear. We proceed from $n-1$ to n. For simplicity we drop the indices ε and write $a(\varphi_\varepsilon^{t_j}) = a_j$ and $a^+(\varphi_\varepsilon^{t_j}) = a_j^+$. Remark, that the a_i and a_j^+ inside normal ordering $:\cdots:$ commute. Then

$$\begin{aligned} &H(t_n)H(t_{n-1})\cdots H(t_1) \\ &\quad = \big(M_1(t_n)a_n^+ + M_0(t_n)a_n^+ a_n + M_{-1}(t_n)a_n\big) \sum_{\{J_1,\ldots,J_p\}\in\mathfrak{P}_{n-1}} \mathbb{O}_t{:}L(J_1)\cdots L(J_p){:} \end{aligned}$$

$$= \mathbb{O}_t \sum_{\{J_1,\dots,J_p\}\in\mathfrak{P}_{n-1}} \big({:}\big(M_1(t_n)a_n^+ + M_0(t_n)a_n^+a_n + M_{-1}(t_n)a_n\big)L(J_1)\cdots L(J_p){:}$$
$$+ \big(M_0(t_n)a_n^+ + M_{-1}(t_n)\big)\big[a_n, {:}L(J_1)\cdots L(J_p){:}\big]\big)$$
$$= \mathbb{O}_t \sum_{\{J_1,\dots,J_p\}\in\mathfrak{P}_{n-1}} \Bigg({:}L\big(\{n\}\big)L(J_1)\cdots L(J_p){:}$$
$$+ \big(M_0(t_n)a_n^+ + M_{-1}(t_n)\big)$$
$$\times \sum_{j=1}^{p} {:}L(J_1)\cdots L(J_{j-1})\big[a_n, L(J_j)\big]L(J_{j+1})\cdots L(J_p){:}\Bigg).$$

Now

$$\big(M_0(t_n)a_n^+ + M_{-1}(t_n)\big)\big[a_n, L\big(\{t_1,\dots,t_l\}\big)\big]$$
$$= \big(M_0(t_n)a_n^+ + M_{-1}(t_n)\big)\big(M_0(t_l)\cdots M_0(t_2)M_1(t_1)$$
$$+ M_0(t_l)\cdots M_0(t_2)M_0(t_1)a_1\big)k(t_n - t_l)k(t_l - t_{l-1})\cdots k(t_2 - t_1)$$
$$= L\big(\{t_n, t_l, \dots, t_1\}\big).$$

So

$$H(t_n)H(t_{n-1})\cdots H(t_1)$$
$$= \mathbb{O}_t \sum_{\{J_1,\dots,J_p\}\in\mathfrak{P}_{n-1}} \Bigg({:}L\big(\{n\}\big)L(J_1)\cdots L(J_p)$$
$$+ \sum_{j=1}^{p} L(J_1)\cdots L(J_{j-1})L\big(J_j + \{n\}\big)L(J_{j+1})\cdots L(J_p){:}\Bigg).$$

As any partition of $[1, n]$ contains either n as a singleton, or is contained in an element of a partition of $[1, n-1]$, one obtains the result. □

Lemma 10.2.2 *Write, for $I \subset [1, n]$, $\#I = l$,*

$$P(I)(t) = \frac{1}{l!}\mathbb{O}_t \int_0^t \mathrm{d}t_1 \cdots \mathrm{d}t_l \, L(I)(t_l, \dots, t_1)$$

then, if $\#I_i = n_i$,

$$\int_{0<t_1<\cdots<t_n<t} \mathrm{d}t_1 \cdots \mathrm{d}t_n \, H_\varepsilon(t_n)\cdots H_\varepsilon(t_1)$$
$$= \sum_{\{I_1,\dots,I_p\}\in\mathfrak{P}_n} \frac{n_1!\cdots n_p!}{n!}\mathbb{O}_t{:}P(I_p)(t)\cdots P(I_1)(t){:}.$$

Proof As

$$\sum_{\{I_1,\dots,I_p\}\in\mathfrak{P}_n} \mathbb{O}_t{:}L(I_1)\cdots L(I_p){:} = \sum_{\{I_1,\dots,I_p\}\in\mathfrak{P}_n} \mathbb{O}_t{:}\big(\mathbb{O}_t\big(L(I_1)\big)\cdots\mathbb{O}_t\big(L(I_p)\big)\big){:}\,.$$

is symmetric in $t_1,\dots,t_n$ we obtain

$$\int_{0<t_1<\cdots<t_n<t} \mathrm{d}t_1\cdots\mathrm{d}t_n\, H_\varepsilon(t_n)\cdots H_\varepsilon(t_1)$$
$$= \frac{1}{n!}\int_0^t\cdots\int_0^t \mathrm{d}t_1\cdots\mathrm{d}t_n\,\mathbb{O}_t \sum_{\{I_1,\dots,I_p\}\in\mathfrak{P}_n} {:}\mathbb{O}_t\big(L(I_1)\big)\cdots\mathbb{O}_t\big(L(I_m)\big){:}.$$

□

We split partitions as

$$\mathfrak{P} = \mathfrak{P}_n' + \mathfrak{P}_n'',$$

where $\mathfrak{P}_n'$ is the set of non-overlapping partitions, and $\mathfrak{P}_n''$ is the set of overlapping partitions.

Lemma 10.2.3 *We have*

$$\int_{0<t_1<\cdots<t_n<t} \sum_{\{I_1,\dots,I_p\}\in\mathfrak{P}_n'} {:}L(I_1)\cdots L(I_p){:}$$
$$= \sum_{n_1+\cdots+n_p=n}\int_{0<s_1<\cdots<s_p<t} \mathrm{d}s_1\cdots\mathrm{d}s_p\,{:}F_{n_p}(s_p,s_{p-1})$$
$$\times F_{n_{p-1}}(s_{p-1},s_{p-2})\cdots F_{n_1}(s_1,0){:}$$

with

$$F_l(s,r) = \int_{r<t_1<\cdots<t_{l-1}<s} \mathrm{d}t_1\cdots\mathrm{d}t_{l-1}\, L\big(\{s,t_{l-1},\dots,t_1\}\big).$$

Proof If $\{I_1,\dots,I_p\}\in\mathfrak{P}_n'$, then it is of the form

$$I_1 = [1,n_1],\dots,I_p = [n_1+\cdots+n_{p-1}, n_1+\cdots n_p = n].$$

Put

$$t_1 = t_{11},\dots,t_{n_1-1} = t_{1,n_1-1},\qquad t_{n_1} = s_1,$$
$$t_{n_1+1} = t_{2,1},\dots,t_{n_1+n_2-1} = t_{2,n_2-1},\qquad t_{n_1+n_2} = s_2,$$
$$\vdots$$
$$t_{n_1+\cdots+n_{p-1}+1} = t_{p,1},\dots,t_{n-1} = t_{p,n_p-1},\qquad t_n = s_p,$$

and obtain the result.

□

Lemma 10.2.4 *Assume, that the M_i are independent of the t_i, then for $\varepsilon \downarrow 0$*

$$\int_{0<t_1<\cdots<t_n<t} \sum_{\{I_1,\ldots,I_p\}\in\mathfrak{P}'_n} :L(I_1)\cdots L(I_p):$$

$$\to \sum_{n_1+\cdots+n_p=n} \int_{0<s_1<\cdots<s_p<t} \mathrm{d}s_1\cdots \mathrm{d}s_p :G_{n_p}(s_p)G_{n_{p-1}}(s_{p-1})\cdots G_{n_1}(s_1):$$

with

$$G_l(s) = \kappa^{l-1}\big(M_0^{l-1}M_1\gamma a^+(s) + M_0^l|\gamma|^2 a^+(s)a(s)$$
$$+ M_{-1}M_0^{l-1}\overline{\gamma}a(s) + M_{-1}M_0^{l-2}M_1\mathbf{1}\{l\geq 2\}\big)$$

and

$$\gamma = \int \mathrm{d}t\, \varphi(t), \qquad \kappa = \int_0^\infty k(t)\mathrm{d}t = \int_0^\infty \mathrm{d}t \int \mathrm{d}s\, \overline{\varphi}(s-t)\varphi(s).$$

Proof

$$F_l(s,r) = \int_{r<t_1<\cdots<t_{l-1}<s} \mathrm{d}t_1\cdots \mathrm{d}t_{l-1}\, L\big(\{s,t_{l-1},\ldots,t_1\}\big)$$
$$= \int_{r<t_1<\cdots<t_{l-1}<s} \mathrm{d}t_1\cdots \mathrm{d}t_{l-1}\big(M_0^{l-1}M_1 a^+(\varphi_\varepsilon^s) + M_0^l a^+(\varphi_\varepsilon^s)a(\varphi_\varepsilon^{t_1})$$
$$+ M_{-1}^{l-1}a(\varphi_\varepsilon^{t_1})$$
$$+ M_{-1}M_0^{l-2}M_1\mathbf{1}\{l\geq 2\}\big)k_\varepsilon(s-t_{l-1})k_\varepsilon(t_{l-1}-t_{l-2})\cdots k_\varepsilon(t_2-t_1).$$

This converges for $\varepsilon \downarrow 0$ to

$$G_l(s) = \kappa^{l-1}\big(M_0^{l-1}M_1\gamma a^+(s) + M_0^l|\gamma|^2 a^+(s)a(s)$$
$$+ M_{-1}M_0^{l-1}\overline{\gamma}a(s) + M_{-1}M_0^{l-2}M_1\mathbf{1}\{l\geq 2\}\big).$$ □

Lemma 10.2.5 *For $\varepsilon \downarrow 0$, the contribution of the overlapping intervals converges*

$$\int_{0<t_1<\cdots<t_n<t} \sum_{\{I_1,\ldots,I_p\}\in\mathfrak{P}''_n} :L(I_1)\cdots L(I_p): \to 0.$$

Proof We obtain an upper estimate if we replace M_i by $\|M_i\|$ and φ by $|\varphi|$. In order to simplify the notation, let us assume that the $M_i \geq 0$ and $\varphi \geq 0$. We have $\gamma \geq 0$, and $\kappa = \gamma^2/2$. In Lemma 10.2.1 we need the operators $\mathbb{O}_t$ only to arrange the M_i. As the M_i commute, we may forget the $\mathbb{O}_t$ and can assume that the M_i are independent of t. We put, for $\#I = l$,

$$P_l(t) = P(I)(t).$$

Then

$$\int_{0<t_1<\cdots<t_n<t} \mathrm{d}t_1 \cdots \mathrm{d}t_n\, H_\varepsilon(t_n) \cdots H_\varepsilon(t_1)$$
$$= \sum_{(n_1,\ldots,n_p):n_1+\cdots+n_p=n} \frac{1}{p!} {:}P_{n_p}(t) \cdots P_{n_1}(t){:}$$

since the number of partitions of n into subsets of $n_1, \ldots, n_p$ elements is

$$\frac{n!}{p!n_1! \cdots n_p!}.$$

Going back to Lemma 10.2.2, we observe that

$$P_l(t) = \int_{0<t_1<\cdots t_l<t} \mathrm{d}t_1 \cdots \mathrm{d}t_l\, L(t_1, \ldots, t_l) \to \int_0^t \mathrm{d}s\, G_l(s)$$

and

$$\int_{0<t_1<\cdots<t_n<t} \mathrm{d}t_1 \cdots \mathrm{d}t_n\, H_\varepsilon(t_n) \cdots H_\varepsilon(t_1)$$
$$= \int_{0<t_1<\cdots<t_n<t} \mathrm{d}t_1 \cdots \mathrm{d}t_n \sum_{\{I_1,\ldots,I_p\}\in\mathfrak{P}_n} {:}L(I_1) \cdots L(I_m){:}$$
$$= \sum_{(n_1,\ldots,n_p):n_1+\cdots+n_p=n} \frac{1}{p!} {:}P_{n_1}(t) \cdots P_{n_p}(t){:}$$
$$\to \frac{1}{p!} \sum_{n_1+\cdots+n_p=n} {:}\int_0^t \mathrm{d}s_p G_{n_p}(s_p) \int_0^t \mathrm{d}s_{p-1} G_{n_{p-1}}(s_{p-1}) \cdots \int_0^t \mathrm{d}t_1\, G_{n_1}(s_1){:}.$$

Hence

$$\int_{0<t_1<\cdots<t_n<t} \sum_{\{I_1,\ldots,I_p\}\in\mathfrak{P}''_n} {:}L(I_1) \cdots L(I_p){:}$$
$$= \int_{0<t_1<\cdots<t_n<t} \sum_{\{I_1,\ldots,I_p\}\in\mathfrak{P}_n} {:}L(I_1) \cdots L(I_p){:}$$
$$- \int_{0<t_1<\cdots<t_n<t} \sum_{\{I_1,\ldots,I_p\}\in\mathfrak{P}'_n} {:}L(I_1) \cdots L(I_p){:}$$
$$\to 0.$$

□

Proof of the theorem We have the iterative solution

$$U_\varepsilon(t) = 1 + \sum_{n=1}^{\infty} \int_{0<t_1<\cdots<t_n<t} \mathrm{d}t_1 \cdots \mathrm{d}t_n\, H_\varepsilon(t_n) \cdots H_\varepsilon(t_1).$$

We go back to Lemma 10.2.2, and assume that $M_i \geq 0$ and $\varphi \geq 0$. Then

$$\begin{aligned}\langle f | U_\varepsilon(t) | g \rangle &= \langle f | 1 + \sum_{n=1}^{\infty} \sum_{n_i \geq 1 : n_1 + \cdots + n_p = n} \frac{1}{p!} {:}P_{n_1}(t) \cdots P_{n_p}(t){:} | g \rangle \\ &= \langle f | 1 + \sum_{p=1}^{\infty} \frac{1}{p!} {:}P(t)^p{:} | g \rangle\end{aligned}$$

with

$$\begin{aligned}P(t) = \sum_{l=1}^{\infty} P_l(t) &= \sum_{l=1}^{\infty} \frac{1}{l!} \int_0^t \mathrm{d}t_1 \cdots \mathrm{d}t_l\, L(t_1, \ldots, t_l) \\ &= \sum_{l=1}^{\infty} \int_{0<t_1<\cdots t_l<t} \big(M_0^{l-1} M_1 a^+(\varphi_\varepsilon^{t_l}) + M_0^l a^+(\varphi_\varepsilon^{t_l}) a(\varphi_\varepsilon^{t_1}) \\ &\quad + M_{-1} M_0^{l-1} a(\varphi_\varepsilon^{t_1}) + M_{-1} M_0^{l-2} \mathbf{1}\{l \geq 2\} M_1\big) \\ &\quad \times k_\varepsilon(t_l - t_{l-1}) \cdots k_\varepsilon(t_2 - t_1).\end{aligned}$$

Recall

$$\gamma = \int \mathrm{d}t \varphi(t), \qquad \kappa = \int_0^\infty k(t) \mathrm{d}t = \int_0^\infty \mathrm{d}t \int \mathrm{d}s\, \overline{\varphi}(s-t) \varphi(s).$$

As we assumed $\varphi \geq 0$, we have $\gamma \geq 0$ and $\kappa = \gamma^2/2$. Assume f and g are two continuous functions of compact support with $0 \leq f, g \leq 1$, and denote by $\mathrm{e}(f)$ the function

$$\mathrm{e}(f) : \mathfrak{R} \to \mathbb{C}, \qquad \big(\mathrm{e}(f)\big)(t_1, \ldots, t_n) = f(t_1) \cdots f(t_n).$$

Then

$$\big\langle \mathrm{e}(f) \big| U_\varepsilon(t) \big| \mathrm{e}(g) \big\rangle = \mathrm{e}^{\langle f | g \rangle} \left(1 + \sum_{p=0}^{\infty} P_{fg}(t)^p\right)$$

where $P_{fg}(t)$ arises from $P(t)$ by replacing $a^+(\varphi_\varepsilon^t)$ by $\langle f, \varphi_\varepsilon^t \rangle$ and $a(\varphi_\varepsilon^t)$ by $\langle \varphi_\varepsilon^t | g \rangle$. So

$$P_{fg}(t) = \sum_{l=1}^{\infty} \int_{0<t_1<\cdots t_l<t}$$

$$\big(M_0^{l-1} M_1 \langle f | \varphi_\varepsilon^{t_l} \rangle + M_0^l \langle f | \varphi_\varepsilon^{t_l} \rangle \langle \varphi_\varepsilon^{t_1} | g \rangle + M_{-1} M_0^{l-1} \langle \varphi_\varepsilon^{t_1} | g \rangle$$

$$
\begin{aligned}
&+ M_{-1}M_0^{l-2}\mathbf{1}\{l\geq 2\}M_1)k_\varepsilon(t_l - t_{l-1})\cdots k_\varepsilon(t_2 - t_1)\\
&\leq t\sum_{l=1}^{\infty}\kappa^{l-1}\big(\gamma M_0^{l-1}M_1 + \gamma^2 M_0^l + \gamma M_{-1}M_0^{l-1} + M_{-1}M_0^{l-2}\mathbf{1}\{l\geq 2\}M_1\big)\\
&= t\frac{1}{1-\kappa M_0}\big(\gamma M_1 + \gamma^2 M_0 + \gamma M_{-1}M_0^{l-1} + \kappa M_{-1}M_1\big).
\end{aligned}
$$

So for $\kappa M_0 = M_0 \int \mathrm{d}t\, \varphi(t)^2/2 < 1$,

$$
\big\langle \mathrm{e}(f)\big|U_\varepsilon(t)\big|\mathrm{e}(g)\big\rangle < \infty.
$$

Remark that any continuous function ≥ 0 with compact support on $\mathfrak{R}$ can be majorized by a function of the type $\mathrm{e}(f)$.

We proceed now to the general case. Consider for $f, g \in \mathscr{K}$, the expression $\langle f|U_\varepsilon(t)|g\rangle$. It can be majorized by replacing M_i by $\|M_i\|$, and φ by $|\varphi|$, f by $\|f\|$ and g by $\|g\|$. The preceding discussion implies that, for

$$
\|M_0\|\int \mathrm{d}t\big|\varphi(t)\big|^2/2 < 1,
$$

we may take $\varepsilon\downarrow 0$ behind the sum and the integrals and obtain, by Lemmata 10.2.4 and 10.2.2,

$$
\begin{aligned}
\langle f|U_\varepsilon(t)|g\rangle &= \langle f|1 + \sum_{n=1}^{\infty}\sum_{n_i\geq 0, n_1+\cdots+n_p=n}\int_{0<s_1<\cdots<s_p<t}\mathrm{d}s_1\cdots\mathrm{d}s_p\\
&\quad\times :K_{n_p}(s_p)K_{n_{p-1}}(s_{p-1})\cdots K_{n_1}(s_1):|g\rangle\\
&= \sum_{p=0}^{\infty}\langle f|\int_{0<s_1<\cdots<s_p<t}\mathrm{d}s_1\cdots\mathrm{d}s_p :K(s_p)K(s_{p-1})\cdots K(s_1):|g\rangle
\end{aligned}
$$

with

$$
\begin{aligned}
K(s) &= \sum_{l=1}^{\infty}G_l(s)\\
&= \sum_{l=1}^{\infty}\kappa^{l-1}\big(M_0^{l-1}M_1\gamma a^+(s) + M_0^l|\gamma|^2 a^+(s)a(s)\\
&\quad + M_{-1}M_0^{l-1}\overline{\gamma}a(s) + M_{-1}M_0^{l-2}M_1\mathbf{1}\{l\geq 2\}\big)\\
&= \frac{\gamma}{1-\kappa M_0}M_1 a^+(s) + \frac{|\gamma|^2 M_0}{1-\kappa M_0}a^+(s)a(s) + M_{-1}\frac{\overline{\gamma}}{1-\kappa M_0}a(s)\\
&\quad + M_{-1}\frac{\kappa}{1-\kappa M_0}M_1.
\end{aligned}
$$

□

Remark 10.2.1 If one makes the replacements

$$
\begin{aligned}
M_i &\Rightarrow -\mathrm{i}M_i, \\
\gamma &\Rightarrow 1, \\
\kappa &\Rightarrow 1/2
\end{aligned}
$$

one obtains the essential part of the formula in Theorem 8.8.1.

Remark 10.2.2 Similar calculations involving overlapping and non-overlapping partitions have been performed in an old paper by P.D.F. Ion and the author [26].

References

1. L. Accardi, Noise and dissipation in quantum theory. Rev. Math. Phys. **2**, 127–176 (1990)
2. L. Accardi, Y.-G. Lu, N. Obata, Towards a non-linear extension of stochastic calculus, in *Publications of the Research Institute for Mathematical Sciences, Kyoto*, ed. by N. Obata. RIMS Kôkyûroku, vol. 957 (1996), pp. 1–15
3. L. Accardi, Y.-G. Lu, I.V. Volovich, White noise approach to classical and quantum stochastic calculus. Preprint 375, Centro Vito Volterra, Università Roma 2 (1999)
4. L. Accardi, Y.-G. Lu, I.V. Volovich, *Quantum Theory and Its Stochastic Limit* (Springer, Berlin, 2002)
5. G.E. Andrews, R. Askey, D. Roy, *Special Functions* (Cambridge University Press, Cambridge, 1999)
6. S. Attal, Problemes d'unicité dans les représentations d'opérateurs sur l'espace de Fock, in *Séminaire de Probabilités XXVI*. Lecture Notes in Math., vol. 1526 (Springer, Berlin, 1992), pp. 617–632
7. S. Attal, Non-commutative chaotic expansions of Hilbert-Schmidt operators on Fock space. Commun. Math. Phys. **175**, 43–62 (1996)
8. F.A. Berezin, *The Method of Second Quantization* (Academic Press, New York, 1966)
9. N. Bourbaki, *Intégration, chap. 6* (Gauthier-Villars, Paris, 1959)
10. N. Bourbaki, *Intégration, chap. 1–4* (Gauthier-Villars, Paris, 1965)
11. N. Bourbaki, *Intégration, chap. 5* (Gauthier-Villars, Paris, 1965)
12. N. Bourbaki, *Algebra I* (Springer, Berlin, 1989)
13. N. Bourbaki, *Lie Groups and Lie Algebras. Chapters 1–3* (Springer, Berlin, 1989)
14. A.M. Chebotarev, Quantum stochastic differential equation is unitary equivalent to a symmetric boundary problem in Fock space. Infin. Dimens. Anal. Quantum Probab. **1**, 175–199 (1998)
15. G.V. Efimov, W. von Waldenfels, R. Wehrse, Analytical solution of the non-discretized radiative transfer equation for a slab of finite optical depth. J. Spectrosc. Radiat. Transf. **53**, 59–74 (1953)
16. F. Fagnola, On quantum stochastic differential equations with unbounded coefficients. Probab. Theory Relat. Fields **86**, 501–516 (1990)
17. I. Gelfand, N. Vilenkin, *Generalized Functions*, vol. 1 (Academic Press, New York, 1964)
18. I. Gelfand, N. Vilenkin, *Generalized Functions*, vol. 4 (Academic Press, New York, 1964)
19. J. Gough, Causal structures of quantum stochastic integrators. Theor. Math. Phys. **111**, 563–575 (1997)
20. J. Gough, Noncommutative Ito and Stratonovich noise and stochastic evolution. Theor. Math. Phys. **113**(N2), 276–284 (1997)
21. M. Gregoratti, The Hamiltonian operator associated with some quantum stochastic evolutions. Commun. Math. Phys. **222**(1), 181–200 (2001)

W. von Waldenfels, *A Measure Theoretical Approach to Quantum Stochastic Processes*, Lecture Notes in Physics 878, DOI 10.1007/978-3-642-45082-2,

22. M. Gregoratti, The Hamiltonian associated to some quantum stochastic differential equations. Thesis, Milano, 2000
23. H. Haken, *Laser Theory*. Handbuch der Physik, vol. XXV/2c (Springer, Berlin, 1970)
24. E. Hille, R.S. Phillips, *Functional Analysis and Semigroups* (Amer. Math. Soc., Providence, 1968)
25. R.L. Hudson, P.D.F. Ion, The Feynman-Kac formula for a canonical quantum mechanical Wiener process, in *Random Fields, Vols. I, II*, Esztergom, 1979. Colloq. Math. Soc. Janos Bolyai, vol. 27 (North Holland, Amsterdam, 1981), pp. 551–568
26. P.D.F. Ion, W. von Waldenfels, Zeitgeordnete Momente des weissen klassischen und des weissen Quantenrauschens, in *Probability Measures on Groups. Proceedings*, Oberwolfach, 1981. Lecture Notes in Math., vol. 928 (Springer, Berlin, 1982)
27. J.M. Jauch, M. Rohrlich, *The Theory of Photons and Electrons* (Springer, Berlin, 1976)
28. L.D. Landau, E.M. Lifschitz, *Lehrbuch der theoretischen Physik IV. Quantenelektodynamik* (Akademie-Verlag, Berlin, 1986)
29. J.M. Lindsay, Quantum and non-causal stochastic calculus. Probab. Theory Relat. Fields **97**, 65–80 (1993)
30. J.M. Lindsay, H. Maassen, An integral kernel approach to noise, in *Lecture Notes in Math.*, vol. 1303 (1988), pp. 192–208
31. H. Maassen, Quantum Markov processes on Fock space described by integral kernels, in *Lecture Notes in Math.*, vol. 1136 (1985), pp. 361–374
32. L. Mandel, E. Wolf, *Optical Coherence and Quantum Optics* (Cambridge University Press, Cambridge, 1995)
33. K. Matthes, J. Kerstan, J. Mecke, *Infinitely Divisible Point Processes* (Wiley, New York, 1978)
34. P.A. Meyer, *Quantum Probability for Probabilists*. Lecture Notes in Mathematics, vol. 1538 (Springer, Berlin, 1993)
35. N. Obata, *White Noise Calculus and Fock Space*. Lecture Notes in Mathematics, vol. 1577 (Springer, Berlin, 1994)
36. K.R. Parthasarathy, *An Introduction to Quantum Stochastic Calculus* (Birkhäuser, Basel, 1992)
37. L. Schwartz, *Théorie des distributions I* (Herrmann, Paris, 1951)
38. W. von Waldenfels, Continous Maassen kernels and the inverse oscillator, in *Séminaire des Probabilités XXX*. Lecture Notes in Mathematics, vol. 1626 (Springer, Berlin, 1996)
39. W. von Waldenfels, Continuous kernel processes in quantum probability, in *Quantum Probability Communications*, vol. XII (World Scientific, Singapore, 2003), pp. 237–260
40. W. von Waldenfels, Description of the damped oscillator by a singular Friedrichs kernel, in *Quantum Prob. and Rel. Fields* (2003)
41. W. von Waldenfels, Symmetric differentiation and Hamiltonian of a quantum stochastic process. Infin. Dimens. Anal. Quantum Probab. Relat. Top. **8**, 73–116 (2005)
42. W. von Waldenfels, The Hamiltonian of a simple pure number process, in *Quantum Probability and Infinite Dimensional Analysis*, vol. 18 (World Scientific, Singapore, 2005), pp. 518–525
43. W. von Waldenfels, White noise calculus and Hamiltonian of a quantum stochastic process. arXiv:0806.3636 (2008), 72 p.
44. W. von Waldenfels, The Hamiltonian of the amplified oscillator. Arch. Math. **92**, 538–548 (2009)
45. H. Weyl, *Gruppentheorie und Quantenmechanik* (Leipzig, 1931)

Index

W. von Waldenfels, *A Measure Theoretical Approach to Quantum Stochastic Processes*, Lecture Notes in Physics 878, DOI 10.1007/978-3-642-45082-2,